S0-CAH-419

EXCITED STATES IN ORGANIC CHEMISTRY AND BIOCHEMISTRY

THE JERUSALEM SYMPOSIA ON QUANTUM CHEMISTRY AND BIOCHEMISTRY

Published by the Israel Academy of Sciences and Humanities, distributed by Academic Press (N.Y.)

1st JERUSALEM SYMPOSIUM: *The Physicochemical Aspects of Carcinogenesis* (October 1968)
2nd JERUSALEM SYMPOSIUM: *Quantum Aspects of Heterocyclic Compounds in Chemistry and Biochemistry* (April 1969)
3rd JERUSALEM SYMPOSIUM: *Aromaticity, Pseudo-Aromaticity, Antiaromaticity* (April 1970)
4th JERUSALEM SYMPOSIUM: *The Purines: Theory and Experiment* (April 1971)
5th JERUSALEM SYMPOSIUM: *The Conformation of Biological Molecules and Polymers* (April 1972)

Published by the Israel Academy of Sciences and Humanities, distributed by D. Reidel Publishing Company (Dordrecht and Boston)

6th JERUSALEM SYMPOSIUM: *Chemical and Biochemical Reactivity* (April 1973)

Published and distributed by D. Reidel Publishing Company (Dordrecht and Boston)

7th JERUSALEM SYMPOSIUM: *Molecular and Quantum Pharmacology* (March/April 1974)
8th JERUSALEM SYMPOSIUM: *Environmental Effects on Molecular Structure and Properties* (April 1975)
9th JERUSALEM SYMPOSIUM: *Metal-Ligand Interactions in Organic Chemistry and Biochemistry* (April 1976)

VOLUME 10

EXCITED STATES IN ORGANIC CHEMISTRY AND BIOCHEMISTRY

PROCEEDINGS OF THE TENTH JERUSALEM SYMPOSIUM ON QUANTUM CHEMISTRY AND BIOCHEMISTRY HELD IN JERUSALEM, ISRAEL, MARCH 28/31, 1977

Edited by

BERNARD PULLMAN

Université Pierre et Marie Curie (Paris VI)
Institut de Biologie Physico-Chimique
(Fondation Edmond de Rothschild), Paris, France

and

NATAN GOLDBLUM

The Hebrew University, Hadassah Medical School
Jerusalem, Israel

D. REIDEL PUBLISHING COMPANY

DORDRECHT-HOLLAND / BOSTON-U.S.A.

Library of Congress Cataloging in Publication Data

Jerusalem Symposium on Quantum Chemistry and Biochemistry, 10th, 1977.
Excited states in organic chemistry and biochemistry.

(The Jerusalem symposia on quantum chemistry and biochemistry; v. 10)
Bibliography: p.
Includes index.
1. Excited state chemistry–Congresses. 2. Chemistry, Physical organic–Congresses. 3. Biological chemistry–Congresses. I. Pullman, Bernard, 1919- II. Goldblum, Natan. III. Title. IV. Series.
QD461.5.J47 1977 547'.1'28 77-21896
ISBN 90-277-0853-3

Published by D. Reidel Publishing Company,
P.O. Box 17, Dordrecht, Holland

Sold and distributed in the U.S.A., Canada, and Mexico
by D. Reidel Publishing Company, Inc.
Lincoln Building, 160 Old Derby Street, Hingham,
Mass. 02043, U.S.A.

Printed in The Netherlands

PREFACE

We are living since such a long time now in a world governed in its many aspects by the decimal system, that the 10th anniversary of any significant event represents an event in itself, in particular for those who have been implicated in its birth and development. It is also a landmark at which one feels necessary to stop for a while and think, make a balance of the value and significance of the efforts expanded.

The inaugural session of this Symposium, presided by Professor Ephraim Katzir, the President of the State of Israël, in the presence of Professor A. Dvoretzky, President of the Israël Academy of Sciences and Humanities, served such a purpose. I hope not to betray the general feeling by saying that, on their modest scale, the Jerusalem Symposia, called in Quantum Chemistry and Biochemistry but which in fact have gone far beyond the quantum aspects of these disciplines seem to have been a significant event in a number of their aspects. The different themes discussed at the ten meetings were among the frontier subjects of present day scientific research in Chemistry and Biochemistry. The Symposia contributed, I believe, in a very positive way to scientific exchanges and contacts and, I hope, also, to the progress of science.

The 10th Symposium was also an occasion to express our appreciation to all those who contributed to their establishment, growth and success. A particular tribute was paid to the generosity and understanding of the Baron Edmond de Rothschild without whose help these meetings would not have been possible. The Baron de Rothschild was presented with two beautiful scrolls, from the Israel Academy of Sciences and Humanities and from the Hebrew University of Jerusalem, expressing their deep appreciation for the good work accomplished. The memory and contribution of Professor Ernst Bergmann, one of the creators of these Symposia and coorganizer of the first eight of them was recalled with emotion.

May I thank again all those who contributed to the success of this meeting : the authorities of the Israel Academy of Sciences and Humanities and in particular its President Professor A. Dvoretzky and Mrs. Agigaèl Hyam and Miriam Yogev, Professor Natan Goldblum, Vice-President of the Hebrew University who carried the heavy burden of local arrangements and the Baron Edmond de Rothschild for his renewed and everlasting generosity. The support of the European Research Office is also gratefully acknowledged.

Bernard Pullman

TABLE OF CONTENTS

Preface *(Bernard Pullman)* V

List of Participants XI

P. Vigny and J.P. Ballini / Excited states of nucleic acids at 300 K and electronic energy transfer 1

M.D. Sevilla / Mechanisms for radiation damage in DNA constituents and DNA 15

R.O. Rahn / Influence of Hg^{2+} on the excited states of DNA: photochemical consequences 27

Shi Yi Wang / A "hot" ground state intermediate in the photohydration of pyrimidines 39

Th. Montenay-Garestier / Excited state interactions and energy transfer between nucleic acid bases and amino acid side chains of proteins 53

C. Hélène / Mechanisms of quenching of aromatic amino acid fluorescence in protein-nucleic acid complexes 65

J. Sperling and A. Havron / Specificity of photochemical cross-linking in protein-nucleic acid complexes 79

J. Hüttermann / Excitation and ionization of 5-halouracils: ESR and ENDOR of single crystals 85

M.F. Maestre, J. Greve, and J.Hosoda / Optical studies on T4 gene product 32 protein DNA interaction 99

E. Hayon / The chemistry of excited states of aromatic amino acids and peptides 113

Th.M. Hooker, Jr., and W.J. Goux / Chiroptical probes of protein structure 123

M. Iseli, R. Geiger, and G. Wagnière / Description of the chiroptic properties of small peptides by a molecular orbital method 137

G. Laustriat, D. Gerard, and C. Hasselmann / Influence of 3-substitution on excited state properties of indole in aqueous solutions 151

L. Salem / The sudden polarization effect 163

J.J. Wolken / Photoreceptors and photoprocesses in the living cell 175

L.J. Dunne / Electron-electron interactions and resonant optical spectral shifts in photoreceptor molecules 187

S. Boué, D. Rondelez, and P. Vanderlinden / Classical and non-classical decay paths of electronically excited conjugated dienes 199

C.A. Bush / Far ultraviolet circular dichroism of oligosaccharides 209

Th. Kindt and E. Lippert / Adiabatic photoreactions in acidified solutions of 4-methylumbelliferone 221

D.B. McCormick / Spectral and photochemical assessments of interactions of the flavin ring system with amino acid residues 233

S.P. McGlynn, D. Dougherty, T. Mathers, and S. Abdulner / Photoelectron spectroscopy of carbonyls. Biological considerations 247

M.S. Gordon and J.W. Caldwell / Excited states of saturated molecules 257

J. Joussot-Dubien, R. Bonneau, and P. Fornier de Violet / Evidence and reactivity of a twisted form of medium size cyclo-alkene rings presenting a double bond past orthogonality 271

J. Wirz / Electronic structure and photophysical properties of planar conjugated hydrocarbons with a 4n-membered ring 283

G. Snatzke and G. Hajós / Excited states of chiral pyrazines 295

G. Köhler, C. Rosicky, and N. Getoff / Wavelength dependence of Q(F) and Q(e^-_{aq}) of some aromatic amines in aqueous solution 303

F.C. de Schryver, J. Huybrechts, N. Boens, J.C. Dederen, and M. Irie / Intramolecular excited state interactions in 1,3-di (2-anthryl) propane 313

Th.J. de Boer, F.J.G. Broekhoven, and Th.A.B.M. Bolsman / Behaviour of excited c-nitroso compounds in the presence and absence of oxygen 323

P. Politzer and K.C. Daiker / Some possible products of the reactions of O(^{1}D) and O_2($^1\Delta$) with unsaturated hydrocarbons 331

H.H. Seliger and J.P. Hamman / Chemical production of excited states: adventitious biological chemiluminescence of carcinogenic polycyclic aromatic hydrocarbons 345

J. Michl, A. Castellan, M.A. Souto, and J. Kolc / Higher excited states and vibrationally hot excited states: how important are they in organic photochemistry in dense media? 361

G.G. Hall and C.J. Miller / Solvent effects on excited states 373

M.B. Rubin / Photochemistry of vicinal polyketones 381

U.P. Wild / Fluorescence from upper excited singlet states 387

J.C. Lorquet, C. Galloy, M. Desouter-Lecomte, M.J. Decheneux, and D. Dehareng / Non-adiabatic interactions in the unimolecular decay of polyatomic molecules 397

E.S. Pysh / Measurement of circular dichroism in the vacuum ultraviolet. A new challenge for theoreticians 409

R. Janoschek / Non empirical calculations of excited states of large molecules by the method of improved virtual orbitals 419

Index of Subjects 431

Index of Names 436

LIST OF PARTICIPANTS

Boué, S.G., Université Libre de Bruxelles, Faculté des Sciences, Avenue F.D. Roosevelt 50, 1050 Bruxelles, Belgium

Bush, C.A., Illinois Institute of Technology, Lewis College of Science and Letters, Department of Chemistry, Chicago, Illinois 60616, USA

Daniels, M., Oregon State University, Radiation Center, Corvallis, Oregon 97331, USA

De Boer, Th. J., Universiteit van Amsterdam, Laboratorium voor Organische Scheikunde, Nieuwe Achtergracht 129, Amsterdam, The Netherlands

De Schryver, F.C., Universiteit te Leuven, Department Scheikunde, Celestijnenlaan 200F, 3030 Heverlee, Belgium

Dunne, L.J., Chelsea College, University of London, Department of Mathematics, Manresa Road, London SW3 6LX, England

Getoff, N., Institut für Theoretische Chemie und Strahlenchemie, Universität Wien, 1090 Wien, Währinger Strasse 38, Austria

Gordon, M.S., North Dakota State University of Agriculture and Applied Sciences, Department of Chemistry, Fargo, North Dakota 58102, USA

Hall, G.G., The University of Nottingham, Department of Mathematics, Nottingham NG7 2RD, England

Hayon, E., Department of the Army, U.S. Army Natick, Research and Development Cmd., Natick, Massachusetts 01760, USA

Hélène, C., C.N.R.S., Centre de Biophysique Moléculaire, Av. de la Recherche Scientifique, 45045 Orléans Cedex, France

Hooker, T.M., Jr., University of California, Santa Barbara, Department of Chemistry, Santa Barbara, California 93106, USA

Hüttermann, J., Universität Regensburg, Fachbereich Biologie und Vorklinische Medizin, Institut für Biophysik und Physikalische Biochemie, 8400 Regensburg, Universitätsstrasse 31, Germany

Janoschek, R., Universität Stuttgart, Institut für Theoretische Chemie, Pfaffenwaldring 55, 7 Stuttgart 80, Germany

Jortner, J., Tel-Aviv University, Institute of Chemistry, 61390 Ramat-Aviv, Tel-Aviv, Israel

Joussot-Dubien, J., Université de Bordeaux I, Unité de Chimie, Laboratoire de Chimie Physique, 351 Cours de la Libération, 33405 Talence, France

Laustriat, G., Université Louis Pasteur U.E.R. des Sciences Pharmaceutiques, Laboratoire de Physique, 3 rue de l'Argonne, 67083 Strasbourg-Cedex, France

Lippert, E., Iwan N. Stranski-Institut für Physikalische und Theoretische Chemie der Technischen Universität Berlin, 1 Berlin 12, Strasse des 17 Juni 112, Ernst-Reuter-Haus, West Germany

Lorquet, J.C., Université de Liege, Institut de Chimie, Department de Chimie Générale et de Chimie Physique, Sart-Tilman B. 4000 par Liege, Belgium

Maestre, M.F., University of California, Space Sciences Laboratory, Berkeley, California 94720, USA

McCormick, D.B., Cornell University, Section of Biochemistry, Molecular and Cell Biology, Division of Biological Sciences, Savage Hall, Ithaca, New York 14853, USA

McGlynn, S.P., Louisiana State University and Agricultural and Mechanical College, College of Chemistry and Physics, Baton Rouge, Louisiana 70803, USA

Michl, J., The University of Utah, Department of Chemistry, Chemistry Building, Salt Lake City 84112, USA

Montenay-Garestier, T., Museum National d'Histoire Naturelle, Chaire de Biophysique, 61 rue Buffon, 75005 Paris, France

Politzer, P., University of New Orleans, Lake Front, Department of Chemistry, New Orleans, Louisiana 70122, USA

Pullman, A., Institut de Biologie Physico-Chimique, 13 rue P. et M. Curie, Paris 5e, France

Pullman, B., Institut de Biologie Physico-Chimique, 13 rue P. et M. Curie, Paris 5e, France

Pysh, E.S., Brown University, Department of Chemistry, Providence, Rhode Island 02912, USA

Rahn, R., Oak Ridge National Laboratory, Union Carbide Corp. Nuclear Div., P.O.B.Y. Oak Ridge, Tennessee 37830, USA

Rosenfeldt, T., The Hebrew University of Jerusalem, Department of Physical Chemistry, Jerusalem, Israel

Rubin, M., Department of Chemistry, Technion, Haifa, Israel

Salem, L., Laboratoire de Chimie Theorique, Batiment 490, Centre d'Orsay, 91405 Orsay, France

Seliger, H.H., The Johns Hopkins University, Mergenthaler Laboratory for Biology, Baltimore, Maryland 21218, USA

Sevilla, M.D., Oakland University, Department of Chemistry, Rochester, Michigan 48063, USA

Snatzke, G., Lehrstuhl für Strukturchemie, Ruhruniversität, 4630 Bochum 1, Postfach 10 21 48, W. Germany

Sperling, J., The Weizmann Institute of Science, Department of Organic Chemistry, Rehovot, Israel

Vigny, P., Fondation Curie, Institut du Radium, Laboratoire Curie, 11 rue P. et M. Curie, 75231 Paris - Cedex 05, France

Wagnière, G., Physikalisch-Chemisches Institut der Universität Zürich, 8001 Zürich, Ramistrasse 76, Switzerland

Wang, S.Y., The Johns Hopkins University School of Hygiene and Public Health, Department of Biochemical and Biophysical Sciences, 615 North Wolfe Street, Baltimore, Maryland 21205, USA

Wild, U., Eidgenössische Technische Hochschule Zürich, Laboratorium für Physikalische Chemie, 8006 Zürich, Universitätsstrasse 22, Switzerland

Wirz, J., Physikalisch-Chemisches Institut der Universität Basel, 4056 Basel, Klingelbergstrasse 80, Switzerland

Wolken, J.J., Carnegie-Mellon University, Biophysical Research Laboratory, Schenley Park, Pittsburgh, Pennsylvania 15213, USA

EXCITED STATES OF NUCLEIC ACIDS AT 300K AND ELECTRONIC ENERGY TRANSFER.

Paul VIGNY and Jean Pierre BALLINI

Institut du Radium, Laboratoire Curie
11, rue Pierre et Marie Curie
75231 PARIS CEDEX 05, France.

I. INTRODUCTION

Investigation of the Excited States of Nucleic Acids appears to be a major step in the understanding of the photochemical changes induced in DNA by ultraviolet radiation. The details of the mechanisms initiated by the absorption of a photon by a base and ending with the formation of a photoproduct on the same or on another base cannot be understood without a knowledge of their excited states. This, together with the amount of information which can be obtained on their ground states as well, certainly accounts for the fact that many luminescence studies have been carried out on nucleic acids for the last ten years. However the main feature of these molecules is that the systems are quenched to a high degree under physiological conditions. The fluorescence quantum yields are so weak that until recently nucleic acid bases were simply considered not to fluoresce at room temperature. In this respect, they differ from many aromatic compounds for which internal conversion from the first excited singlet state is unimportant. Therefore, most of the work has been performed either at extreme pH values where the nucleic bases exhibit measurable fluorescence emission or at 77K in glasses where the quantum yields are of the order of 10^{-1} or 10^{-2}, thus permitting normal recording of the luminescence spectra. Under such conditions a good understanding of the lowest excited singlet and triplet states has been thus achieved (for a review, see for example Guéron et al (1). Although many interesting results were obtained, the major question which has been constantly raised (1) (2) is whether conclusions obtained in a rigid medium at 77K can be extrapolated to fluid aqueous

B. Pullman and N. Goldblum (eds.), Excited States in Organic Chemistry and Biochemistry, 1-13.

solutions at room temperature, specially in view of the drastic temperature effect on the quantum yields. Other questions such as the existence of interactions between bases in the excited states or the ability of electronic energy to be transfered from one base to another, in nucleic acids under physiological conditions, were unresolved.

The experimental problem of the detection of nucleic bases at room temperature has been therefore reinvestigated independently by Daniels in Oregon (3) and by our group in Paris (4) leading to the identical conclusion that nucleic bases do weakly fluoresce at 300K (fluorescence quantum yield of the order of 10^{-4}). Our earliest experiments were rather crude. They have been further refined and extended to such complex structures as dinucleotides, polynucleotides and nucleic acids. The aim of the present contribution is to summarize our recent results which should still be considered as a preliminary approach to the large field of the excited states of nucleic acids at room temperature.

II. EXPERIMENTAL CONSIDERATIONS.

Several difficulties are encountered when trying to study the fluorescent properties of compounds with very low quantum yields. They will be briefly discussed. Of course the first requirement lies in a highly sensitive spectrophotofluorometer. It is not intended to discuss here the apparatus which was used for these studies since it has already been described (5). Its sensitivity is partly due to the photon-counting method which allows an increase of the signal-to-noise ratio by increasing counting time, and partly to the optical components. Two kinds of improvements have been performed as compared to the above mentioned apparatus i) a pdp11 computer is now used for accumulation which allows automatic corrections of the spectra and ii) a more recent model of the instrument is now operating, which is somewhat more sensitive.

Rather high concentrations (10^{-3} M - 10^{-4} M) were used in our experiments. Important corrections had thus to be operated and their validity to be carefully checked (5). The improved sensitivity of our new apparatus will now allow us to use more dilute solutions and to avoid most of these corrections. As an example Figure 1 shows a recently recorded fluorescence spectrum of Adenine at concentration 10^{-5} M. At present, a concentration ten times lower is therefore our limit.

Sample purity is an important limitation since traces of a highly fluorescent impurity can give rise to perturbation in the fluorescence spectrum. A number of commercially available bases, nucleosides and

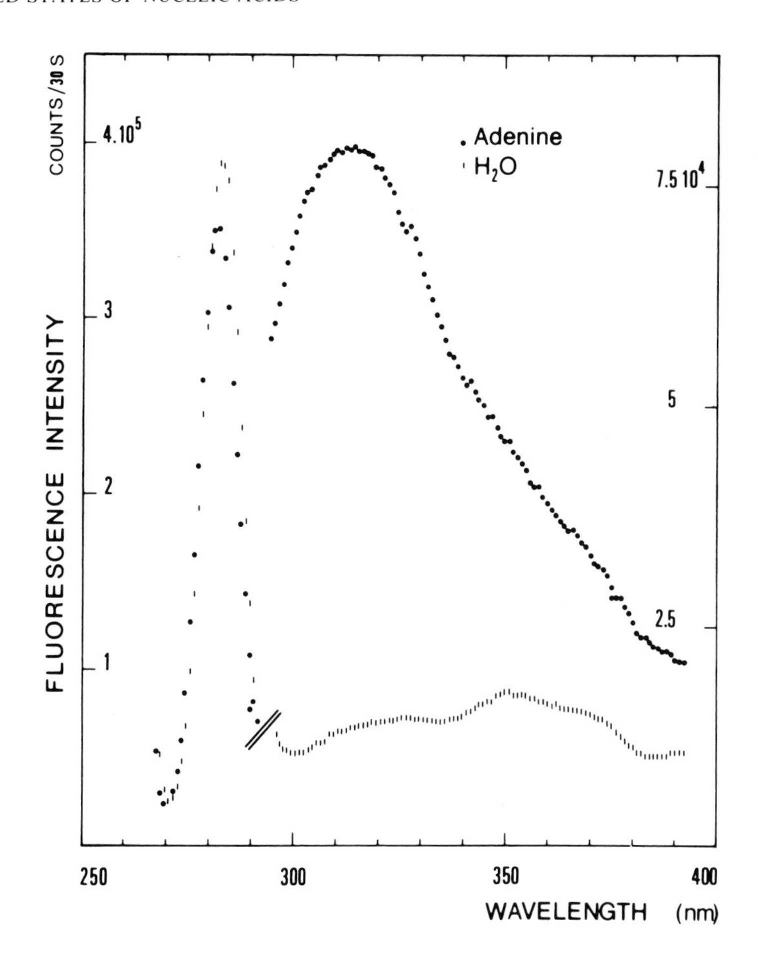

Figure 1. Room temperature fluorescence of Adenine at 10^{-5} M in water. The left part of the figure shows the Raman scattering on a different intensity scale. ($\lambda_{exc.}$=255nm,$\Delta\lambda_{exc.}$=6.0nm, $\Delta\lambda_{em.}$=3.0nm, counting time=30s per point).

nucleotides (Merck, Sigma, Calbiochem, Schwartz Bioresearch, Nutritional Biochemicals Corporation) have therefore been tested, some of which have been shown to be unsuitable for fluorescence measurements. Most of the reported fluorescence spectra are issued from products purchased from Calbiochem (A grade). Suprasil quartz cells are carefully selected and the water is triple distilled from K MnO_4 and $Ba(OH)_2$.

Polynuclenotides were purchased from Miles Laboratories. They can be more easily purified by extensive dialysis. However, due to their structure, they may undergo photochemical reactions giving rise to fluorescent adducts either during their preparation or during the recording of the spectra. The identification of the fluorescence spectrum of a polynucleotide may therefore be troublesome.

III. EXCITED STATES OF MONONUCLEOTIDES

The corrected room temperature fluorescence spectra of the five common nucleotides are given in Figure 2. As compared to the low-temperature spectra, they are broader and structureless but not very different. Except for GMP, the red-shifts when going from rigid samples

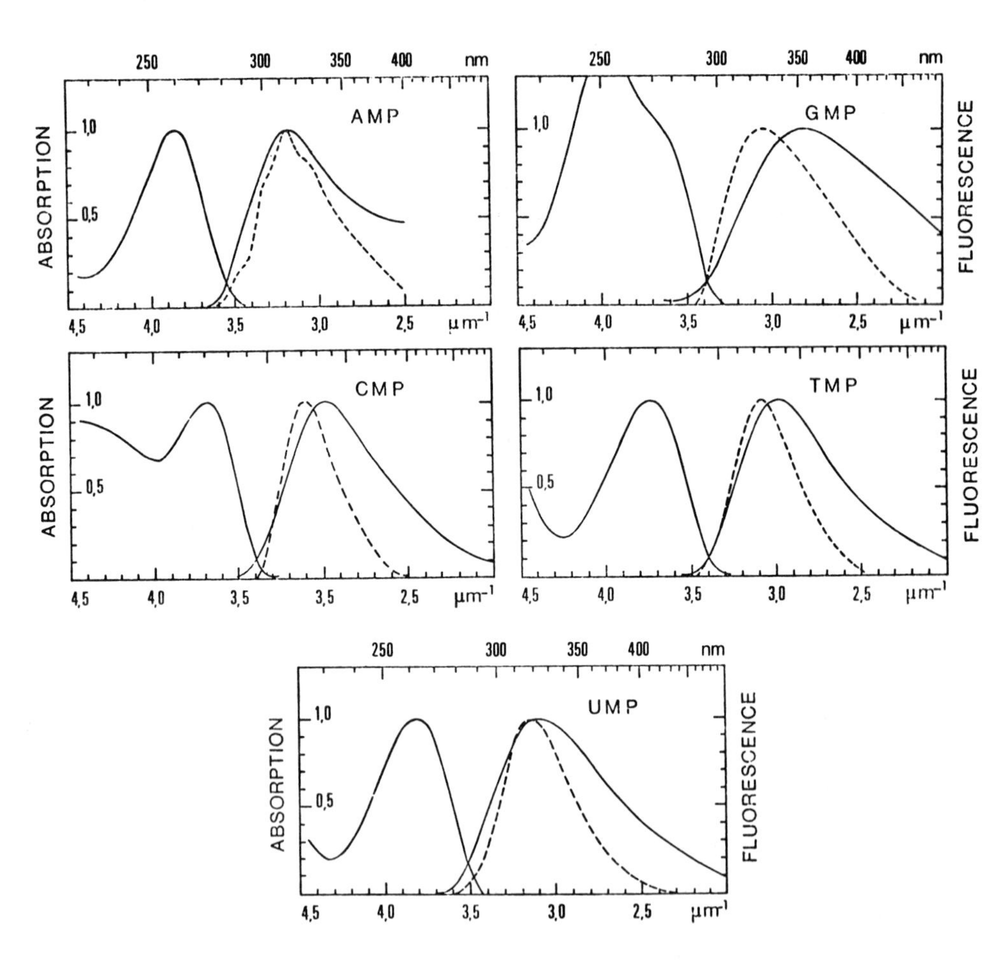

Figure 2. Absorption and fluorescence spectra of the common nucleotides at 300K. A comparison is made with fluorescence data obtained at 77K (----) by Guéron et al (1) (Our experimental conditions $C=10^{-4}M$, $\lambda_{exc.}=248nm$, $\Delta\lambda_{exc}=4.2nm$, $\Delta\lambda_i=3.2nm$).

to fluid solutions are small (CMP, TMP) or negligible (AMP, UMP). The most important feature lies in the quantum yields. Their values are between 0.3 10^{-4} (UMP) and 1.2 10^{-4} (CMP and TMP) (Table 1). It is of interest to notice that addition of the ribose and phosphate group leaves the quantum yields of C, T and U unchanged, whereas those of A and G are decreased by a factor of five.

To interpret the difference between 77° and 300°K, it is necessary to postulate a very efficient $S_1 \rightarrow S_0$ internal conversion since other deactivation processes cannot quantitatively explain the low quantum yields observed at room temperature (6). It is not possible to state whether this quenching is intra or intermolecular or -more likely- have both origins.

Another interesting point about nucleotides is the knowledge of their fluorescence lifetimes. Calculations derived from the room temperature data and assuming that their entire low-energy absorption band is responsible for emission lead to singlet lifetimes of 10^{-12}s for bases (3) and nucleotides (7), in agreement with experiments involving energy transfer to Eu^{+} (8). No doubt that direct experimental determination of these lifetimes in the future would be an important contribution in this field.

IV. EXCITED STATE INTERACTIONS IN POLYNUCLEOTIDES

Bases are brought together in polynucleotides so that interactions may occur. In addition to the well-known ground state interactions, can excited state interactions also occur at room temperature ? Such exciplexes and excimers have been proposed at 77K to explain the red-shift observed in their emission spectra (see reference (1) for a review).

Beside the monomer-like emission, the room temperature emission spectrum of the dinucleotide ApA shows a new broad band at ~420nm (9). This emission can be thought to arise from an excimer formed between two stacked bases. According to what is known about excimer emission, its intensity should be more or less intense, depending on the stacking of the two bases. At room temperature ApA is supposed to be in a stacked conformation. Moreover this stacking is very temperature dependent and becomes less important when temperature is increased. Part a of Figure 3 shows that the second emission band is effectively temperature dependent and notably increased when the temperature is lowered to 4°C. The same interpretation has been proposed for $C_{5'pp5'}C$ (10), whose second emission band ($\lambda_{em.}^{max.}$ =410nm) is strongly increased when ionic strength is increased (Figure 3, part 3).

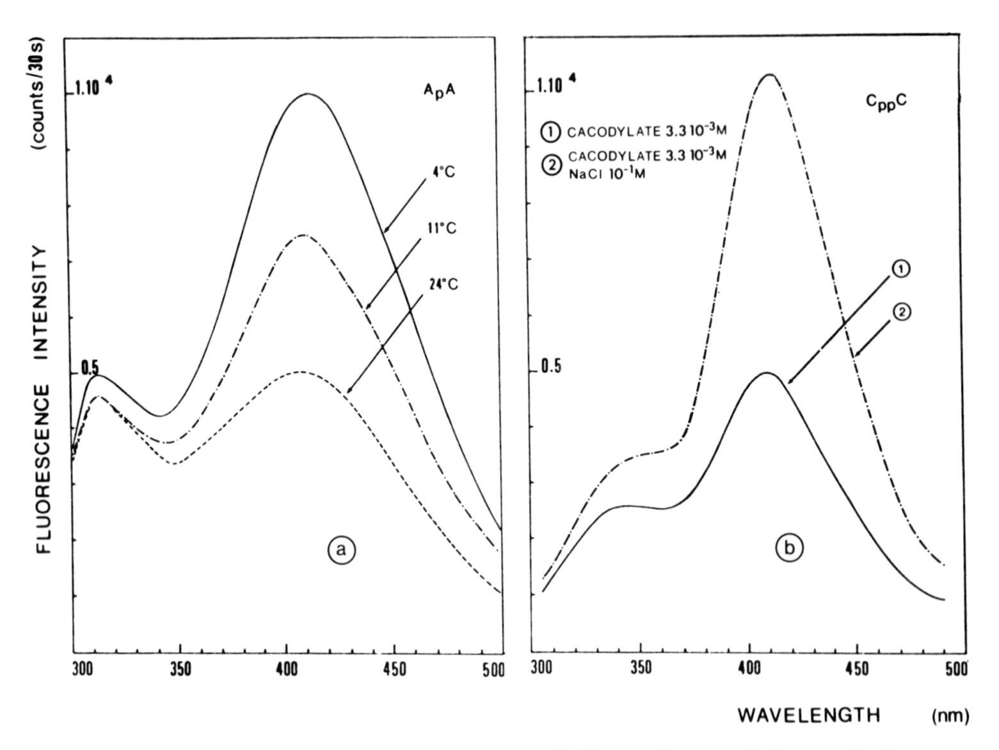

Figure 3. Temperature and ionic strength dependence of the emission spectra of dinucleotides. (experimental conditions $\lambda_{exc.}=248$ nm, $\Delta\lambda_{exc.}=4.2$nm, $\Delta\lambda_i=6.4$nm, concentration of ApA 1.5×10^{-4}M in monomer in phosphate buffer 10^{-2}M, concentration in $C_{5'pp5'}C$ 2×10^{-4}M in monomer. Uncorrected spectra).

A good example of excimer emission in polynucleotides at room temperature is given by PolyC whose emission spectrum is strongly dependent on the polymeric structure. At pH7 where PolyC is known to be in a random coil, the emission spectrum is monomer-like ($\lambda^{max.}_{em.}=343$nm) with a weak contribution above 400nm. At pH4 on the other hand, PolyC is known to be in a double stranded helix. The monomer-like emission is then very weak whereas an intense emission is observed at 410nm with an excitation spectrum superimposable on the absorption spectrum. Figure 4 shows other polynucleotides which can be thought to form excimers. PolyA is known to have a locally organized structure and shows a second emission band at 395nm which is strongly temperature dependent. Such is also the case of Poly d (A-T) whose second emission band ($\lambda^{max.}_{em.}=415$nm) is absent at 80°C when the double stranded polymer is melted, a phenomenon which is reversible.

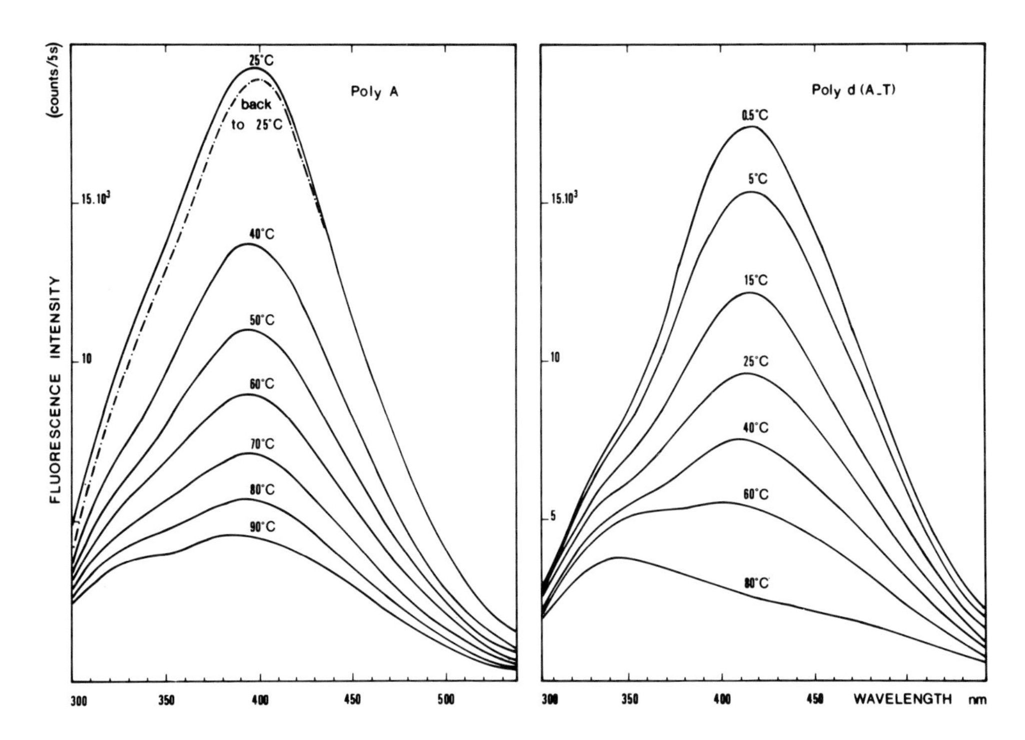

Figure 4. Temperature dependence of the emission spectra of dinucleotides. (optical density 6.6 at 260nm, in phosphate buffer 0.15M. Other experimental conditions are identical to those of Fig. 3).

A systematic study of the room temperature emission of all the common polynucleotides clearly shows that all the observed second emission bands cannot be understood in terms of excimers. From results summarized in Table 1, three classes of polynucleotides are to be distinguished

i) class I contains those, already discussed, which are thought to form excimers (PolyA, Poly d (A-T) and acidic PolyC)

ii) class II contains those whose second emission band must be ascribed to the fluorescence of photo-adducts that can be formed between residues in well defined stacked positions. In these polynucleotides, the emission is not related to the polymeric structure but appears to be dependent on irradiation time. That excimer emission may also be present cannot be totally excluded ; an attractive idea would be that the excimer is a common intermediate in both radiative and photochemical deactivation processes. Most of the polynucleotides belonging to this class are pyrimidine derivatives (namely Poly dT, PolyU, Poly dG.Poly dC). However, in addition to the excimer-like emission of PolyA, Poly dA appears to show a photoproduct emission ($\lambda_{em.}^{max.}$ 345-360nm). This finding, already

	pH	max. (nm)	$\varnothing_f$	E_{0-0} (cm^{-1})	probable origin of the 2nd emission band
AMP	7	312	0.5 10^{-4}	35550	
ApA	7	315 and 420	1.4 10^{-4}	35300	excimer
PolyA	7	325 and 395	3 10^{-4}	34800	excimer
GMP	7	340	0.8 10^{-4}	33800	
GpG	7	350	1.3 10^{-4}	33400	
PolyG	7	342	4.7 10^{-4}	32700	
CMP	7	330	1.2 10^{-4}	34000	
CpC	7	335	1.4 10^{-4}	34100	
CppC	7	330 and 410	2.7 10^{-4}	34000	excimer
PolyC	7	343 and ~400 (weak)	1.3 10^{-4}	33500	excimer
	4	~330 (weak) and 415	8 10^{-4}		excimer
TMP	7	330	1.2 10^{-4}	34100	
Poly dT	7	328 and 400	10^{-3}	34200	adduct(s)
UMP	7	320	0.3 10^{-4}	35100	
PolyU	7	322 and 380 (weak)	0.4 10^{-4}	35000	adduct(s)
Poly d(A-T)	7	~330 (shoulder) and 415	1.8 10^{-4}	34400	excimer
Poly dG. Poly dC	7	335 and 395	1.3 10^{-4}	33800	adduct(s)

Table 1. Fluorescent properties of nucleotides and polynucleotides at 300K. (The fluorescence quantum yields have been estimated with reference to Adenine $\varnothing_f$=2.6 10^{-4} (3) with an excitation at 248nm. For nucleotides, the values are somewhat higher than the previously reported ones (4), which were obviously underestimated. For polynucleotides the whole spectrum is taken into account. Therefore the quantum yield of polynucleotides which present a fluorescence due to adduct(s) is overestimated. The 0-0 energy has been determined by the absorption emission intersection).

mentioned in our previous work on PolyA (9) is probably related to the specific photoreaction in Poly dA observed by means of other techniques (11) (12)

iii) class III contains polymers which only show the monomer-like emission spectrum. Only PolyG belongs to this class and one can wonder if this observation can be related to the peculiar properties of Guanosine already discussed (13).

V. ROOM TEMPERATURE LUMINESCENCE OF DNA

Going on with our investigation on polynucleotides, a characterization of the DNA emission was tempted. It was not hoped to get a complete understanding of such a complicated system containing four bases in more or less random fashion. A number of questions should be elucidated at the monomeric and polymeric level before thinking to reach this ultimate goal. Even at 77K is the DNA luminescence reported to be a difficult study. Under such conditions, DNA quantum yield is about one tenth that of an equimolar mixture of the four constituent nucleotides. Although the emission is not well characterized, what comes out is that G-C base pairs probably introduce quenching while the emission itself is mainly from exciplexes involving A and T (1) (14).

Difficulties considerably increase at 300K since quantum yields are two or three orders of magnitude lower. Highly purified samples are needed and attention should be paid to fluorescent adducts that can be formed by U.V. irradiation of DNA (15). A number of commercially available DNAs have been extensively dialysed against phosphate buffer and their fluorescence spectra have been recorded. All tested samples, extracted respectively from Calf Thymus, Calf Spleen, Salmon Sperm, Chicken Blood (Calbiochem. A grade) or from Calf Thymus, Micrococcus Lysodeikticus (Sigma), show a maximum emission between 330 and 335 nm. Some of them also showed an emission at higher wavelength (around 400nm). This last observation, however, was not reproducible. Quantum yields, relative to Adenine, were estimated between 0.6 and 0.8×10^{-4}, depending on the sample. These results are in agreement with those reported by Daniels (16). Unfortunately no excitation spectrum was given by this author . We were surprised to find for the above mentioned DNAs excitation spectra with maxima around 280nm, thus very different from the absorption spectra. Before trying to give an explanation of this phenomenon, one must therefore ask the question whether commercial DNA is suitable for refined fluorescence measurements.

We would prefer to focus our attention on the data obtained from a highly purified DNA, extracted from Mouse Skin for other experiments requiring very pure DNA (17). Its fluorescence characteristics are shown in Figure 5. As in commercial DNA, the emission has a maximum at 335 nm, but a lower quantum yield has been found

$$\phi_f \simeq 3 \; 10^{-5}.$$

Such a low value, lower than that of most nucleotides and polynucleotides (Table 1) allows us to think that all excited bases in DNA do not emit. Whether the observed emission is issued from only one or from several

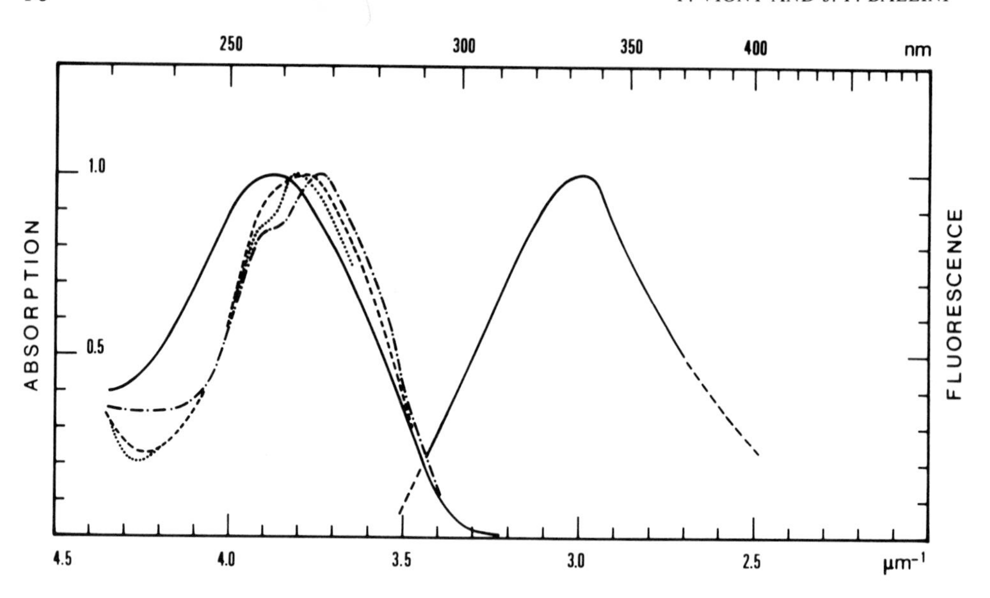

Figure 5. Fluorescence characteristics of Mouse Skin DNA at 300K. The fluorescence spectrum (right part of the figure) is obtained with an excitation wavelength at 260nm ($\Delta\lambda_{exc.}$ =6. 0nm, $\Delta\lambda_{em.}$ =1. 5nm). Corrected excitation spectra are respectively monitored at emission wavelength 350nm (—·—·—), 330nm (----) and 310nm (····). pH7 tris NaCl 10^{-2}M buffer is used and the optical density is 3. 21 at 257. 5nm.

of the four bases is an important question. A first indication that the four bases are probably not present is found in the fact that emission appears to be less broad, specially in the red side region, than that of a mixture of the four nucleosides at the same concentration. No answer can be drawn from the position of the maximum emission. 335nm could correspond to T residue, although C and G maxima, which are red-shifted in polymers to respectively 343 and 342nm, cannot be excluded. Finally A residue which emits at 312nm in aqueous solution is shifted to 325nm in PolyA and should also be considered. This idea is corroborated by the 0-0 transition energy value

$$E_{0-0} \simeq 34\,400\text{cm}^{-1}$$

derived from Figure 5, a value which is near those of PolyA (34 800cm^{-1}), Poly d (A-T) (34 400cm^{-1}) and Poly dT (34 200cm^{-1}). It has to be noticed also that the blue-side shape of the emission spectrum of these polynucleotides is very close to that of DNA. More striking is the situation of the excitation spectra, clearly different from DNA absorption. The fact that they depend on the monitoring wavelength emission is another argument in favour of the contribution of several residues to DNA emission. Considering DNA as a sum of individual residues, one can compare the excitation spectra to the absorption spectra of the four bases. C and T residues seem then to be involved. On the other hand taking DNA as an arrangement of A-T and G-C base pairs, one can compare the excitation spectra to the absorption spectrum of the heteropolynucleotides Poly d (A-T) and Poly dG. Poly dC. The excitation spectra clearly resemble that of Poly d (A-T) absorption spectrum, the observed shifts being related to the amount of A and T measured at different emission wavelengths. If this was true, A-T base pairs would be more important in DNA emission than the G-C pairs, a situation which would not be distant from that observed at low temperature (1). However, further work is still needed to identify with certainty the residues involved in the room temperature DNA emission.

VI. ENERGY TRANSFER IN NUCLEIC ACIDS

At this stage, the question of electronic energy transfer in nucleic acids under physiological conditions may be reinvestigated. From this point of view, what comes out of our DNA study is somewhat disappointing since G which among the four residues has the lowest excited singlet state (Table 1) and should act as an efficient energy trap in the case of an important energy transfer, does not seem to play an important role in DNA emission. However in view of DNA complexity, transfer studies should be undertaken on simpler models such as di- and oligonucleotides. No doubt that their interpretation will be difficult due to the overlap between the fluorescence spectra of the four bases.

Other nucleic acids such as tRNA are probably more suitable for energy transfer studies because of spectroscopic and structural reasons. tRNAs are much smaller molecules whose sequences are known nowadays. For some of them, the crystallographic structure has been recently established. On the other hand, they often possess odd nucleosides which may have completely different spectroscopic properties and may therefore be distinguished from the common bases. Such is the case of 4-Thiouridine which is present in position 8 of 70% of E. Coli tRNA. Its absorption spectrum ($\lambda^{max.}_{abs.}$ 335nm) is shifted as compared to the normal nucleosides whereas it emits an unusual weak emission at 510nm in tRNA (18). Moreover it can undergo a specific photoreaction which can be monitored by the fluorescence of the reduced form of the product (19). In collaboration with A. Favre and G. Thomas, we have recently determined the luminescence excitation spectrum in the range 230-380nm. The two spectra are identical but present a new peak around 260nm. At this wavelength they are amplified by a factor of nine as compared with the absorption and excitation spectra of the free nucleoside in aqueous solution. A detailed discussion of the possible origins of this peak led us to conclude that electronic energy transfer does occur in native tRNA at room temperature, from the common bases to the 4-Thiouridine residue (20). Moreover, from the sets of atomic coördinates obtained on Yeast $tRNA^{Phe}$ crystals a satisfactory account of this phenomenon can be obtained assuming a singlet-singlet transfer.

Singlet-singlet energy transfer also occurs in tRNAs in which the 8-13 link has been photochemically introduced. The acceptor is not the 4-Thiouridine in position 8 but the reduced 8-13 link. (to be published, in collaboration with A. Favre and G. Thomas). Work is now in progress on this subject following two directions i) a further investigation of the transfer mechanism ii) the use of transfer properties as a tool in the study of tRNA structure in aqueous solution, since significant differences between the tRNA species are observed.

The last example shows that the understanding of the Excited States of Nucleic Acids at 300K can be of help not only for photochemical and photobiological problems but also for applications to the ground state properties, in the field of Molecular Biology. From the photobiological point of view, however, it is clear that such an understanding is far from being solved and needs further exhaustive investigations. At the monomeric level a direct determination of the fluorescence lifetimes would be an important contribution. At the polymeric level it is now important to know whether the electronic energy transfer evidenced in tRNAs does also occur between the common bases of DNA.

Acknowledgements - The authors wish to acknowledge Prof. M. Duquesne for his help and encouragements in this work.

REFERENCES

1. GUERON, M., J. EISINGER and A.A. LAMOLA in Basic Principles in Nucleic Acid Chemistry. P.O.P. Ts'o Ed. Academic Press (1974).
2. EISINGER, J., A.A. LAMOLA, J.W. LONGWORTH and W.B. GRATZER Nature, 226, 113 (1970).
3. DANIELS, M. and W. HAUSWIRTH Science, 171, 675 (1971), HAUSWIRTH, W. and M. DANIELS. Photochem.Photobiol. 13, 157 (1971).
4. VIGNY, P., C.R. Acad. Sc. Paris D272, 2247 (1971), VIGNY, P., C.R. Acad. Sc. Paris D272, 3206 (1971), VIGNY, P., Proceedings of the 5th Jerusalem Symposium : The Purines, theory and experi-

ment, 4-8 April (1971).
5. VIGNY, P., and M. DUQUESNE. Photochem. Photobiol. 20, 15 (1974).
6. HAUSWIRTH, W. and M. DANIELS. Chemical Physics Letters 10, 140 (1971).
7. VIGNY, P. and M. DUQUESNE in Excited States of Biological Molecules. J.B. Birks Ed.
8. EISINGER, J. and A.A. LAMOLA. Biochim. Biophys. Acta 240, 299 (1971).
9. VIGNY, P., C.R. Acad. Sc. Paris D277, 1941 (1973).
10. VIGNY, P. and A. FAVRE. Photochem. Photobiol. 20, 345 (1974).
11. PÖRSCHKE, D., Proc. Natl. Acad. Sc. USA 70, 2683 (1973).
12. RAHN, R.O., Abstracts 3rd Annual Meeting of the American Society for Photobiology, 73 (1975).
13. GUSHLBAUER, W., Proceedings of the 5th Jerusalem Symposium. The Purines, theory and experiment, 4-8 April (1971).
14. HELENE, C., M. PTAK and R. SANTUS. J. Chim. Phys. 65, 160 (1968).
15. HAUSWIRTH, W. and S.Y. WANG. Biochem. Biophys. Res. Comm. 51, 819 (1973).
16. DANIELS, M., in Physico-Chemical Properties of Nucleic Acids. J. DUCHESNE Ed. Academic Press, 99 (1973).
17. DAUDEL, P., M. DUQUESNE, P. VIGNY, P.L. GROVER and P. SIMS. FEBS Letters 57, 250 (1975).
18. FAVRE, A., Photochem. Photobiol. 19, 15 (1974).
19. FAVRE, A. and M. YANIV. FEBS Letters 17, 236 (1971).
20. BALLINI, J.P., P. VIGNY, G. THOMAS and A. FAVRE. Photochem. Photobiol. 24, 321 (1976).

MECHANISMS FOR RADIATION DAMAGE IN DNA CONSTITUENTS AND DNA

Michael D. Sevilla

Department of Chemistry
Oakland University
Rochester, Michigan

The initial effects of all forms of ionizing radiation on DNA are the production of positive ions, negative ions and excited states of the DNA bases. These species may later react to produce biologically significant damage. Ion radicals in γ-irradiated DNA have been reported.[1-3] For example Gräslund, Ehrenberg, Rupprecht, and Ström suggest an anion radical on thymine and a cation radical on guanine in γ-irradiated oriented DNA at 77 K.[1] The reactions of these ion radicals produced individually in DNA bases have been recently investigated with some success.[4-10] In this discussion we will concentrate on the initial events immediately after ionization i.e. the fate of the electron and positive hole in irradiated DNA.

There are a number of techniques which have been used to produce ion radicals in DNA bases for ESR studies. They include radiolysis, electrolysis and photolysis. For the most part our studies have employed photolytic techniques. Below we describe these techniques and several recent studies on the ion radicals of DNA constituents.

A. The formation of DNA Base π-Cation* Radicals by Photoionization.

In early work by Helene, Santus, and Douzou the cation radicals of several purines were reported to be produced through

*The terms π-cation and π-anion refer to the loss or gain of one electron from the π electron system and do not refer to the charge on the molecule.

B. Pullman and N. Goldblum (eds.), Excited States in Organic Chemistry and Biochemistry, 15-25.

photolysis of frozen aqueous solutions at low temperatures.[11] The mechanism was shown to be photolysis of the metastable triplet state of these species. However, for the pyrimidine compounds, e.g., thymine and cytosine, cation radicals are not produced at pH 7. In agreement with these results Shulman and coworkers found through both optical and esr methods that the neutral thymine and cytosine molecules did not show phosphorescence or appreciable population of the triplet state.[12-13] By increasing the pH to 12 where thymine loses its N_3 photon, phosphorescence and esr signals due to the triplet were observed by these workers with a decay time of 0.60 sec. From this previous work it is reasonable to expect that under conditions of high pH and low temperature thymine should photoionize from its excited tripled state.

ESR studies of the uv photolysis of thymine have shown that thymine does photoionize at 77 K in alkaline aqueous glasses such as 5 M K_2CO_3, 8 M NaOD or basic 8 M $NaClO_4$.[14,15] Perhaps unexpected from the previous work the thymine π-cation is produced by photolysis in an acid glass (5 M D_3PO_4) as well.

In Table I the triplet lifetimes at 77 K reported by Shulman and Rahn for a number of DNA constituents are shown.[12]

Table I. Triplet Lifetimes (τ) at 77 K

Compound	τ(sec)[a]	pH
Thymine	0.6	12
Thymidine	0.6	12
TMP	0.5	12
Cytosine	-	7,12
GMP	1.26	7
AMP	2.4	7

In general it has been found that the DNA bases, nucleosides or nucleotides can be photoionized at 254 nm in aqueous glasses if they have triplet lifetimes on the order of those shown in the table and significant intersystem crossing.

The aqueous glasses that have been employed in these studies are prepared from 8 M NaOD, 5 M D_3PO_4, 8 M $NaClO_4$, 12 M LiCl, 5 M K_2CO_3 or 50% glycerol in H_2O by cooling these solutions to 77 K.[14,15] Inorganic glasses are preferred since matrix radicals are less likely to form. Glasses prepared from 8 M $NaClO_4$ have proven most useful in the study of cations due to the fact that the photoejected electron is scavenged by ClO_4^- as in reaction 1

$$ClO_4^- + e^- \rightarrow ClO_3^- + O^- \qquad (1)$$

The broad ESR signal from O^- can be easily subtracted from the π-cation signal with computer techniques.

B. The Formation of DNA Base Anion Radicals by Electron Attachment

Esr studies of the reaction of electrons with DNA bases and nucleotides in alkaline and neutral aqueous glasses have been helpful in elucidating the role of the electron in radical production.[7-9,14,16] In our work electrons are produced by the uv photolysis (254 nm) of $K_4Fe(CN)_6$(10^{-2}M) in a number of aqueous glasses at 77 K.[17] The DNA bases are kept at 10^{-3}M or lower to prevent photoionization. The photo-oxidation of of ferrocyanide produces ferricyanide. Ferrocyanide is not paramagnetic and ferricyanide does not interfere with the g=2 region of magnetic field that is of interest. The glasses we have found most useful in studies of electron attachment reactions are prepared from 8 M NaOH and 12 M LiCl. Matrix radicals, chiefly Cl_2^-, are produced in LiCl by photolysis only if the concentration of $K_4Fe(CN)_6$ is low. Deuterated solvents are employed to reduce depolar broadening and improve resolution.

Immediately after photolysis the sample is photobleached with light from an incandescent lamp to mobilize electrons trapped in the glass. The photoproduced and photobleached electrons either react with the solute or the ferricyanide formed during the photolysis.

This technique has been employed successfully in a number of studies. In the following section we describe perhaps the most significant of these studies, the investigation of electron reactions with dinucleoside phosphates.[7]

C. π-Anions of Dinucleoside Phosphates

An ESR and pulse radiolysis study of the reaction of electrons with dinucleoside phosphates (DNPs) was begun to better understand the role of the electron in DNA radiolysis. Specifical goals were 1. to ascertain on which DNA base the electron localized and 2. to determine what reactions occur after formation of the DNP anion.[7]

It is important to the first goal to learn whether DNA bases in DNPs are stacked so as to make electron transfer to the more electron affinic base possible in the rigid glass used in the ESR study or in the aqueous solution used in the pulse radiolysis study. There are a number of investigations

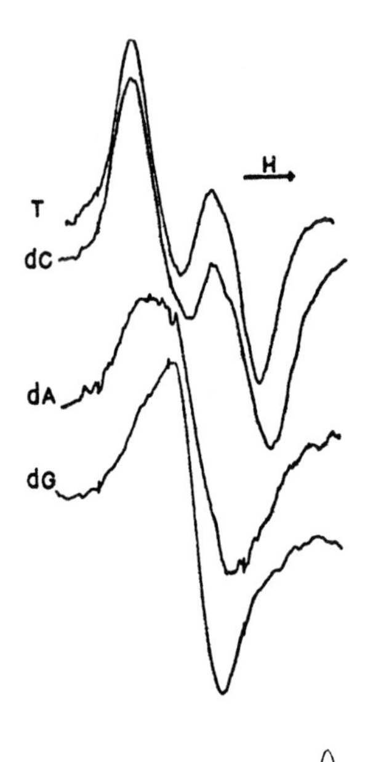

Figure 1. First derivative ESR spectra of the nucleoside anions in 12 M LiCl-D_2O at 110 K: (T) thymidine anion (dC) deoxycytidine anion, (dA) deoxyadenosine anion, (dG) deoxyguanosine anion. The distance in magnetic field between the markers in the spectrum is 13.0 G. The central marker is at g = 2.0056.

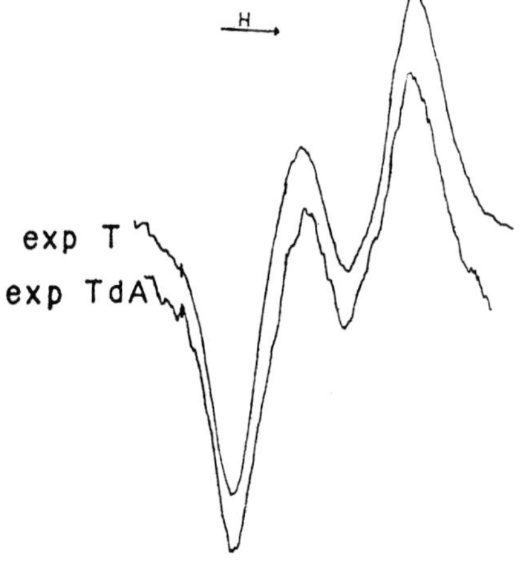

Figure 2. The ESR spectrum (lower spectrum) of the TpdA anion in 12 M LiCl-D_2O at 110 K. The ESR spectrum (upper spectrum) found for T anion. Computer simulations which summed the dA anion with the T anion gave a best fit for pure T anion.

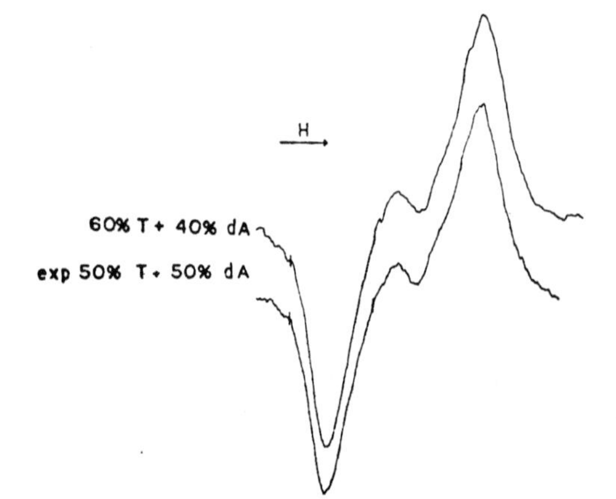

Figure 3. The ESR spectrum (lower spectrum) found after the reaction of electrons with an equal molar mixture of T and dA in 12 M LiCl at 110 K. Computer simulation (upper spectrum) of the lower curve summing T anion and dA anion in the ratio of areas 6:4.

which have shown that the DNA bases in dinucleotides, or free solution, will tend to stack.[18-20] The stacking has been found to be temperature dependent with stacking favored at lower temperatures. Purine-purine stacking is found to be favored somewhat over pyrimidine-purine stacking. In this ESR work the room temperature solutions of DNPs in 12 M LiCl are cooled to 77 K. But the solution remains a liquid upon cooling to approximately 190 K where the solution becomes a glass. Thus, in this system stacking should be greatly favored. The high salt concentration in the solution might be expected to affect the stacking; however, Brahms, Maurizot and Michelson have shown in work with a number of DNPs in 5 M KI that the ionic strength had no important effect on the stacking.[20]

The ESR spectra of the anion radicals of the four DNA nucleosides in 12 M LiCl-D_2O at 110°K produced by uv photolysis of $K_4Fe(CN)_6$ are shown in Figure 1. These spectra were used to simulate the spectra of the DNP anions discussed below. The analyses of the T, dC, adenine and guanine anion spectra have been reported elsewhere.[9,14,21,22]

In Figure 2 we show the result of electron attachment to TpdA. Computer simulations show the best fit to be 100% T anion (shown in Figure 2). Simulations with as little as 10% dA anion were distinguishable. This results suggests that the DNA bases are stacked and the electron is transferred from dA to T. Another possibility is that they are not stacked and the electron reacts much more readily with T than dA in the DNP. This possibility is not in accord with the kinetics of electron reaction with thymine and adenine in aqueous solution.[23] These results show the rates of reaction to be comparable.

Since it could be argued that the rates may differ in a frozen glass of high ionic strength a test of the second explanation was performed. In these experiments equal molar mixtures of T and dA (1×10^{-3}M) were prepared. Electron addition gave the spectrum shown in Figure 3. The computer simulation gave a best fit for 60% T anion and 40% dA anion. This result is quite reasonable in light of the previous kinetic studies. The combined results for TpdA and T + dA strongly suggests that electron transfer to T is occuring with in the DNP.

Electron reactions with TdG and T plus dG mixtures also suggest electron transfer from the purine, dG, to the pyrimidine, T. Electron reactions with TpdC and dApdC were not as conclusive, but show that T and dC share the excess electron equally in TpdC anion whereas the electron resides principally on dA in dApdC anion.

The pulse radiolytic work confirmed that electron transfer occurs from dA to T in TpdA anion.

Our results show that electron transfer occurs in the DNPs from the purine to the pyrimidine base at 77 K. This initial unpaired electron distribution in the DNPs is a consequence of the relative attraction of the DNA bases for an excess electron. Thus the ESR results found for the DNP anions can be used to order the four nucleosides in terms of their electron affinity in an aqueous solution. The results for TpdA and TpdG clearly show that the electron affinity of T is greater than dG or dA. Although the relative ordering of the electron affinity of dC was not unequivocably determined, the results suggest that is is intermediate between T and the purine nucleosides. Thus, the following order of electron affinities for all the DNA nucleosides in indicated:

$$T \simeq dC > dA \simeq dG$$

The ordering of the electron affinities of the free DNA bases would be expected to be the same as found in the DNPs.

Our results are in agreement with theoretical calculations of the electron affinity of the RNA bases.[24,25] These calculations predict the order of electron affinity to be uracil > cytosine >> guanine > adenine. Experiments on the radiation chemistry of DNA or DNA bases also lend some support to our findings.[7]

D. π-Cations in DNA Constituents and DNA

π-cation radicals of DNA bases have been identified in γ-irradiated crystalline thymine[26], thymidine[27], cytosine[28] as well as DNA[1] itself. The π-cation radicals of all the DNA bases save cytosine have been produced by photoionization in aqueous glasses.[9,11,14,15,21] The purine cations show relatively unresolved spectra[9] while those of thymine and 5-methyl cytosine show well resolved ESR spectra which have been fully interpreted.[15] Recent work[21] with π-cations of TMP, and thymidine has resulted in spectra which show a resolved coupling to the ribose group. In Figure 4A the spectrum of the TMP π-cation in 8 M $NaClO_4$ is shown. The parameters used in the reconstruction given in Figure 4B are $a(CH_3) = 21.3G$, $a(\text{ribose } \beta\text{-proton}) = 8.3G$, $A_{\parallel}(N_1) = 13.1$ G, $g_{\parallel} = 2.0022$ and $g_{\perp} = 2.0040$. The methyl group and nitrogen splittings are not found to vary significantly with the substituent at N_1. In fact the thymine π-cation itself shows values of these splittings very near those reported above.[15]

In our most recent work we have performed a study of various dinucleoside phosphate and DNA π-cations. Photolysis of these molecules in 8 M $NaClO_4$ at 77 K has resulted in the finding that principally the guanine π-cation is stabilized in DNA or in DNP's containing guanine. In the following we illustrate our findings for DNA. Figure 5A shows the spectrum of DNA in 8 M $NaClO_4$ immediately after photolysis. The low field signal is due to O^- formed by reaction of the photoejected electron with ClO_4^- (reaction 1). In Figure 5B we show the spectrum of O^- itself. The subtraction of B from A is shown in Figure 5C. Double integration of A and B showed that about 50% of the radical was due to O^-. This suggests the signal observed in Figure 5C is due only to the DNA π-cation. In Figure 5D we show the guanine π-cation from the photoionization of GpG. The photoionization of dG produced an identical spectrum. The spectrum in Figure 5D is virtually identical to that found for DNA as evidenced by the subtraction of 5D from 5C shown in Figure 5E. The above evidence combined with the fact that the spectra found for the π-cations of T, dC (from single crystal work) and dA are distinguishable from that of dG leads us to the conclusion that the π-cation in DNA is localized largely on guanine.

We believe there are two possible interpretations of the results found for DNA. First the observation of the guanine π-cation may simply be due to its selective photoionization. In agreement with this interpretation we find that guanine is somewhat more easily photoionized than adenine whereas the pyrimidines can not be photoionized in neutral glasses. However, under basic conditions where thymine can be photoionized still only the π-cation of guanine is observed. Interestingly the signal due to the π-cation is much more intense in basic solutions. The second possibility is that photoionization from guanine, adenine and thymine is followed by hole transfer to the DNA base with the lowest ionization potential (guanine). Since hole transfer through the stacked DNA bases is likely to occur and since on the average a guanine base should be only a few DNA bases from the original hole site, this mechanism has the most appeal.

The report of Shulman and Rahn that the phosphoresence from thymine containing dinucleotides and DNA is from the thymine base alone is pertinent to this point.[12] They suggest that the triplet excitation is transferred through the DNA chain to the base with the lowest triplet energy (thymine). The build up of the thymine triplet state is of course a prerequisite to the photoionization of thymine. From the above it seems a major fraction of the cation radicals should be produced on thymine. However this was not found to be the case.

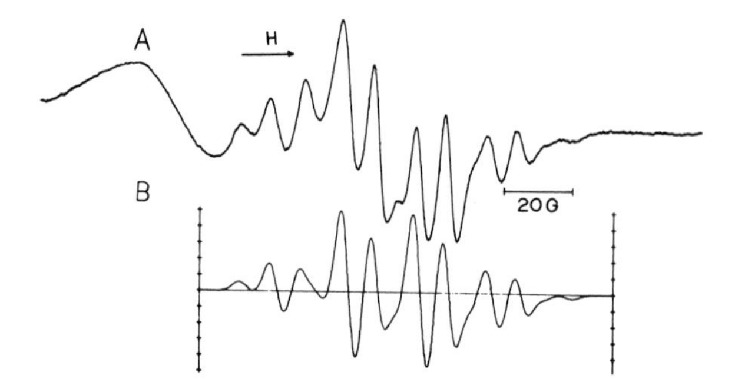

Figure 4. (A) The first derivative ESR spectrum of the TMP (thymidine-5'-monophosphate) π-cation radical in a basic 8M $NaClO_4$-D_2O glass at 115 K, produced by photoionization. The O^- spectrum accounts for the broad low field and the central underlying components. (B) Computer simulation of the π-cation spectrum in A employing parameters in text.

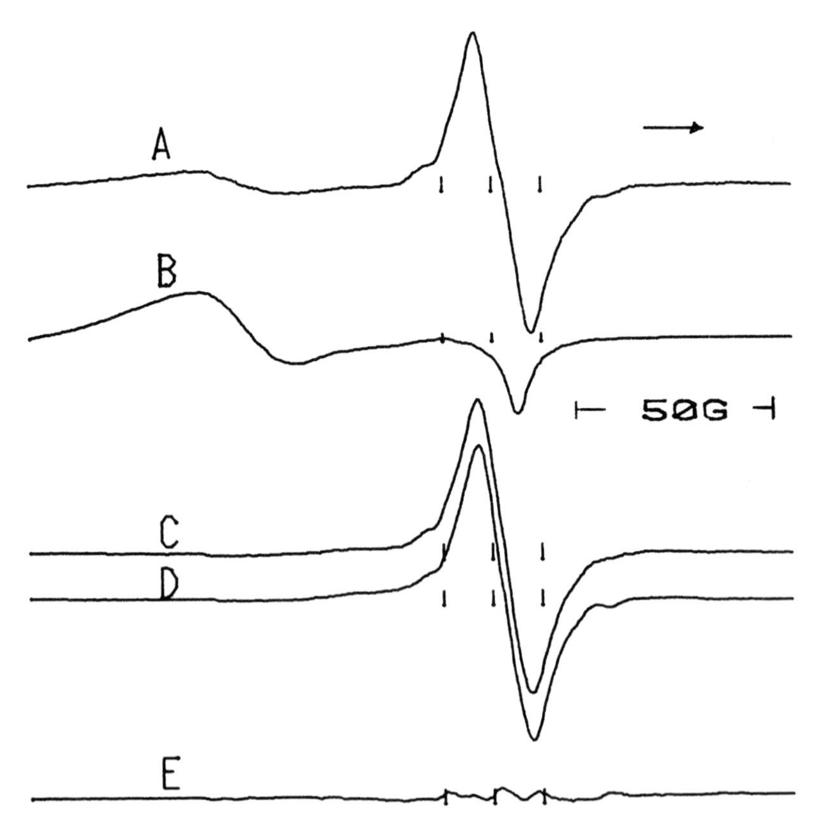

Figure 5. (A) ESR spectrum of salmon sperm DNA after photoionization in 8 M $NaClO_4$ (B) The spectrum of O^- in 8 M $NaClO_4$ (C) A-0.5 B (D) The spectrum of the π-cation of GpG in 8 M $NaClO_4$ (E) C-D

This combined with the fact that only the guanine π-cation is thought to be present in γ-irradiated DNA, points to hole transfer within the DNA strand.

E. Reactions of the Ion Radicals in DNA and DNA Constituents

The results presented here and those of previous workers suggest that the initial localization of charge on the DNA strand after an ionization event is likely to place the excess electron on thymine and the hole on guanine. The reactions which take place subsequently will of course depend on environment. However results with DNA and DNA constituents have shown the following reactions occur in aqueous matrices at low temperature.

1. Protonation of the thymidine anion at the 6-carbon position to form TH· .[6,10,16]

2. Protonation of the dAMP and dGMP anions at the 8-carbon position to form AH· and GH· .[5]

3. Protonation of solely thymine in dinucleoside phosphate anions containing thymine and the other DNA bases to form TH· .[7]

4. Deprotonation of the thymidine cation methyl group to form T(-H)· .[6]

5. Hydroxyl ion addition to the thymidine cation at the 6-carbon position to form TOH· .[4,8]

6. Hydroxyl ion addition to the dAMP and dGMP cations at the 8-carbon position to form AOH· and GOH· .[5]

In γ-irradiated DNA the TH· radical has been observed by a number of workers and has been shown to have the thymine anion as its precursor.[2] There has been one report of the T(-H) radical in DNA,[29] however its presence is not as well established as TH·. Although the fate of the cation in DNA is unclear, the deprotonation and hydroxylation mechanisms noted above seem most likely to explain the subsequent reactions of the hole in DNA at this time.

Acknowledgement

The author would like to thank the United States Energy Research and Development Administration for support of this research.

References

1. A. Gräslund, A. Ehrenberg, A. Rupprecht and G. Ström, Biochem. Biophys. Acta, 254, 172 (1971).
2. A. Gräslund, A. Ehrenberg, A. Rupprecht, B. Tjälldin and G. Ström, Radiat. Res., 61, 488 (1975).
3. A. Gräslund, A. Rupprecht and G. Ström, Photochem. Photobiol., 21, 153 (1975).
4. S. Gregoli, M. Olast and A. Bertinchamps, Radiat. Res., 65, 202 (1976).
5. S. Gregoli, M. Olast and A. Bertinchamps Radiat. Res., 60, 388 (1974).
6. M. D. Sevilla, C. Van Paemel and G. Zorman, J. Phys. Chem., 76, 3577 (1972).
7. M. D. Sevilla, R. Failor, C. Clark, R. A. Holroyd and M. Pettei, J. Phys. Chem., 80, 353 (1976).
8. M. D. Sevilla and M. Engelhardt, Discussions of the Faraday Society, in press.
9. M. D. Sevilla and P. Mohan, Internat. J. Radiat. Biol., 25, 635 (1974).
10. A.van de Vorst, Int. J. Radiat. Phys. Chem., 6, 143 (1974).
11. C. Helene, R. Santus and P. Douzou, Photochem. Photobiol., 5, 127 (1966).
12. R. G. Shulman and R. O. Rahn, J. Chem. Phys., 45, 2940 (1966).
13. J. W. Longworth, R. O. Rahn, and R. G. Shulman, J. Chem. Phys., 45, 2930 (1966).
14. M. D. Sevilla, J. Phys. Chem., 75, 626 (1971).
15. M. D. Sevilla, J. Phys. Chem., 80, 1898 (1976).
16. R. A. Holroyd and J. W. Glass, Int. J. Radiat. Biol., 14, 445 (1968).
17. P. B. Ayscough, R. G. Collins and F. S. Dainton, Nature, 205, 965 (1965).
18. S. I. Chan and J. H. Nelson, J. Am. Chem. Soc., 91, 168 (1969).
19. P.O.P. Ts'o and S. I. Chan, Biochemistry, 8, 997 (1969).
20. J. Brahms, J. C. Maurizot and A. M. Michelson, J. Mol. Biol., 25, 481 (1967).
21. M. D. Sevilla, C. Van Paemel and C. Nichols, J. Phys. Chem., 76, 3571 (1972).
22. M. D. Sevilla and C. Van Paemel, Photochem. Photobiol., 15, 407 (1972).
23. G. Scholes in "Radiation Chemistry of Aqueous Systems," G. Stein Ed., Interscience, New York, N.Y., 1968.
24. N. Bodor, M. J. S. Dewar and A. J. Harget, J. Am. Chem. Soc., 92, 2929 (1970).
25. H. Berthod, C. Gressner-Prettre, and A. Pullman, Theor. Chim. Acta., 5, 53 (1966).

26. A. Dulcic and J. N. Herak, J. Chem. Phys., 57, 2537 (1972).
27. G. Hartig and H. Dertinger, Int. J. Radiat. Biol., 20, 577 (1971).
28. J. N. Herak and V. Galogaza, J. Chem. Phys., 50, 3101 (1969).
29. G. Hartig and H. Dertinger, Int. J. Radiat. Biol., 20, 577 (1971).

INFLUENCE OF Hg^{2+} ON THE EXCITED STATES OF DNA: PHOTOCHEMICAL CONSEQUENCES*

Ronald O. Rahn

Biology Division, Oak Ridge National Laboratory, Oak Ridge, Tennessee, 37830, U.S.A.

INTRODUCTION

Metal ions have been widely used as probes in excited-state studies of nucleic acids. One of the first demonstrations of triplet energy transfer in poly(A) and DNA was made by Bersohn and Eisenberg (1964), using Mn^{2+} as a triplet state quencher. These observations were extended by Eisinger and Shulman (1966) to include Co^{2+}, Ni^{2+}, and Cu^{2+}, all of which were shown to quench long-range triplet transfer in poly(A) without quenching the fluorescence. Hélène and co-workers (see Hélène, 1973) showed that similar quenching mechanisms operated in adenosine aggregates but that, in addition, Cu^{+2} quenched the adenosine fluorescence by singlet transfer to an adenosine-Cu^{2+} complex. Energy transfer methods, employing europium ions as energy traps have also been used to study fluorescence lifetimes and intersystem crossing yields of nucleic acid monomers at 25°C (Lamola and Eisinger, 1971).

Attempts by Eisinger (1966) and Sutherland and Sutherland (1969a), using Co^{2+} and Ni^{2+}, respectively, failed to show any effect of these metal ions on thymine dimerization in DNA. These experiments were, in part, designed to explore the possibility of dimerization taking place via a triplet mechanism. However, even when acetophenone is used as a triplet sensitizer of thymine in DNA, Mn^{2+} has no effect on dimerization (Rahn, unpublished results). Other metals known to interact with DNA bases but which have little influence on thymine dimerization include Pb^{2+}, Cd^{2+}, and Zn^{2+} (Rahn, unpublished results). Pt^{2+} binds to DNA but not to thymine

*Research supported by the Energy Research and Development Administration under contract with the Union Carbide Corporation.

B. Pullman and N. Goldblum (eds.), Excited States in Organic Chemistry and Biochemistry, 27-37.

and consequently exerts a minimal effect on thymine dimerization (Munchausen and Rahn, 1975).

The first demonstration of a metal ion substantially influencing photochemistry was made by Sutherland and Sutherland (1969a) who found that Cu^{2+} enhanced dimerization when bound to the phosphate backbone but quenched dimerization when attached to the bases (although subsequent unpublished studies by Rahn have failed to show significant dimer reduction by Cu^{2+}). In their paper, the Sutherlands also reported that Ag^{+} significantly enhanced the dimer yield. Rahn and Landry (1973) have subsequently conducted a detailed study of Ag^{+} binding to DNA and have shown that the enhancement of dimerization by Ag^{+} binding in both poly(dT) and DNA is accompanied by a parallel increase in the phosphorescence intensity measured at 77 K. It was proposed, therefore, that Ag^{+} induces a heavy atom effect leading to an increase in intersystem crossing and that dimerization occurs from the more heavily populated triplet state.

In this presentation, the influence of Hg^{2+} on the photochemistry and luminescence of DNA will be discussed. A previous report (Rahn and Landry, 1970) suggested that the binding of Hg^{2+} to bases other than thymine creates energy traps which results in thymine dimerization and phosphorescence being quenched. Additional evidence bearing on this mechanism including some recent room-temperature fluorescence measurements will be presented here.

PHOTOCHEMICAL STUDIES

The influence of Hg^{2+} on the yield of the cis-syn thymine dimer for various thymine containing polynucleotides ranging from native DNA to poly(dT) is given in Table I. Dimerization in poly(dT) is quenched about 2-fold by Hg^{2+} for irradiation at 254 nm, but there is little effect of Hg^{2+} on dimerization done with 280 nm irradiation. Since the absorbance of poly(dT) at 280 remains constant upon binding Hg^{2+} while the absorbance at 254 is reduced nearly 2-fold [similar to the absorbance changes observed for poly(U) by Yamane and Davidson (1962)], it is concluded that the quenching for 254-nm radiation mainly reflects a change in the absorbance and not a change in the quantum yield. Similar results were also obtained for apurinic acid, which is prepared from DNA by removing all the purines.

In contrast to poly(dT) and apurinic acid, the yield of dimers in native and denatured DNA is reduced upon mercuration by > 10-fold and ~4-fold for irradiation at 254 nm and 280 nm, respectively. It is concluded, therefore, that saturating amounts of Hg^{2+} reduce the quantum yield of dimerization at least 4-fold in either single- or double-stranded DNA but do not quench thymine dimerization in polynucleotides void in purines.

Table I

Influence of Hg^{2+} (r = 1) on Yields of cis syn Thymine Dimer in Various Thymine Containing Polynucleotides

Polynucleotide	λex (nm)	Fluence at 254 nm (J/m^2 X 10^{-3})	Yield of $\hat{TT}_1$	
			- Hg^{2+}	+ Hg^{2+}
Poly(dT)	254	2.0	23.1	12.4
	280	0.6	10.8	9.3
Apurinic Acid	254	1.0	5.3	3.0
	280	1.0	3.3	3.2
Native DNA	254	5.0	6.9	0.6
	280	2.5	6.5	1.7
Denatured DNA	254	10.0	9.0	0.8

The variation in the dimer yield as a function of r, the molar ratio of bound Hg^{2+} to nucleotide is shown in Fig. 1 for various polynucleotides. The yield in native DNA for 280-nm irradiation increases slightly at low r values, reaches a maximum at r ≈ 0.1 and then decreases with increasing r. From previous studies by Yamane and Davidson (1961) it is known that Hg^{2+} binds to all of the bases in DNA but preferentially binds to thymine at low r values. Since Hg^{2+} binds to two thymines simultaneously, displacing two protons from each of the N_3 positions, it is expected that thymine binding sites of the form

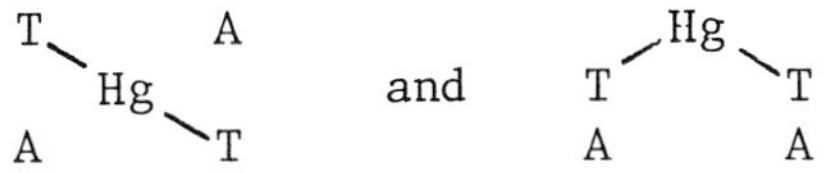

```
T       A                  Hg
  \   /                  /    \
   Hg         and      T        T
  /   \
A       T              A        A
```

will be saturated by r ≈ 0.1. However, only the latter contains thymines suitably arranged to form dimers. It is not understood why there is an increase in the dimer yield at r = 0.1 because the dimer yields in mercurated poly(dT) are unchanged for excitation at 280 nm (Table I). Complexing bases other than thymine with Hg^{2+} (r > 0.1) initiates a quenching of thymine dimerization, an indication that bases other than thymine when mercurated act as energy traps. The strong red shift in the absorbance of poly(A) (Yamane and Davidson, 1962) upon mercuration supports the notion that adenine when mercurated has a lower single state than thymine and can act as a trap.

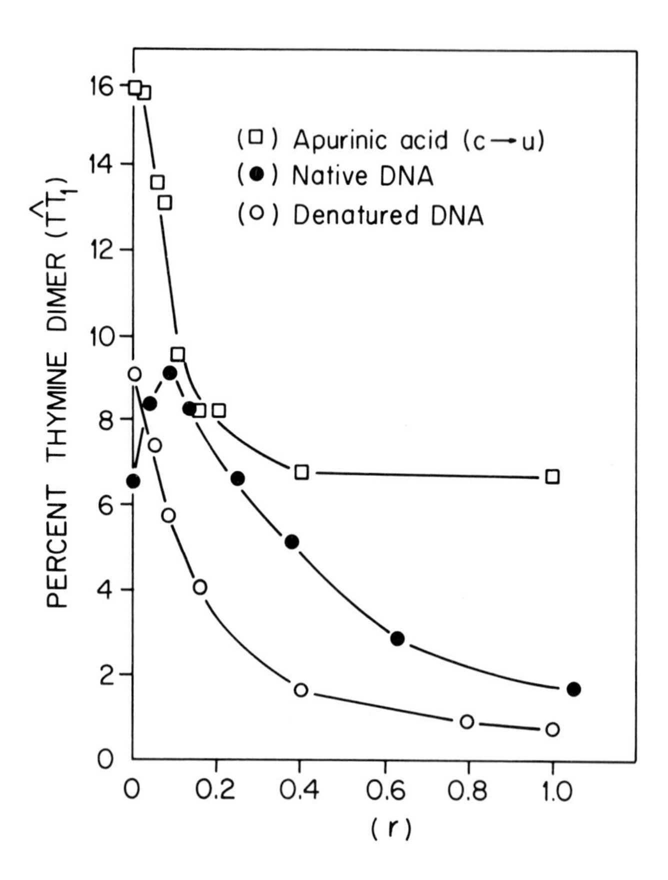

FIG. 1. Variation in the yield of cis-syn thymine dimer as a function of r, the molar ratio of $HgCl_2$ to DNA (E. coli) nucleotide. Native DNA was irradiated at 280 nm with 2500 J/m^2. Both denatured DNA and apurinic acid were irradiated at 254 nm with 10,000 J/m^2. The apurinic acid was deaminated, and the denatured DNA was in 50% ethylene glycol in order to unstack the bases.

In denatured DNA, quenching of dimers by Hg^{2+} occurs continuously for 254-nm irradiation starting from r = 0 without reaching a maximum, although at this irradiation wavelength the maximum is less pronounced for native DNA. The absence of a maximum is consistent with the fact that the irradiation is at 254 nm and the yield of dimers in the absence of other interactions should reflect the decreased absorbance of thymine at this wavelength. At r = 0.1 when all the available dimerizable pairs of the form —TT— are saturated, the yield is, as expected, reduced 2-fold, same as with poly(dT) or apurinic acid (shown in Fig. 1 for comparison). At higher r values, the yield of dimers in denatured DNA continues to

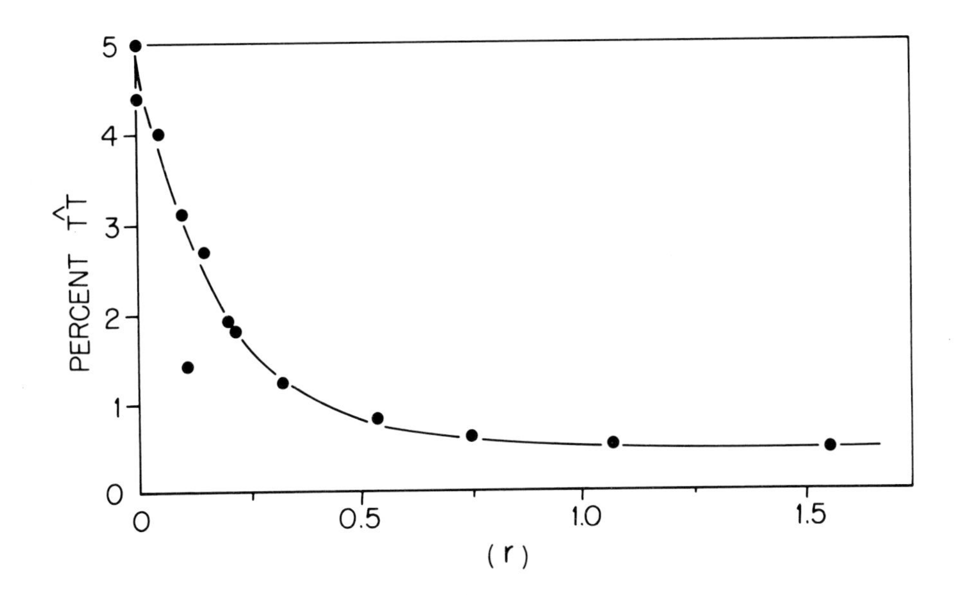

FIG. 2. Influence of Hg^{2+} on the yield of cis-syn thymine dimer (T̂T) in acetophenone sensitized E. coli DNA. Irradiation was 20,000 J/m^2 at 313 nm, in the absence of oxygen.

decrease corresponding to quenching by other mercurated bases acting as energy traps. It is sufficient to assume that Hg^{2+} quenching, in both native and denatured DNA, is probably short range in nature and takes place between a thymine pair and a neighboring mercurated base. The short range nature of the quenching is indicated by the lack of sharpness of the quenching curves.

Since the absorbance of DNA is significantly red shifted upon binding Hg^{2+}, it was of interest to measure the action spectrum for dimerization in completely mercurated DNA in order to determine whether absorption by residues other than thymine, which is only slightly red-shifted by Hg^{2+}, contributes to the yield of dimers. As shown in Fig. 2, the action spectrum of dimerization resembles the absorbance of poly(dT) both for the uncomplexed (Fig. 2a) as well as the complexed polymer (Fig. 2b). It is concluded, therefore, that dimerization only takes in mercurated DNA from photons absorbed by thymine and not by other bases.

In order to determine whether the Hg^{2+} quenching mechanism operates at the triplet level, triplet sensitized dimerization using acetophenone was studied. In this fashion, dimerization occurs from the excited triplet state of thymine, which is populated by energy transfer from an acetophenone triplet molecule. The sensitized dimer yields in poly(dT) were unaffected by mercuration.

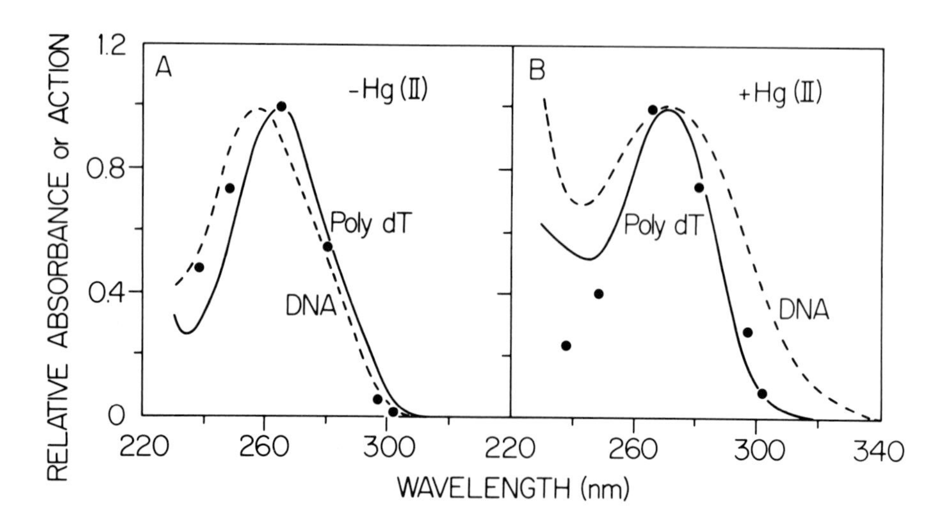

FIG. 3. Relative thymine, dimer yields in E. coli DNA at various irradiation wavelengths (•) as compared with the absorbance of DNA and poly(dT); (A) in the absence of Hg^{2+}, (B) in the presence of Hg^{2+} ($r = 1$).

However, as shown in Fig. 3, the sensitized dimer yields in DNA were quenched up to 5-fold by Hg^{2+}, indicating that mercuration of bases other than thymine results in triplet energy traps.

LUMINESCENCE RESULTS

Binding of Hg^{2+} to all of the polynucleotides studied quenches their fluorescence at 77 K completely. Phosphorescence quenching also occurs for poly(A) and poly(G) but the phosphorescence of poly(dT) and poly(C) is greatly enhanced by Hg^{2+}. A similar enhancement of the phosphorescence occurs with poly(dA-dT) as shown in Fig. 4, but only for r values less than $r = 0.25$ when preferential binding to thymine occurs. Values of r greater than 0.25 quench the phosphorescence, indicating that binding of Hg^{2+} to adenine quenches thymine triplets. Similar studies with $d(Tp)_4$ (Fig. 4) show that no quenching of the phosphorescence of the homotetranucleotide occurs at $r = 1$ and that the phosphorescence is identical to poly(dA-dT) ($r = 0.25$) except for a blue shift of 50 nm.

The phosphorescence of DNA also increases with increasing r (Fig. 5) and after reaching a maximum decreases markedly. As shown, the maximum phosphorescence intensity occurs at higher r values for those DNAs with higher thymine contents. This evidence, which suggests that thymine is the phosphorescent base, is corroborated in Fig. 6 by a comparison of the phosphorescence spectra of

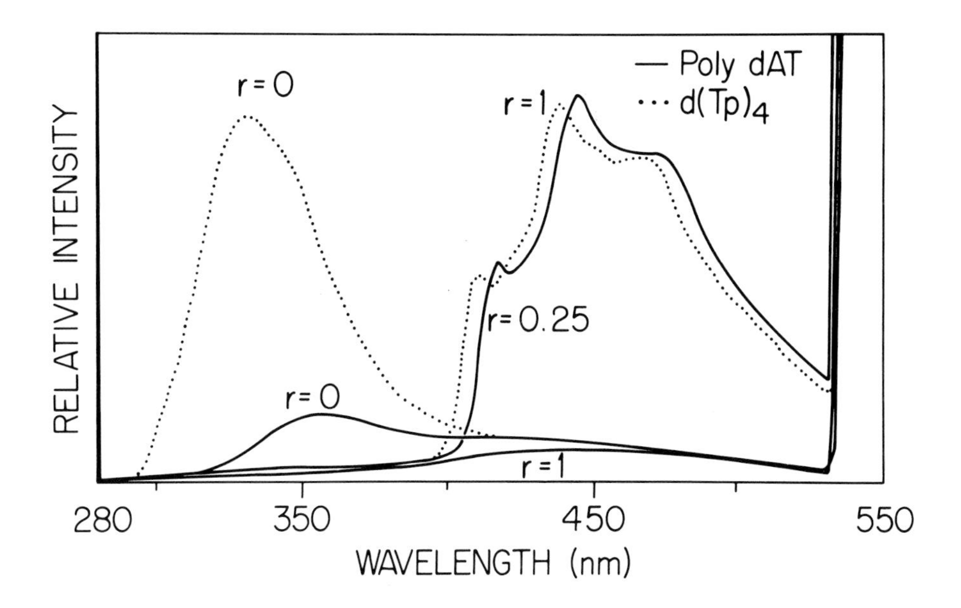

FIG. 4. Total luminescence spectrum at 77 K of poly(dA-dT)-Hg^{2+} complexes for r = 0, 0.25 and 1 as compared with that of $d(Tp)_4$ at r = 0 and r = 1.

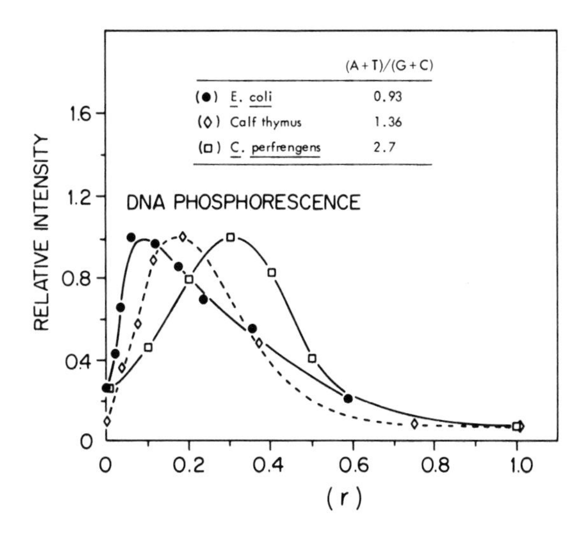

FIG. 5. Variation in the phosphorescence intensity as a function of r for three DNAs differing in their thymine content.

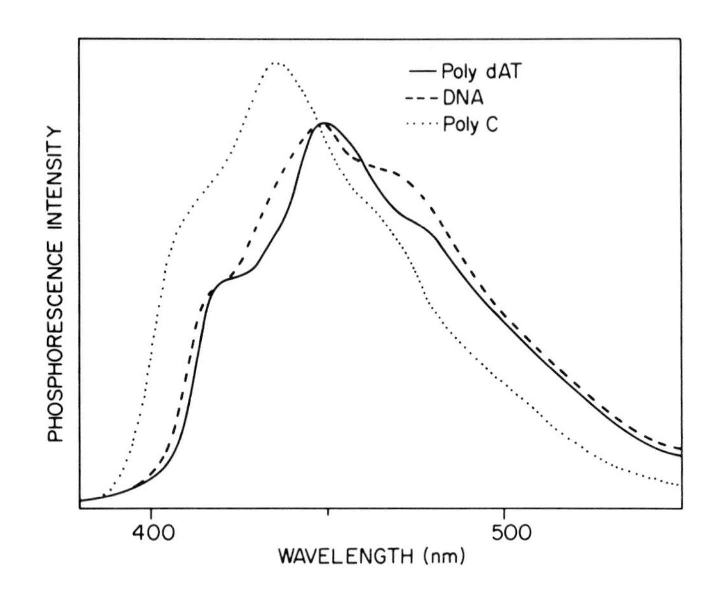

FIG. 6. Comparison of the phosphorescence spectra of poly(C) (r = 1), poly(dA-dT) (r = 0.25) and calf thymus DNA (r = 0.2).

DNA (r = 0.2) with poly(dA-dT) (r = 0.25) and poly(C) (r = 1). Clearly, DNA and poly(dA-dT) show similar spectral features while poly(C) has a distinctly blue-shifted spectrum. Since both A and G in DNA are presumed to have their phosphorescence quenched by Hg^{2+}, it is clear that it is the thymine phosphorescence which is maximized. A comparison of the phosphorescence curve for *E. coli* in Fig. 6 with that for dimerization in Fig. 1 shows that the changes in dimerization and phosphorescence as a function of r occur in a parallel fashion. Similar results were obtained with *H. influenzae* DNA.

Recently, we have examined the influence of Hg^{2+} on the room temperature fluorescence of various thymine derivatives. The purpose of these studies was to determine whether Hg^{2+} was indeed capable of quenching singlet states in these compounds under conditions used to study the photochemistry, i.e., at 25°C in aqueous solution. Work by Chen (1971) has indicated that Hg^{2+} can quench the room temperature fluorescence of tyrosine in proteins. However the lifetimes of aromatic amino acid at 25°C ($\sim 10^{-9}$ sec) are several orders of magnitude longer than those of the nucleic acids ($\sim 10^{-12}$ sec) (Lamola and Eisinger, 1971). Hence, one might anticipate that even if bound Hg^{2+} enhanced intersystem crossing [estimated by Hauswirth (1971) to be 2×10^9 sec^{-1} for thymine] by

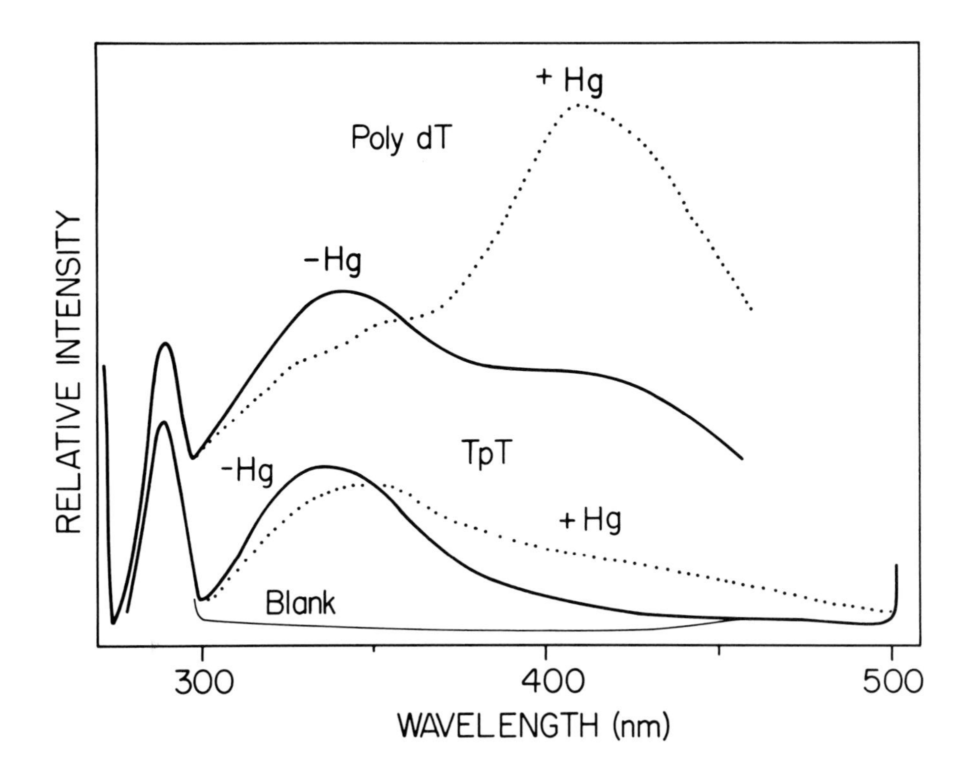

FIG. 7. Room temperature luminescence of TpT and poly(dT) with and without Hg^{2+} (r = 1) for excitation at 260 nm.

several orders of magnitude, there still would not be complete quenching of the singlet energy. Such is the case as indicated in Fig. 7 in which the effect of Hg^{2+} on the fluorescence of TpT and poly(dT) is shown. Only a slight decrease in fluorescence is noted in the mercurated samples. Interestingly, there is an increase in the emission in the region where phosphorescence would appear. Studies are being conducted to determine whether this emission is indeed associated with the thymine triplet state.

DISCUSSION AND SUMMARY

Bound Hg^{2+} quenches thymine dimerization and thymine phosphorescence in DNA only when bases other than thymine are mercurated. Triplet sensitization studies show that this quenching can occur at the triplet level. However, room temperature fluorescence studies indicate that only a fraction of the short-lived ($\sim 10^{-12}$ sec) thymine singlet state is quenched by Hg^{2+} enhanced intersystem crossing. Hence, at room temperature the most likely mechanism for quenching of photodimerization in DNA involves energy trapping at the singlet level by a neighboring mercurated base, presumably adenine which has the largest red shift in its absorbance upon complexing with Hg^{2+}. At 77 K, on the other hand, Hg^{2+} can completely quench the fluorescence of all the bases because of the longer singlet lifetimes. Consequently the quenching of the thymine phosphorescence occurs via triplet transfer, a result substantiated by acetophenone sensitization studies at room temperature

It has been shown (Rahn and Landry, 1971) that removal of the purines from DNA enhances the rate of thymine dimerization 2-fold. Presumably, a mechanism for deactivating thymine excited states exists which involves electronic interactions between thymine and neighboring purines. It is possible that this mechanism involves a charge transfer interaction and is enhanced when Hg^{2+} binds to the purine. The breakdown of this mechanism due to localized distortions of the DNA structure* may account for the maximum at r = 0.1 for 280-nm irradiation of native DNA. The absence of this mechanism at the triplet level may also explain, in part, why no maximum at r = 0.1 is obtained with acetophenone sensitization.

---

*Such distortions due to an axial chain-shift of one base spacing have been proposed by Katz (1963) to account for the cross-linking reaction of Hg^{2+} between two thymines located on opposite strands.

REFERENCES

Bersohn, R. and Isenberg, I. (1963) Biochem. Biophys. Res. Commun., 13, 205

Chen, R. F. (1971) Arch. Biochem. Biophys. 142, 552

Eisinger, J. (1966) in Electron Spin Resonance and the Effects of Radiation on Biological Systems (Snipes, W., ed), p. 76, National Academy of Sciences, Washington

Eisinger, J. and Shulman, R. G. (1966) Proc. Natl. Acad. Sci. U.S.A. 55, 1387

Hauswirth, W. (1971) Thesis, Oregon State University

Hélène, C. (1973) in Physico-chemical Properties of Nucleic Acids, Vol. 1 (Duchesne, J., ed.), p. 119, Academic Press, New York

Katz, S. (1963) Biochim. Biophys. Acta 68, 240

Lamola, A. A. and Eisinger, J. (1971) Biochim. Biophys. Acta 240, 313

Munchausen, L. L. and Rahn, R. O. (1975) Biochim. Biophys. Acta 414, 242

Rahn, R. O. and Landry, L. C. (1970) Proc. Natl. Acad. Sci. U.S.A. 67, 1390

Rahn, R. O. and Landry, L. C. (1971) Biochim. Biophys. Acta 247, 197

Rahn, R. O. and Landry, L. C. (1973) Photochem. Photobiol. 18, 20

Sutherland, B. M. and Sutherland, J. C. (1969a) Biophys. J. 9, 1329

Yamane, T. and Davidson, N. (1961) J. Amer. Chem. Soc. 83, 2599

Yamane, T. and Davidson, N. (1962) Biochim. Biophys. Acta 55, 780

A "HOT" GROUND STATE INTERMEDIATE IN THE PHOTOHYDRATION OF PYRIMIDINES

Shih Yi Wang

The Johns Hopkins University, School of Hygiene and Public Health, Division of Radiation Chemistry, Department of Biochemistry

Because the theme of this symposium concerns the excited state molecular species in organic and biochemistry and because you, the participants, are experts from diversified disciplines, this symposium provides an ideal forum for the discussion of a "hot" ground state intermediate in the photohydration of pyrimidines (Pyr). This is an issue which has confronted our laboratory ever since 1956. The general belief has been that the lifetime of a vibrationally excited or "hot" ground state Pyr ($^{V}Pyr_o$) is too short ($\sim 10^{-12}$ sec) for a photoreaction to take place. However, favorable evidence accumulated from our laboratory, as well as from others, has been considerable. Recently, this proposition has met with approval in the latest comprehensive review of Pyr photohydrates (Fisher and Johns, 1976). Therefore, first, I would like to discuss the nature of this photoreaction, second, the early background that led to my proposal, and third, the experimental findings which may be interpreted as being in favor of this proposal.

NATURE OF PYR PHOTOHYDRATION:

Upon absorption of a photon, a ground-state molecule of Pyr ($^{1}Pyr_o$) can be promoted to a higher energy level, excited singlet state $^{1}Pyr_1$, as shown in the following Scheme. At room temperature, the lifetime of $^{1}Pyr_1$ is only a few picoseconds (Hauswirth and Daniels, 1971) during which time $^{1}Pyr_1$ can undergo intersystems crossing (isc) by spin inversion, resulting in the formation of a triplet state Pyr molecule, $^{3}Pyr_1$. This triplet state has a lifetime of several microseconds, during which $^{3}Pyr_1$ can interact with a $^{1}Pyr_o$ to form a pyrimidine dimer (Pyr<>Pyr). $^{3}Pyr_1$ may also be populated by energy transfer through photosensitization from a donor molecule (sen), such as acetone, excited to its triplet state. When this is done, $^{3}Pyr_1$ is produced without any involvement of $^{1}Pyr_1$. Under such a condition, no Pyr hydrates ($ho^{6}hPyr$) could be detected (Greenstock and Johns, 1968).

B. Pullman and N. Goldblum (eds.), Excited States in Organic Chemistry and Biochemistry, 39-52.

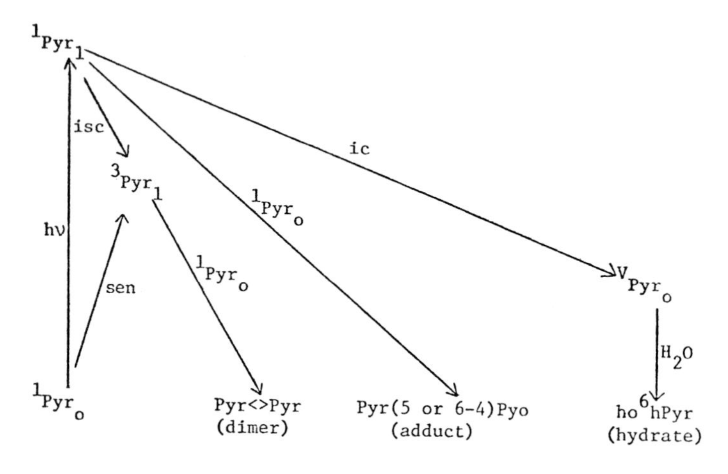

Reaction pathways for an electronically excited pyrimidine

In addition, the insensitivity of hydration rates to the presence or absence of oxygen and photosensitizer (Lamola and Mittal, 1966; Eisinger and Lamola, 1967; Greenstock et al., 1967; Nnadi, 1968) and of triplet quenchers (Burr and Park, 1967) have also been observed. These findings gave evidence to exclude $^3Pyr_1$ as the possible precursor for photohydrates leaving $^1Pyr_1$ as the alternative. Since $^1Pyr_1$ has also been indicated as the precursor for Pyr adducts [Pyr(5-4)Pyo and Pyr(6-4)Pyo; see Wang (1976a) for a review], it may be helpful to gain further knowledge on photohydration reactions by means of the study of Pyr adducts. However, such an approach is unfeasible because photohydrates are usually not formed under the conditions most favorable for the formaion of Pyr adducts, i.e. the irradiation of Pyr in frozen aqueous solutions. This can be explained in terms of solid state reactions in accordance with the "molecular aggregates-puddle formation" hypothesis (Wang, 1961; 1965; for reviews are Montenay-Garestier et al., (1976); Pincock, (1969)). Consequently, the evidence thus far gathered concerning the mechanistic aspects of photohydration has been, in most instances, generated by carrying out reactions in dilute aqueous solutions, as first described by Sinsheimer and Hastings (1949) and is illustrated as follows:

$$\text{Ura } (R_1, R_3 = H) + H_2O \underset{H^+, HO^-, \Delta}{\overset{h\nu}{\rightleftharpoons}} \text{ho}^6\text{hUra}$$

Photohydration of uracil

BACKGROUND FOR THE PROPOSAL OF A ZWITTERIONIC INTERMEDIATE:

In this reaction, Pyr is the only light absorbing substance, and both the solvent (H_2O) and the product (photohydrates) are non-absorbing. Thus, the number of moles reacted per unit time is

$$-d[Pyr]/dt = E\phi, \quad (1)$$

in which E represents the energy absorbed per unit time, and ϕ is the quantum yield. Since

$$E = AI_{abs},$$

in which A is the front surface area of the reaction vessel, and I_{abs} is the incident intensity absorbed per unit area per unit time, and then

$$E = A(I_o - I)$$

$$= AI_o(1-10^{-OD})$$

in which I_o is the incident intensity per unit area per unit time, I is the emergent intensity per unit area per unit time, and OD is the optical density of the solution at the wavelength of the incident radiation. Thus, Eq. (1) may take on the form of

$$-d[Pyr]/dt = AI_o(1-10^{-OD})\phi \quad (2)$$

In Eq. (2), only OD is time dependent. When OD is small, i.e. OD<<1, $1-10^{-OD} \cong OD \ln 10$, Eq. (2) becomes

$$-d[Pyr]/dt = k[OD] = k_1[Pyr]$$

which is linear in OD. Thus, at relatively low concentrations (OD<<1/ln 10), the absorption of incident light must be dependent on the OD or the concentration of the solution, [Pyr]. This absorption decreases steadily during the course of the reaction in proportion to the decreasing [Pyr]. As a result, such a photoreaction should be strictly first order.

When OD is large, i.e. OD>2, $1-10^{-OD} \cong 1$, Eq. (2) is reduced to pseudo zero-order kinetics, and becomes

$$-d[Pyr]/dt = k_2$$

which is independent of OD. At relatively high concentrations, the "complete" absorption of the incident light by the reaction solutions is expected. This, in turn, results in a constant decrease of the reactant and a constant production of the photoproducts during most of the course of the reaction. Such photolysis must yield strictly zero-order kinetics.

If photolysis is carried out with the OD of the solution ranging between 0.4 and 2, the photoreaction should exhibit rate constants between zero- and first-order kinetics. Therefore, it is characteristic for a light-dependent reaction to change its kinetic order of reaction due to the change in concentration (Wang, 1962; 1976b).

Unexpectedly, photohydration reactions of uridine (Urd) and 1,3-dimethyluracil (Me_2Ura) were found to follow first-order kinetics at all concentrations, even with OD>>2 (Wang, 1962; Nnadi, 1968; Wang and Nnadi, 1968). These findings can be interpreted as indicating that the rate-determining step is no longer light dependent. Furthermore, the quantum yields for these reactions were found with a great number of Ura derivatives (Nnadi, 1968; Wang and Nnadi, 1968) to vary with concentration of the solution, light intensity, temperature, etc. These characteristics are generally manifested in <u>dark</u> rather than <u>light-dependent</u> reactions.

Thus, a more detailed reaction scheme has been proposed (Wang and Nnadi, 1968) and is shown in Scheme 1.

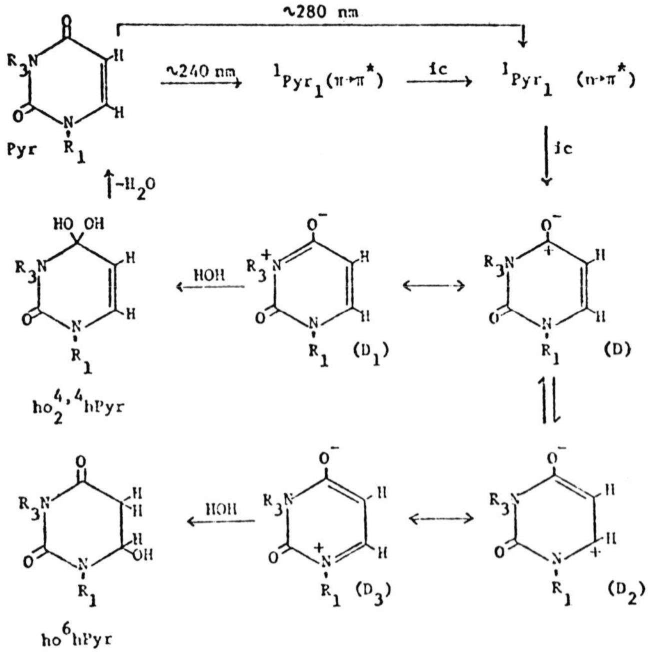

Proposed mechanism for the photohydration of pyrimidines

Upon absorption of 254 nm light, the Pyr molecule is electronically excited to a $(\pi \rightarrow \pi^*)\,^1Pyr_1$ which is rapidly deactivated by internal conversion to the lowest $(n \rightarrow \pi^*)\,^1Pyr_1$ (Kasha, 1960). Our suggestion of the intervention of the $n \rightarrow \pi^*$ state is based on the finding that quantum yields of photohydration are insensitive to the variation of wavelengths from 240-280 nm (Haug and Wang, 1962), as 280 nm light is most likely to promote a $n \rightarrow \pi^*$ excited state. This is supported by the observation that the quantum yields of photohydration of Ura and Urd are also independent of wavelengths between 230-280 nm (Brown and Johns, 1968). Formation of species D or D_2 from $(n \rightarrow \pi^*)\,^1Pyr_1$ is possible and D and D_2 could be further stabilized by resonance of D_1 and D_3. Such a zwitterionic intermediate was proposed (Wang _et al._, 1956; Wang, 1958; Moore, 1959; Wacker _et al._, 1964; Summer, Jr. _et al._, 1973) for these photoreactions and is theoretically justifiable (Zimmerman, 1963). D_1 and D_3 then undergo water addition to yield $ho_2^{4,4}hUra$ and ho^6hUra, respectively. While $ho_2^{4,4}hUra$ is unstable and readily reverts to the starting material, ho^6hUra is isolable and is the well-known photohydrate. In either case, a zwitterionic intermediate is suggested as shown, and a portion of its positive charge may delocalize through N(3)-C(4) or N(1)-C(6). Conceivably, the rate-limiting step could involve a vibrationally excited state species $(^VPyr_0)$ such as D_1 or D_3 for the formation of Pyr photohydrates.

SUPPORTING EVIDENCE:

N- and C-Substitution Effects:

If indeed the rate-limiting step is not light dependent, the general principles of organic chemistry rather than those of photochemistry should be applicable. Thus, the studies of substituent effects on the rates of hydration should yield information pertinent to our mechanistic understanding (Wang, 1962, Wang and Nnadi, 1968). Table 1 shows that the order of the decreasing rate of the four basic compounds is $Me^1Ura > Ura > Me^3Ura > Me^6Ura$ under neutral conditions. This clearly indicates that the rate of hydration is increased by N(1)-Me, decreased by N(3)-Me, and greatly suppressed by C(6)-Me as compared with Ura, the unsubstituted compound. Apparently, all N(1)-substituted Ura have greater rates of hydration than unsubstituted compounds. Table 1 also shows that the rates increase as the N(1)-substituents become more and more electron donating, suggesting that this increase parallels their abilities in stabilizing a positive charge on N(1). In moving from methyl to ribose-phosphate, the rates are 4 to 10 times that of Ura in water. On the other hand, N(3)-substituents decrease their hydration rates in proportion to their ability to stabilize a positive charge on N(3) is shown by the fact that $Me_2^{1,3}Ura$ and Ura-1,3-di-diethylmalonate undergo hydration at rates 14 and 32% slower than Me^1Ura and Ura-1-diethylmalonate, respectively.

Compound[a]	K x 10^3/min[b]	Relative Rate
6-Methyluracil	1.74	0.14
1,3,6-Trimethyluracil	6.1	0.50
3-Methyluracil	8.1	0.66
Uracil	12.2	1
1-Methyluracil	51.3	4.2
Uracil-1-acetic acid	64.4	5.3
Uracil-1-propionic acid	66	5.4
Uracil-1-propionamide	70	5.7
Uracil-1-ethyl acetate	83	6.8
Uracil-1-ethyl proprionate	82.5	6.8
Uracil-1-acetamide	86	7.0
Uracil-1-diethylmalonate	98	8.0
Uridine	121	9.9
Uridine-2'(3')-phosphate	126	10.3
1,3-Dimethyluracil	44	3.6
Uracil-1,3-di-diethylmalonate	66	5.4

[a]The compounds described were identified by elemental analyses, infrared and nuclear magnetic resonance spectroscopy. The hydration products can be isolated in over 85% yields.

[b]Precision within 1.8%.

Effect of substitution on the rate of hydration of uracil derivatives 0.1 mM in distilled water

These observations may be interpreted as that both D_1 and D_3 are the intermediates which compete for the addition of water (Nnadi, 1968; Wang and Nnadi, 1968). D_3 reacts to yield a stable photohydrate, ho^6hUra, and thus, results in a decrease in UV absorption of the solution proportional to the extent of the reaction. D_1, however, produces an unstable photohydrate, $ho_2^{4,4}hUra$ which reverts spontaneously to the starting material and, therefore, no apparent spectral changes can be detected. The latter reaction in competition with D_3 is manifested itself as a slower rate of hydration because the kinetics were analyzed in terms of spectral decreases. Me^6Ura and $Me_3^{1,3,6}Ura$ undergo hydration at a rate about seven times slower than Ura and $Me_2^{1,3}Ura$, respectively (Nnadi, 1968; Wang and Nnadi, 1968). This may be explained solely in terms of the steric hindrance effect operating at the reaction site.

Apparently, a large increase in the rates of hydration was also found for Ura substituted at N(1), such as Et^1Ura, Urd, and 1-cyclohexyl-Ura (Burr and Park, 1968; Burr et al., 1972). This trend holds when one examines the existing data for several frequently studied N(1)-substituted bases, nucleosides, and nucleotides (Sinsheimer and Hastings, 1949; Sinsheimer, 1954; Moore and Thomson, 1955; Shugar and Wierzchowski, 1957; Guschlbauer *et al*., 1965; Brown and Johns, 1968). In addition, N(3)-methylation of Ura, causing a reduction in the rates of photohydration has also been confirmed (Burr *et al*., 1972). These results further suggest the possible importance of a positive charge on N-atoms or of the D_1 and D_3 species in influencing the rates of Pyr photohydration.

pH Effects:

At pH 2, where protonation of N(1) is much favored, the hydration rate for Ura, Me^3Ura, and Me^6Ura was found (Nnadi, 1968; Wang and Nnadi, 1968) to approach that of N(1)-substituted Ura and was six, eight, and two-fold faster than at pH 7 (Table 2). Again, this indicates that quaternary-N(1) intermediates, i.e. D_3, are important. Furthermore, the relative rates at pH 2 are in general agreement with the electron donating and withdrawing abilities of the substituents.

		Rate[c] (10^3min./min.)					
Uracils[a]	H_2O	pH 2	pH 4	pH 6	pH 7	pH 8	Relative rate pH 2
6-Me	1·74	4·1[d]	3·7	1·7	2·1	1·8	d
1,3,6-Me_3	6·1	d	4·2	5·9	5·7	5·4	d
3-Me	8·1	61	18	8·1	8·8	6·0	0·95
Uracil	12·2	64	29	11·5	11·5	11	1·0
1-Me	51·3	52	51	51	51	42	0·80
1-CH_2CO_2H	64·4	87	62	61	56	52	1·36
1-$(CH_2)_2CO_2H$	66	84	70	64	55	50	1·30
1-CH_2CONH_2	70	72	70	70	62	58	1·14
1-$CH_2 \cdot CO_2Et$	83	83·5	83	82	74	68	1·30
Ethyl-1-propionate	82·5	81	83	72	71	63	1·29
1-Acetamide	86	91	88·7	86	83	80	1·42
Diethyl-1-malonate	98	106	105	123	—	120	1·65
Uridine	121	126	123	121	122	117	2·0
Uridylate	126	130	126	125	126	120	2·0
1,3-Me_2	44	d	e	44	44	41	d
Diethyl 1,3-di-malonate	66	e	e	66	89	e	e

[a]These compounds and hydration products (over 86% yields) were identified by elemental analyses, i.r. and n.m.r.

[b]Phosphate buffer system (0.1M) with KCl added to keep [Cl^-] constant. pH 2 buffer was HCl-KCl system.

[c]Precision within 1·8%.

[d]Hydration product is unstable at pH 2.

[e]No determination was made.

Effect of substitution and pH[b] on the rate of hydration of uracil derivatives

At pH 4-7, the rates of hydration for N(1)-substituted Ura were, as would be expected, generally unaltered (Wang and Nnadi, 1968).

Photohydration rates as functions of pH of Me^3Ura, FlUra, Up, pU, and Cyt derivatives have also been critically examined (Burr et al., 1972) and again these functions were found to be sigmoidal (Fig. 1). However, the inflection points of the curves were interpreted as reflecting the pK of the singlet excited state molecules, $^1Pyr_1$, for the loss of N(1)H (pK 4-5) or N(3)H (pK 6-7), respectively and were claimed to agree with the calculated excited state pK values [i.e. $pK^* = pK - 0.625/T(\Delta\nu)$, (Weller, 1961)]. Consequently, it was concluded that whereas the reactive state for photohydration does not resemble the ground state electronically, it does resemble a singlet excited state (Burr et al., 1972). This conclusion is in direct contrast to our proposal (Wang et al., 1956; Wang, 1958; 1962; Wang and Nnadi, 1968).

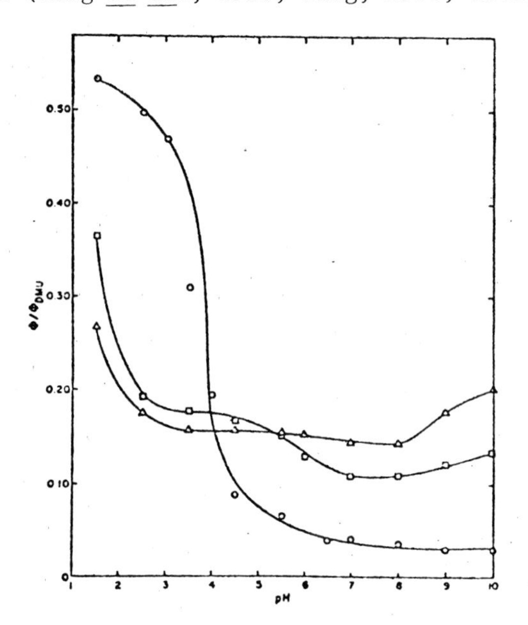

Quantum yields for photohydration of several uracil derivatives as functions of the pH, 10^{-4} M in oxygen-saturated water at 5°; uracil ⊙; uridine 5'-phosphate, ⊡; uridine 2'-phosphate, ⚠. (Burr et al., 1972.)

At the same time, experiments demonstrating that vibrationally excited state species of protonated Ura as possible precursors for photohydration have been reported (Khattak et al., 1972). Since the reaction must be carried out in highly acidic media and since the reversibility of Ura photohydrates at room temperature with pH <2 is well recognized, these studies have to be performed at low temperatures. It is well known that freezing a dilute acid solution results in the formation of concentrated "acid puddles" in the imperfect regions between ice crystals (Wang, 1961; 1965; Bruice and Butler, 1965) and that maintaining such a solution over dry-ice insures a temperature ∿ -70°C. Under such a condition pH 2 solutions resulted in >1000-fold increases in acidity in "acid puddles," UV irradiation of Ura, Me^1Ura, Me^3Ura, and $Me_2^{1,3}Ura$ produced corresponding photoproducts with quantum yields in the range of 0.2 - 0.6 (Khattak *et al.*, 1972). These values are 10 to 400-fold greater than those determined in water and strongly indicate that ground state Ura species with a positive charge on N-atoms are efficient in producing photohydrates.

Concentration Effects:

On the other hand, the studies of the effects of concentration (Fisher and Johns, 1976; a review) have resulted not only in quantitative but also qualitative differences. For a mechanistic understanding, however, only the qualitative aspects are of concern. The points of issue are 1) whether the quantum yield or photohydration rate is concentration dependent and 2) whether first-order kinetics remain at a higher concentration with OD>2 as discussed in B. Burr *et al.*, (1976) critically re-examined the photohydration of Me_2Ura and found that the quantum yield is a sigmoidal function of Me_2Ura concentration over the range of 0.1 - 1 mM. They also showed that the concentration dependence of Me_2Ura loss is not caused by the combination of a low yield photohydration process together with a high yield photodimerization process. Rather, it must represent a genuine concentration dependence in the supposedly first-order photohydration of Me_2Ura. These results confirm our early findings (Wang, 1962; Nnadi and Wang, 1968) and corroborate our proposed mechanism (Scheme 1).

Fluorescence Quenching Study:

Whitten *et al.* (1970) found that the quenching of fluorescence of Me_2Ura by water and a variety of other nucleophiles does not correlate well with the rate of photo-addition to the Pyr. This finding indicates that the fluorescent singlet state $^1Pyr_1$ is not involved in the addition reaction. Since $^3Pyr_1$ has been ruled out as a precursor, vibrationally excited ground state, $^VPyr_o$ remains as the only alternative.

Nucleophilic Center at C(6):

Molecular orbital calculations of the polarization of the 5,6-double bond of Pyr, i.e. the difference between the charge densities on C(5)-C(6), have also been carried out (Danilov, 1967; Malrieu, 1967). As Table 3 shows, a negative value indicates that C(6) is relatively positive, as in D_3, which favors nucleophilic addition or hydration. Hence, a positive value of polarization would be expected to lead to a

Compound	Polarization of 5,6-bond $^{1}Pyr_{1}$	$^{1}Pyr_{o}$	ϕ_H
Ura	-0.193	-0.122	$\sim 10^{-2}$
Cyt	0.083	-0.189	$\sim 10^{-2}$
Thy	0.158	0.052	$\sim 10^{-5}$

Molecular orbital calculations and quantum yields of photohydration

low hydrate yield. As Fisher and Johns (1976) have pointed out, the approximate ϕ_H for neutral Pyr correlates far better with the polarization of the 5,6-bond for $^{1}Pyr_{o}$ than with the values for $^{1}Pyr_{1}$.

Spontaneous and Light Induced Hydration:

In 2M or 8M D_2SO_4, nmr spectral changes showed that 4-methoxy-1-methyl-2-oxopyrimidine gradually converts to $Me^{1}Ura$ and also C(5)-deuterium exchange was observed (Hauswirth <u>et al</u>., 1972). A general mechanism to account for this type of reaction was proposed as shown in Scheme 2. Apparently, "dark" water addition could occur at C(4) or C(6) in highly acidic media parallel to those taking place with light-

Proposed mechanism for spontaneous and light induced hydration of pyrimidine derivatives

induced hydration of Pyr under neutral conditions. As can be seen, the release of CH_3OH results from the conversion of the C(4)-hydrate to $Me^{1}Ura$ and the C(5)-deuterium exchange is brought about by the dehydration of the C(6)-hydrate. While the rates of the CH_3OH release were not influenced by irradiation, the rates of the C(5)-deuterium exchange showed a 2.3 to 5-fold increase. These findings are indicative of the possible importance of ground state protonated species and involvement of a $^{V}Pyr_{o}$ in Pyr photohydration.

Cytidine C(5)-Photoexchange and D_2O Solvent Effects:

Consequently, it should be of interest to learn the exact nature of C(5)-deuterium exchange in the photohydration of Pyr. Previously, tritium release from [5-T]Cyd during irradiation at 265 nm has been reported (DeBoer and Johns, 1970). Recently, this exchange was found to be light-dependent and its quantum efficiency is of the same order of magnitude as that for Cyd photohydration (Hauswirth and Wang, 1977). In addition, the precursor for C(5)-photoexchange was shown to be excited singlet, $^1Pyr_1$, because the quantum yield is insensitive to incident UV wavelength (237 to 285 nm) and the inability to sensitize this exchange using acetone and 313 nm light, or to quench it with O_2 or Mn^{2+} (Hauswirth and Wang, 1974). However, when D_2O was used as a solvent, the quantum yields of both photohydration and photoexchange were reduced to the same extent (Table 4, Hauswirth and Wang, unpublished results). Therefore, both processes must share one and the same intermediate which is derived from $^1Pyr_1$. This intermediate is

Solvent	$\phi_{hyd} \times 10^2$	$\phi_{-T} \times 10^2$	I_T	$\phi_{exch} \times 10^2$
H_2O	1.95	0.53	4.9	2.60
D_2O	0.74	0.20		
decrease %	38	38		

Quantum yields of photohydration and photoexchange of cytidine

probably formed in the step preceding both processes by the reaction of a solvent molecule with the Pyr subsequent to irradiation. To account for the above, a mechanism has been proposed (Hauswirth and Wang, 1977) as shown in Scheme 3. The most likely structure of the intermediate is

C(5)-Photoexchange and Photohydration of Cytidine

also depicted. Such a structure may be unique and has been considered (Wang, 1959) to be connected with the photohydration of Cyt derivatives.

In conclusion, the evidence presented, together with possibly other findings, seems to indicate not only that $^{V}Pyr_{o}$ may be involved, but also that various resonance forms of D_3 may serve as the intermediate in the photohydration of Pyr derivatives.

ACKNOWLEDGEMENT:

I am most grateful to my coworkers who undertook this contraversial yet fascinating research problem with understanding, diligence, and courage. I also wish to thank USPHS-NIH, and USERDA-AEC for financial support during the past two decades.

REFERENCES:

I.H. Brown, and H.E. Johns (1968) Photochem. Photobiol. 8, 273.
T.C. Bruice, and A.R. Butler (1965) Fed. Proc., Fed. Amer. Soc. Exp. Biol. 24, S-45.
J.G. Burr, and E.H. Park (1967) Radiat. Res. 31, 547.
J.G. Burr, and E.H. Park (1968) Photochem. Photobiol. 81, 418.
J.G. Burr, E.H. Park, and A. Chan (1972) J. Amer. Chem. Soc. 94, 5866.
J.G. Burr, C. Gilligan, and W.A. Summers (1976) Photochem. Photobiol. 24, 483.
V.I. Danilov (1967) Photochem. Photobiol. 6, 233.
G. DeBoer, and H.E. Johns (1970) Biochim. Biophys. Acta 204, 18.
J. Eisinger, and A.A. Lamola (1967) Biochem. Biophys. Res. Commun. 28, 558.
G.J. Fisher, and H.E. Johns (1976) in "Photochemistry and Photobiology of Nucleic Acids. Chemistry" Vol. 1, Chap. 4, ed. S.Y. Wang, Academic Press, New York, N.Y. p.169.
C.L. Greenstock, I.H. Brown, J.W. Hunt, and H.E. Johns (1967) Biochem. Biophys. Res. Commun. 27, 431.
C.L. Greenstock, and H.E. Johns (1968) Biochem. Biophys. Res. Commun. 30, 21.
W. Guschlbauer, A. Favre, and A.M. Michelson (1965) Z. Naturforsch. B20, 1141.
A. Haug, and S.Y. Wang (1962) Experiments carried out at the University of Köln, Germany.
W.W. Hauswirth, and M. Daniels (1971) Chem. Phys. Lett. 10, 140.
W.W. Hauswirth, and S.Y. Wang (1977) Photochem. Photobiol. (in press).
W.W. Hauswirth, and S.Y. Wang (1974) 2nd Annual Meeting of the Amer. Soc. Photobiol., Vancouver, B.C., Canada, July 1974.
W.W. Hauswirth, B.S. Hahn, and S.Y. Wang (1972) Biochem. Biophys. Res. Commun. 48, 1614.
M. Kasha (1960) in "Comparative Effects of Radiation" ed. M. Burton, J.S. Kirby-Smith, and J.L. Magee, Wiley, N.Y. p.72.

M.N. Khattak, W.W. Hauswirth, and S.Y. Wang (1972) Biochem. Biophys. Res. Commun. 48, 1622.
A.A. Lamola, and J.P. Mittal (1966) Science 154, 1560.
J.P. Malrieu (1967) C.R. Acad. Sci., Ser. D264, 662.
T. Montenay-Garestier, M. Charlier, and C. Helene (1976) in "Photochemistry and Photobiology of Nucleic Acids. Chemistry" Vol. 1, Chap. 8, ed. S.Y. Wang, Academic Press, N.Y. p.382.
A.M. Moore (1959) Can. J. Chem. 37, 1281.
A.M. Moore, and C.H. Thomson (1955) Science 122, 594.
J.C. Nnadi (1968) Ph.D. Thesis, Johns Hopkins University, Baltimore, Maryland.
R.E. Pincock (1969) Account. Chem. Res. 2, 97.
D. Shugar, and K.L. Wierzchowski (1957) Biochem. Biophys. Acta 9, 199.
R.L. Sinsheimer (1954) Radiat. Res. 1, 505.
R.L. Sinsheimer, and R. Hastings (1949) Science 110, 525.
W.A. Summers, Jr., C. Enwall, J.G. Burr, and R.L. Letsinger (1973) Photochem. Photobiol. 17, 295.
A. Wacker, H. Dellweg, L. Träger, A. Kornhauser, E. Lodemann, G. Türck, R. Selzer, P. Chandra, and M. Ishimoto (1964) Photochem. Photobiol. 3, 369.
S.Y. Wang (1958) J. Amer. Chem. Soc. 80, 6196.
S.Y. Wang (1959) Nature (London) 184, 184.
S.Y. Wang (1961) Nature (London) 190, 690.
S.Y. Wang (1962) Photochem. Photobiol. 1, 135.
S.Y. Wang (1965) Fed. Proc., Fed. Amer. Soc. Exp. Biol. 24, S-71.
S.Y. Wang (1976a) in "Photochemistry and Photobiology of Nucleic Acids. Chemistry", Vol. 1, Chap. 6, ed. S.Y. Wang, Academic Press, N.Y., p.295.
S.Y. Wang (1976b) ibid., Chap 1, p.1.
S.Y. Wang, and J.C. Nnadi (1968) Chem. Commun. p.1160.
S.Y. Wang, M. Apicella, and B.R. Stone (1956) J. Amer. Chem. Soc. 78, 4180.
A. Weller (1961) Progr. React. Kinet., 1, 189.
D.G. Whitten, J.W. Hopp, G.L.B. Carlson, and M.T. McCall (1970) J. Amer. Chem. Soc. 92, 3499.
H.E. Zimmerman (1963) Adv. Photochem. 1, 183.

DISCUSSION

MICHL :

I would like to know in more details how the "hot ground state intermediate" mechanism yields a kinetic scheme which accomodates the concentration dependence of the reaction rate in the limit of high optical densities. Also, I would like to know whether a thermally equilibrated ground as possibly excited state intermediate such as perhaps a protonated pyrimidine could not fit such a kinetic scheme equally well. I would like to add that I see no fundamental physical objection to the occurrence of hot ground state reactions in dense media ; certainly, hot excited state reactions have been documented (e.g., R.V. Carr, B. Kim, J.K. McVey, N.C. Yang, W. Gechartz, and J. Michl, Chem. Phys. Lett. 39, 57 (1976)).

WANG :

First, I must say that I am delighted in receiving your supporting statement and in knowning of the documentation of hot excited state reactions. As for the more detailed kinetic treatments, I suspect that both treatments may lead to a similar equation in the form of

$$- d[Pyr]/dt = k_1 AI_o (1 - 10^{-OD}) [H^+][Pyr]$$

at high concentration, OD > 2, the equation will take on the form of

$$- d[Pyr]/dt = k_2 I_{abs} [Pyr]$$

EXCITED STATE INTERACTIONS AND ENERGY TRANSFER BETWEEN NUCLEIC ACID BASES AND AMINO ACID SIDE CHAINS OF PROTEINS

THERESE MONTENAY-GARESTIER

Muséum National d'Histoire Naturelle de Paris
Laboratoire de BIOPHYSIQUE 61, Rue Buffon 75005 PARIS France

and ENSIA au CERDIA 91305 MASSY - France

A precise knowledge of interactions in nucleic acid protein complexes requires a good understanding of excited-state properties and energy transfer processes in these systems. Weak interactions between identical or different organic species may be greatly enhanced by freezing aqueous solutions. Rapid cooling of an aqueous solution induces the formation of a microcrystalline ice structure. Solute molecules are excluded from the growing ice crystals and accumulate in the interstices of solvent (ice) crystallites where they form *aggregates*. Aggregate formation is responsible for many different phenomena involving constituents of proteins and nucleic acids.

- Photodimerization of thymine upon UV irradiation occurs very efficiently in frozen aqueous solutions whereas it is quite inefficient in fluid medium (Wang, 1961).
- Spectroscopic properties of nucleic acid derivatives in aggregates are markedly affected as compared to the properties of dispersed molecules in glassy media and are concentration dependent (Montenay-Garestier, 1973 a).
- Interactions of an electron donor-acceptor type occurring between two different ionic species of bases such as cytidine or adenosine and their cations or thymidine and its anion (Montenay-Garestier and Hélène, 1970 and 1973 b).
- Efficient energy transfers from nucleic acid constituents to energy traps such as metallic cations or aromatic amino acids which allowed us to determine their energy levels in this particular environment (Montenay-Garestier and Hélène, 1973 c).

The exact structure of aggregates is unknown but many arguments favor the idea of a microcrystalline state (Montenay-Garestier *et al.*, 1976). Energy transfer studies allowed us to

B. Pullman and N. Goldblum (eds.), Excited States in Organic Chemistry and Biochemistry, 53-64.

estimate the size of the aggregates. It was demonstrated that in the case of adenosine aggregates energy transfers at the triplet level to metallic cations or dyes involved up to 400 molecules (Hélène, 1973a). Therefore, molecules are *stacked* in aggregates induced by freezing down aqueous solutions. This proximity favors electronic interactions which cannot be observed in fluid medium. This particular and interesting behavior of solute molecules in frozen aqueous solutions was used in our studies to provide evidence for weak and specific interactions occurring between nucleic acid constituents and aromatic amino acids which could not be detected earlier in fluid medium. These electronic interactions might affect the properties of the ground and/or excited states and lead to different phenomena depending on their interaction energy : i) If strong interactions occur, absorption and emission modifications will be observed. ii) If weak interactions are involved, only emission properties will be affected.

We shall limit this review to the interactions of nucleic acid constituents with tryptophan, tyrosine and disulfide bridges. Results concerning phenylalanine or histidine can be found elsewhere (Montenay-Garestier and Hélène 1973 b).

INTERACTIONS BETWEEN NUCLEIC ACID BASES AND AROMATIC AMINO ACIDS: EXCITED-STATE PROPERTIES OF COMPLEXES.

- Tryptophan

Upon freezing an aqueous solution containing an equimolar mixture of tryptophan and one of the nucleosides absorption and emission properties markedly changed as compared to those of the separated components (Montenay-Garestier and Hélène, 1971). A new absorption and a new fluorescence appeared at longer wavelengths. These effects were much more important with pyrimidine derivatives than with purine ones. The microcrystalline state of the samples prevented us to measure their absorbance by standard transmission techniques. Absorption spectra were deduced from reflectance spectra. Upon addition of increasing concentrations of nucleoside or tryptophan, the change in absorbance at a given wavelength allowed us to determine the stoichiometry of complex formation. A 1 : 1 stoichiometry was found in neutral medium and a 2 : 1 complex (2 tryptophan molecules for one nucleoside) was formed in acidic medium when the nucleoside was protonated. At room temperature, only tryptophan emits fluorescence with a rather high quantum yield. Nucleic acids are only weakly fluorescent. At liquid nitrogen temperature in rigid medium, however, both tryptophan and nucleic acid derivatives exhibit fluorescence emissions in the same wavelength range.

Typical emission spectra obtained for mixture of tryptophan with four different nucleosides are presented on Fig. 1. Tryptophan and nucleoside fluorescences were strongly quenched in the mixture. A weak and broad emission appeared at longer wavelengths. The quantum yield of this new fluorescence is higher with pyrimidines than with purines. The red-shift is also larger with pyrimidines than with purines. This indicates that the lowest excited state of the complex is lower than that of the individual

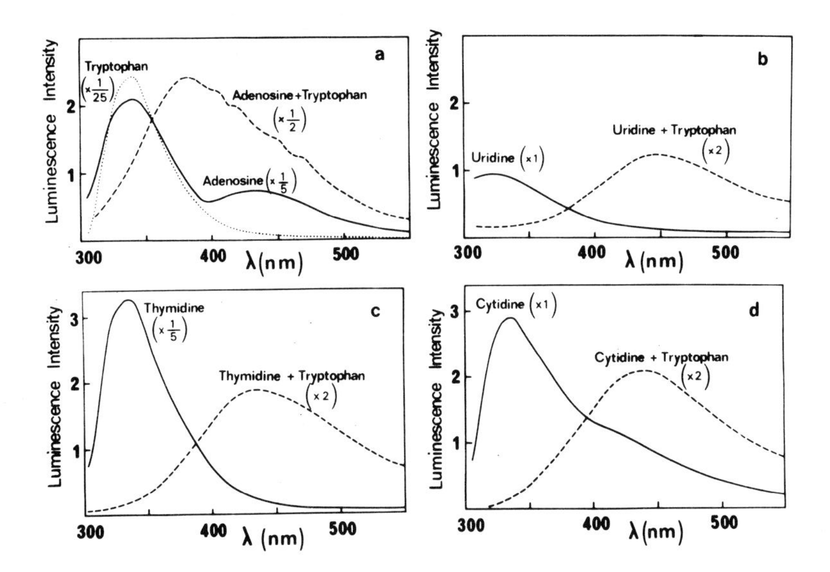

Figure 1 : *Luminescence spectra of frozen aqueous solutions (5mM) at 77 K of a : tryptophan (····), a : adenosine, b : uridine, c : thymidine, d : cytidine (——), and their equimolar mixture at pH 7(— — —). Excitation wavelength was at 280 nm.*

components. The complex of tryptophan with protonated cytidine or adenine did not emit fluorescence at all. Only phosphorescence was emitted. To determine the respective role of the indole ring and the amino acid chain, a number of derivatives were investigated. All the experimental results are consistent with the involvment of the indole ring as electron donor and the base moiety as acceptor in the EDA complex formed. Neither the phosphate nor the ribose are involved. As expected (Mataga and Murata, 1969) the charge-transfer interaction between the two molecules is much more important in their excited state than in their ground state. Thus fluorescence is always more affected than absorption.

From the value of the fluorescence maximum wavelength and from the red-edge of the reflectance spectra, the electron affinities of bases were shown to increase in the following order

$$\text{Gua} < \text{Ade} < \text{Cyt, Ura, Thy} < \text{Gua H}^+ < \text{Ade H}^+ < \text{Cyt H}^+$$

As expected electron transfer to the protonated bases is enhanced (in acidic medium). It is important to emphasize that these results have been thoroughly confirmed by NMR and absorption studies of complex formation between tryptophan and nucleic acid derivatives in concentrated fluid aqueous solutions (Dimicoli and Hélène, 1971). When interactions could be detected in frozen dilute aqueous mixtures interactions of the same nature could be demonstrated in fluid medium at much higher concentrations. Due to the property of frozen aqueous solutions of inducing aggregate formation, much lower concentrations are required as compared to fluid solutions.

Fluorescence studies performed on tryptophan-uridine mixtures at different temperatures in a fluid ternary hydroalcoholic mixture also provided evidence for a new charge-transfer fluorescence band *above* the freezing point (Hui Bon Hoa and Douzou, 1970).

Thus, when tryptophan is stacked with a nucleic acid component, fluorescence emission of both components is quenched at low temperature.

- Tyrosine

When an equimolar mixture of thymidine and tyrosine was frozen, fluorescence of both tyrosine and thymidine was strongly quenched. No red-shift could be observed. In equimolar mixtures with purines, tyrosine emission was quenched while fluorescence of purine nucleosides was not markedly affected. It should be noted that purine derivatives have much higher extinction coefficients than tyrosine in the same wavelength range so that their screening effect may contribute greatly to the apparent quenching of tyrosine fluorescence.

Methylation of the hydroxyl group of tyrosine led to the appearance of a new fluorescence at longer wavelengths in presence of pyrimidine nucleosides. With purine nucleosides, the fluorescence of p-methoxyphenetylamine was quenched whereas purine emission was only slightly affected. All the foregoing results have been ascribed to the formation of *EDA complexes* between phenol and pyrimidine rings as previously proposed to explain the quenching in the case of tryptophan (Hélène *et al.* 1971). The different behavior of pyrimidine and purine bases is consistent with the fact that pyrimidines are better electron acceptors than purines.

- Disulfide group

It seemed to us interesting to investigate interactions of nucleic acid bases with compounds containing disulfide group (such as cystamine) as models of cystine residues of proteins. These base-disulfide interactions could be involved in the photochemical behavior of protein-nucleic acid complexes. At 77 K, upon addition of increasing concentrations of cystamine to adenine aqueous solutions, a *quenching of adenosine phosphorescence* was observed, together with a slight red-shift of the fluorescence maximum. To elucidate the respective contributions of the disulfide bridge and of the electrostatic interactions, the emission behavior of adenosine aggregates in presence of hexamethylenediamine (HMDA) was investigated (hexamethylenediamine has a skeleton analogous to that of cystamine but with two methylene groups replacing the disulfide bridge). A marked increase of the phosphorescence was observed instead of the quenching with cystamine and the vibrational structure was retained (Montenay-Garestier *et al.*, 1976).

Interactions of cystamine with poly(A) could also be demonstrated in mixtures of ethylene glycol and water (1 : 1 v/v) which gave a transparent glass at 77 K. Addition of cystamine to the solution of poly(A) before freezing led to modifications of poly(A) luminescence emission similar to those observed in in the case of adenosine aggregates : strong phosphorescence quenching, slight red-shift of the fluorescence together with a loss of vibrational structure. Addition of 0.1 M NaCl to the cystamine-poly(A) solutions before freezing restored the vibrational structure of the fluorescence band, the maximum position and the initial intensity of the phosphorescence spectrum. This result demonstrated the important role played by electrostatic interactions in the cystamine-poly(A) complex. From the initial slope of the phosphorescence quenching curves, we calculated that one added cystamine molecule was able to quench 50 adenine bases in poly(A) and 14 in adenosine aggregates. Since the amount of bound cystamine molecules remained unknown these values represent a lower limit of the quenching efficiency of cystamine.

Cystamine has a higher singlet state energy than the triplet state of adenosine so triplet-singlet transfer can be ruled out. Its triplet state energy is not known so a triplet-triplet energy transfer could possibly explain the poly(A) phosphorescence quenching. But disulfides are known to act as electron donors or electron acceptors. Any of these mechanisms requires a close proximity of disulfide groups and nucleic acid bases and can explain the observed interaction.

ENERGY TRANSFER PROCESSES

Together with modifications of emission spectra, electronic interactions between molecules may lead to energy migration. As already noticed in the case of adenosine aggregates this process can occur over a large number of molecules. At the singlet level as well as at the triplet level, excitation energy transfer will proceed according to decreasing energies of the excited states.

For aromatic amino acids and nucleic acid components the energy levels are reported in table 1.

TABLE 1

	Singlet energy (cm^{-1})	Triplet energy (cm^{-1})
Tyrosine	37,000	29,000
AMP*	35,200	26,700
UMP*	34,900	-
TMP*	34,100	26,300
GMP*	34,000	27,200
CMP*	33,700	27,900
Tryptophan	34,400	25,000

*from Guéron *et al.* 1967.
Energy of the lowest excited singlet and triplet states of nucleotides and aromatic amino acids determined in a water-ethylene glycol glass at 77 K.

Predictions can be made from the values given in table 1. Tyrosine should be able to transfer its excitation energy to nucleic acid bases both at the singlet and at the triplet level. Tryptophan has the lowest triplet state and bases should be able to transfer their excitation energy at the triplet level to tryptophan.

Two main different mechanisms of energy transfer from a donor to an acceptor molecule may take place :
- *A long range* process according to Förster theory (Förster, 1965) involving the singlet states and-*a short range* process (Dexter 1952) involving an exchange mechanism at the triplet level. This latter process requires a very close proximity of molecules since partial overlap of the electronic clouds of both partners is needed. To fullfil this condition, donor and acceptor molecules have to be stacked. If many molecules are stacked, e.g., in aggregates, this transfer can proceed step by step (hoping model) and lead to energy migration over a large number of molecules.

Förster's theory (1965) allows us to evaluate a distance R_o at which the rate of energy transfer from the donor D to the acceptor A is equal to the sum of rates of all other modes of deactivation

$$R_o^6 = 8.9 \times 10^{-25}\ K^2 \cdot n^{-4} \cdot \phi_D \cdot J$$

where J is the overlap integral

$$\int_o^{\infty} F(\bar{\nu}) \cdot \varepsilon(\bar{\nu}) \cdot d\bar{\nu}/\bar{\nu}^4$$

$F(\bar{\nu})$ is the fluorescence quantum spectrum of the donor normalized to unity on a wavenumber scale (m^{-1}), $\varepsilon(\bar{\nu})$ is the molar decadic extinction coefficient of the acceptor, n is the index of refraction of the medium intervening between the donor and the acceptor, K is the dipole-dipole orientation factor and ϕ_D the donor emission quantum yield. The index of refraction is taken equal to 1.333 for aqueous solutions at room temperature and to 1.359 for ethanolic glasses at liquid nitrogen temperature. All the results given below (table 2) have been calculated for an average value of $\overline{K}^2$ equal to 2/3. A discussion of appropriate values of $\overline{K}^2$ depending on the relative expected orientations of the donor and the acceptor has been published (Eisinger and Dale 1974, Dale and Eisinger, 1974). For these calculations (Montenay-Garestier, 1975), several conditions are assumed :

- The interaction between the donor and the acceptor can be classified as the "very weak coupling" case of Förster theory.
- The dipole approximation is valid.
- Vibrational relaxation to the lowest excited state of the donor is fast compared with the energy transfer (*after-relaxation transfer*, Guéron *et al.* 1967). However in the case of tryptophan solvent reorientation may take place in the excited state which leads to a broadened and red-shifted fluorescence spectrum in water at room temperature. So critical distances for transfer from tryptophan to nucleosides were also evaluated *before* solvent *relaxation* (table 2, 1rst column) using the fluorescence spectrum of tryptophan at low temperature in an ethanolic glass to calculate the overlap integral J.

Several cases must be distinguished depending on the environment, lifetime and fluorescence quantum yield of tryptophan
i) if solvent reorientation can take place around the excited tryptophan before transfer the calculated Förster distances are quite small : Table 2, 2nd column.
ii) if an isolated tryptophan tranfers its excitation energy to a nucleic acid base *before relaxation* the lifetime of the non-relaxed state has to be used. This lifetime is not known but the rotational correlation time for H_2O molecules (in the range 10^{-11} sec) may represent a good approximation. The relaxed state of tryptophan has a lifetime of about 4.10^{-9} sec. Since the overlap integral is two orders of magnitude *larger* before than after relaxation and the lifetime (and therefore the quantum yield ϕ_D) two orders of magnitude *smaller* , the critical distances R_o corresponding to this case are quite similar to those given in table 2, 2nd column in the case of *after relaxation* transfer.
iii) if the tryptophan residue is involved in a protein-nucleic acid complex where it cannot relax in its lowest excited state (i.e. it is not accessible to solvent molecules) then the actual lifetime or quantum yield must be used to evaluate R_o. In the first column of table 2, R_o has been calculated using a ϕ_D value of 0.13. From comparison of the first column and the second one it can be seen that energy transfer from tryptophan to bases will

NUCLEOSIDES	TRYPTOPHAN			TYROSINE	
	Room temperature (ϕ = 0.13)		Low Temperature (ϕ = 0.68)	Room temperature (ϕ = 0.14)	Low temperature (ϕ = 0.425)
	Before relaxation	After relaxation			
Guanosine	9.6	5.1	13.75	15.7	23.6
Cytidine	9	4.8	11.4	15.3	22.2
Thymidine	8.4	4.3	9.2	14.6	21.2
Uridine	negligible overlap integral			12.1	20.1
Adenosine				10.3	20

Förster critical distances (in Å) for energy transfer from tryptophan and tyrosine to nucleosides in water at room temperature and in ethanolic solution at 77 K.
The fluorescence quantum yields of tryptophan and tyrosine used are Chen's values at room temperature(1967) and Longworth's values at low temperature (1971).

be more probable before than after relaxation.
iV) in most proteins, the environment of tryptophan is such that partial relaxation of the excited state is possible. Therefore the emission spectrum will be intermediate between the fully relaxed emission spectrum at room temperature and the non relaxed emission spectrum observed in rigid medium at low temperature; thus R_0 values will be intermediate between those given in 1st and 2nd column of Table 2.

At low temperature, the overlap between the fluorescence spectrum of tyrosine and the absorption spectrum of nucleosides is more important than at room temperature. Förster distances are expected to increase accordingly (Table 2). Values for critical energy transfer distances from tyrosine to nucleosides at low temperature are higher than the corresponding distance between tryptophan and tyrosine previously calculated by Eisinger (1971) at 77 K.

TABLE 3

	Guanosine	Cytidine	Thymidine	Adenosine
ϕ_D	0.13	0.05	0.16	0.01
Tryptophan	12.5	13.9	15.2	9.5
Tyrosine	4.65	8.5	8.8	6

Förster critical distances (in Å) for singlet energy transfer from nucleosides to tryptophan and tyrosine at 77 K.
The fluorescence quantum yields used for nucleosides are values given by Eisinger and Lamola (1971).

Table 3 gives critical distance values for energy transfer from bases to tryptophan and tyrosine at 77 K. (Calculations relative to the transfer from bases to tyrosine lead to non-negligible values). Though these distances are within a factor of one-half of the transfer distances from tyrosine to bases, the transfer from tyrosine to bases is expected to be much more efficient since the transfer probability increases as the sixth power of the critical distance.

Due to the low quantum yield (10^{-4} - 10^{-5}) of nucleosides at room temperature (Vigny, 1977), the critical distances for energy transfer from bases to tyrosine will be negligible under these conditions.

Experimental investigations did not give any evidence for *singlet tranfer* from *tryptophan to nucleosides* in aggregates. In mixed aggregates transfer occurs either from tryptophan or from nucleosides to the EDA complexes (Montenay-Garestier and Hélène,1971).

When small amounts of nucleosides were added to an aqueous solution of *tyrosine* (10^{-3} M) before freezing the fluorescence emission was quenched whatever the nature of the *nucleoside*. Analysis of the fluorescence quenching curves led to the conclusion that a single nucleoside molecule was able to quench the fluorescence of about 70 tyrosine molecules stacked in aggregates. Although the excited-state properties of nucleoside-tyrosine complexes did not allow us to observe sensitized fluorescence of the nucleosides in tyrosine aggregates, it appeared likely that the quenching of tyrosine fluorescence observed at low nucleoside-tyrosine ratio ($r < 0.1$) was due to energy transfer from tyrosine to nucleosides. Förster critical distances for tyrosine singlet energy transfer are quite different from one nucleoside to the other (Table 2). Therefore it appears reasonable to assume that singlet excitation energy is migrating in tyrosine aggregates from tyrosine to tyrosine until it is trapped by an acceptor which can be a nucleoside interacting with its two nearest neighbors (nucleosides or tyrosine) rather than an isolated nucleoside. As previously described, the excited states of pyrimidines in tyrosine complexes are more affected than purine ones. Since fluorescence quenching of tyrosine is similar with pyrimidines and purines, it seems likely that in this last case singlet transfer from tyrosine to bases is the main process rather than transfer to tyrosine-nucleoside complex. At room temperature in systems where polynucleotides or nucleic acids are binding oligopeptides containing a tyrosine residue fluorescence quenching of the tyrosine residues is very likely due to a transfer at the singlet level from tyrosine to nucleic acid bases (see Hélène, this Symposium).

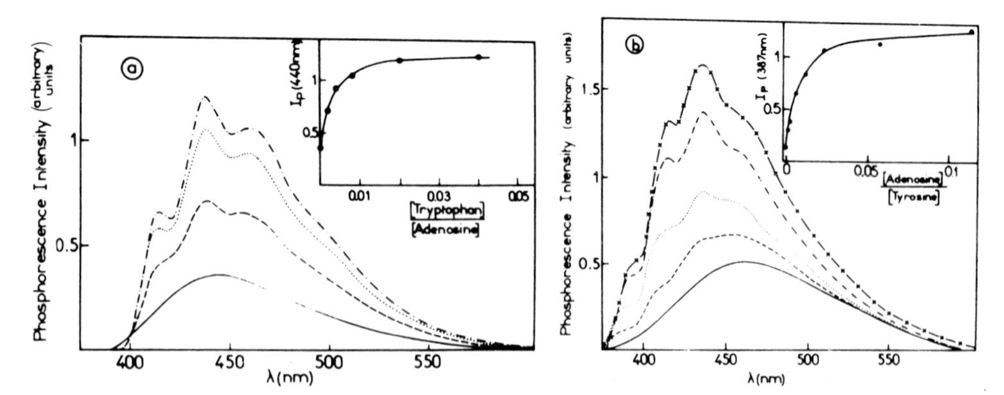

Figure 2 : *a) Phosphorescence spectra of Ado-tryptophan mixtures in aqueous solutions frozen at 77 K. The ratios of tryptophan and Ado concentration are 0 (—), 0.2% (---), 0.8% (···), and 2% (—·—·). Ado concentration is 2mM.* Inset : *change in phosphorescence intensity at 440 nm vs. the ratio of tryptophan and Ado concentration* (Hélène 1973 b).

b) Phosphorescence spectra of tyrosine-Ado mixtures in aqueous solutions frozen at 77 K. The ratios of Ado and tyrosine concentration are 0 (——), 0.2% (---), 1% (···), 5% (-·-·), and 10% (x-x-x). *Tyrosine concentration is 2mM.* Inset : *change in phosphorescence intensity at 387 nm vs. the ratio of Ado and tyrosine concentration (in the low concentration range).*

At the *triplet level*, very efficient transfers from *tyrosine to bases* and from *bases to tryptophan* should be expected. They both have been demonstrated in aggregates as shown on Figure 2. Excitation energy transfer to tryptophan involves about 200 adenosine molecules (Hélène, 1973a). Energy transfer to adenosine involves about 200 tyrosine molecules.

These transfers have also been demonstrated at low temperature in oligopeptide - nucleic acid systems (Hélène, 1973b). In the particular case of the poly(A) complex of luliberin, a decapeptide containing one tyrosine and one tryptophan residues an efficient triplet transfer from adenine bases to tryptophan takes place involving at least 60 adenine bases.

In the case of cystamine - poly(A) complexes, phosphorescence quenching of adenine can be either attributed to an energy transfer from adenine bases to cystamine or to an electron transfer from adenine to cystamine which is known as a good electron acceptor.

Aromatic amino acids (and cystamine) interact with nucleic acid bases. They form stacked complexes the spectral properties of which (and in particular excited state properties) are markedly different from those of the isolated components. Study of aggregates has been shown to be a useful tool to gain a better understanding about the behavior of these complexes.

In these systems efficient energy transfers may take place at the singlet or at the triplet level. Energy transfers in nucleoprotein complexes may occur from tyrosine to bases at the singlet level or from bases to tryptophan at the triplet level and these processes can be superimposed upon energy transfers taking place in each of these macromolecules. The knowledge of excited state properties is an essential step in the study of these complexes and is also fundamental to the understanding of the effects of UV radiation upon living systems.

- Chen R.F. (1967) Anal. Letters 1, 35-42.
- Dale R.E. and J. Eisinger (1974) Biopolymers 13, 1573-1605.
- Dexter D.L. (1952) J. Chem. Phys. 21, 836-850.
- Dimicoli J.L. and C. Hélène (1971) Biochimie 53, 331-345.
- Eisinger J. and A.A. Lamola (1971) in "*Excited States of Proteins and Nucleic Acids*" (Edited by R.F. Steiner and I. Weinryb) pp. 107-198. Plenum Press. New York.
- Eisinger J. and R.E. Dale (1974) J. Mol. Biol. 84, 643-647.
- Förster Th. (1965) in "*Modern Quantum Chemistry*" (Edited by O. Sinanoğlu) pp. 93-137. Academic Press. New York and London.
- Guéron M., J. Eisinger and R.G. Shulman (1967) J. Chem. Phys. 47, 4077-4091.
- Hélène C. (1973 a) in "*Physico Chemical properties of Nucleic acids*" (Edited by J. Duchesne) vol. 1, pp. 119-142. Academic Press. New York.
- Hélène, C. (1973b)Photochem. Photobiol. 18, 255-262.
- Hélène, C., T. Montenay-Garestier and J.L. Dimicoli (1971) Biochim. Biophys. Acta 254, 349-365.
- Hui Bon Hoa G. and P. Douzou (1970) J. Chim. Phys.Biol. 67 suppl. 197-203.
- Longworth, J.W. (1971) in "*Excited States of Proteins and Nucleic Acids*" Edited by R.F. Steiner and I. Weinryb), pp. 319-484. Plenum Press. New York.
- Mataga N. and Murata Y. (1969) J. Amer. Chem. Soc. 91, 3144-3152.
- Montenay-Garestier T. (1973 a) J. Chim. Phys. 70, n°10, 1379-1384.
- Montenay-Garestier T. (1975) Photochem. Photobiol. 22, 3-6.
- Montenay-Garestier T. and C. Hélène (1970) Biochem. 9, 2865-2870.
- Montenay-Garestier T. and C. Hélène (1971) Biochem. 10, 300-306
- Montenay-Garestier T. and C. Hélène (1973 b) J. Chim. Phys. 70, n°10, 1385-1390.
- Montenay-Garestier T. and C. Hélène (1973 c) J. Chim. Phys. 70, n°10, 1390-1399.
- Montenay-Garestier T., M. Charlier and C. Hélène in "*Photochemistry and Photobiology of Nucleic Acids*" (1976)(Edited by S.Y. Wang) vol. 1, pp. 381-417. Academic Press New York.
- Montenay-Garestier T., F. Brun and C. Hélène (1976) Photochem. Photobiol. 23, 87-91.
- Vigny P. (1977) This Symposium.
- Wang S.Y. (1961) Nature 190, 690-694.

MECHANISMS OF QUENCHING OF AROMATIC AMINO ACID FLUORESCENCE IN PROTEIN-NUCLEIC ACID COMPLEXES

Claude HELENE

Centre de Biophysique Moléculaire
45045 Orléans Cedex (France)

Introduction

Fluorescence spectroscopy has been widely used to obtain information on protein structure (1) as well as to study protein-ligand interaction (2). Both the intrinsic fluorescence of the aromatic amino acids and the extrinsic fluorescence of covalently or non-covalently bound labels have been used.

The specific recognition of nucleic acids by proteins is a central problem of molecular biology since this phenomenon is involved at every step of genetic expression. In most cases where intrinsic fluorescence has been used to investigate protein-nucleic acid interactions a quenching of the protein fluorescence has been observed. Some typical examples are given in table 1 (Fluorescence emission of nucleic acids is very weak at room temperature and does not interfere with that of aromatic amino acids (see P. Vigny, this Symposium). Most often proteins have tryptophyl residues and their fluorescence is dominated by tryptophan emission. Fluorescence investigations yield information on the behavior of these residues. In the case of staphylococcal nuclease, however, a quenching of tyrosine fluorescence by the inhibitor thymidine-3', 5'-diphosphate has been observed in the presence of the single tryptophan emission (which is not affected) (10). To our knowledge there is not a single case of phenylalanine fluorescence investigation for a protein interacting with a nucleic acid.

B. Pullman and N. Goldblum (eds.), Excited States in Organic Chemistry and Biochemistry, 65-78.

Table 1
Fluorescence quenching of Trp and Tyr residues in protein-nucleic acid complexes

Protein	Ligand	Observed Quenching	Ref.
Val tRNA ligase (E. coli)	$tRNA^{Val}$	Trp 20 %	(3)
Ser tRNA ligase (yeast)	$tRNA^{Ser}$	Trp 33 %	(4)(5)
Phe tRNA ligase (E. coli)	$tRNA^{Phe}$	Trp 12-15 %	(6)(7)
Glu tRNA ligase (E. coli)	$tRNA^{Glu}$	Trp 22 %	(8)
Arg tRNA ligase (B. stearothermophilus)	$tRNA^{Arg}$	Trp 12 %	(9)
Nuclease from Staphylococcus Aureus	pTp	Tyr 50 %	(10)
RNase T 1	3'-Gp	Tyr 50 % Trp 50 %	(11)
Gene 32 protein from phage T 4	poly(A) denatured DNA	Trp 20.5 % Trp 36 %	(12,13)
Gene 5 protein of phage fd	poly(dT) fd DNA }	Tyr 60-75 %	(14)
DNA binding protein (E. coli)	single stranded DNA	Trp 70 %	(15)

The purpose of this report is to analyze the different phenomena which are responsible for the quenching of aromatic amino acid fluorescence when a protein interacts with a nucleic acid. The binding of small oligopeptides to nucleic acids provides a convenient way of testing the different paths of excited state deactivation.

1. Conformational change in the protein and/or modification of protein-solvent interactions induced by nucleic acid binding

A change in the protein conformation is very likely to occur when a ligand binds to such a macromolecule. A conformational change in the protein may result in a modification of the environment of the aromatic amino acid even when it is far removed from the binding site of the ligand. Interactions of aromatic residues with neighboring groups in the protein is known to alter their fluorescence quantum yield. Therefore any modification in these local interactions will result in a change of fluorescence quantum yield. Also if ligand binding affects the accessibility of aromatic residues to the solvent, changes in fluorescence quantum yields may occur. However, there is no reason a priori to expect a decrease rather than an enhancement of the fluorescence quantum yield of aromatic residues. All possible situationscan be found in the litterature (enhancement, quenching or no change).

It will be beyond the scope of this report to list all the changes in fluorescence quantum yields which have been observed upon ligand binding to proteins. As mentionned above, however, quenching appears to be a general rule when the ligand is a nucleic acid. This is particularly striking in the case of aminoacyl-tRNA ligases interacting with cognate tRNAs where many different systems from different origins have been investigated (see table 1). In all cases, the tryptophan fluorescence of the protein is quenched upon tRNA binding. If this appears to be a general phenomenon, then tryptophan fluorescence quenching must be due to a more specific phenomenon than just a change in the protein conformation.

Different mechanisms can be proposed to explain fluorescence quenching of aromatic amino acids

i- direct interaction between the aromatic amino acid residue and one of the constituents of the nucleic acid (base, sugar, phosphate) : this may result from stacking with the bases, hydrogen bonding, electrostatic interactions...

ii- energy transfer from the aromatic residue to the nucleic acid bases which are only very weakly fluorescent in aqueous solutions at room temperature.

Each of these possibilities will be examined below for the different aromatic amino acids.

2. Quenching of tryptophan fluorescence

a) Stacking interactions with bases

Fluorescence quenching induced by stacking of tryptophan with bases was first demonstrated in frozen aqueous solutions (16, 17). In these frozen systems, mixed aggregates are formed whose minimum size was determined by utilizing the properties of triplet-triplet energy transfer (18). Solute molecules are stacked in these aggregates. When a tryptophan-base aqueous mixture is frozen the fluorescence of both components is quenched and a new fluorescence band is observed at longer wavelengths (17). A similar observation was also made for tryptophan-uridine mixtures in alcoholic solutions at low temperature (19). In model systems containing the indole ring and nucleic acid base covalently linked through a trimethylene bridge stacking was also shown to lead to indole fluorescence quenching at room temperature without any new fluorescence band appearing at longer wavelengths (20).

The quenching of tryptophan fluorescence due to stacking with nucleic acid bases was also shown to occur in tryptamine and oligopeptide - nucleic acid complexes (21-25). Proton magnetic resonance was used to demonstrate that stacking occured in such complexes (26). Stacking was clearly favored in single-stranded nucleic acids (single-stranded polynucleotides or single-stranded regions in tRNAs). Analysis of fluorescence data and comparison with PMR data revealed that fluorescence quenching was related to the concentration of stacked complexes .

The results obtained in aggregates led us to propose that tryptophan fluorescence quenching induced by stacking with bases was due to electron donor-acceptor (EDA) interactions. The indole ring is clearly the electron donor and the base the electron acceptor (17-18). This was confirmed by the observation that the new fluorescence band characteristic of the EDA complexes in aggregates was more red-shifted with protonated bases which are expected to be better acceptors than neutral bases.

b) Interactions with phosphate groups

It has already been reported that phosphate monoanions ($H_2PO_4^-$) are weak quenchers of tryptophan fluorescence while dianions ($HPO_4^{=}$) have no effect (27). These observations have been correlated with the electron scavenging properties of these anions. The indole ring is a good electron donor and solvated electrons are produced from the singlet state of the excited molecule.

In order to determine whether phosphate groups of nucleic acids could be possible candidates for the quenching of tryptophan fluorescence we have investigated the quenching efficiency of dimethylphosphate $(CH_3O)_2PO_2^-$ which is much more similar to the phosphodiester linkage in nucleic acids than phosphate monoanions ($H_2PO_4^-$). Dimethylphosphate has no effect on the fluorescence of N-Acetyltryptophanamide at pH 4. 5. At higher pHs (8. 0) a weak quenching is observed but this effect is always smaller than that due to $H_2PO_4^-$ at pH 4. 5 (J. J. Toulmé, unpublished results).

If proton transfer from the indole NH group was involved in phosphate anion quenching one would expect that phosphate dianions would be more efficient quenchers than monoanions (as observed in the case of tyrosine ; see below). This is not the case (27). It can be concluded that the phosphodiester group in nucleic acids is not expected to be an efficient quencher of tryptophan fluorescence in protein-nucleic acid complexes.

c) Hydrogen bonding interactions

Hydrogen bonding might take place between the NH group of the indole ring and any of the acceptor sites on the nucleic acid (bases, sugar, phosphate) although there is no available evidence for this interaction. The indole NH group has a high ground state pK ($\sim$ 16) and is not expected to be a strong hydrogen donor. However the excited state pK of indole is certainly much lower than that of the ground state. Hydroxyl ions (OH^-) do quench tryptophan fluorescence in a diffusion-controlled reaction leading to loss of the NH proton in the excited state (28-29). The indolate ion is only weakly fluorescent under these conditions. The quenching of tryptophan fluorescence by anions has been related to their electron affinity and not to their ability to abstract proton in the excited state (27). For example, acetate ions or phosphate dianions have no effect whereas phosphate monoanions have a small effect (see above). From these studies it does not appear likely that hydrogen bonding of the indole NH to any of the nucleic acid acceptor groups would lead to proton transfer in the excited state and thus to fluorescence quenching.

d) Energy transfer from tryptophan to nucleic acid bases

Due to the poor overlap between the fluorescence spectrum of tryptophan and the absorption spectrum of nucleic acid bases, energy transfer at the singlet level from Trp to bases is very unlikely. The Förster critical distances which have been calculated (30) are so small that under such conditions it would be expected that electron transfer would then be as likely to explain fluorescence quenching as energy transfer. However if tryptophan is able to transfer its excitation energy before relaxation of its excited singlet state, then the contribution of energy transfer to fluorescence quenching might become non negligible. This would be particularly true in a protein-nucleic acid complex if the tryptophyl residue is not accessible to relaxing solvent molecules.

When the nucleic acid contains a modified base, which absorbs at longer wavelengths than usual bases then energy transfer could be an efficient source of quenching of tryptophan fluorescence. This phenomenon could be important in the complexes formed by amino acyl-tRNA ligases with tRNAs which often contain modified nucleosides such as 4-thiouridine in several E. coli tRNAs, the Y base of $tRNA^{phe}$ from yeast...

3. Tyrosine fluorescence quenching

a) Stacking with bases

In the stacked aggregates formed in frozen aqueous solutions the fluorescence of tyrosine is quenched by nucleic acid bases (31). The fluorescence of pyrimidine bases is also quenched whereas that of purine bases is not markedly affected. In the former case, it appears that electron donor-acceptor interactions are responsible for fluorescence quenching even though these complexes do not show any new fluorescence band at long wavelengths (as observed, e.g., in the case of tryptophan complexes). In the latter case, the stacking interaction itself or energy transfer from tyrosine to the purine base in the stacked complex are equally likely to explain tyrosine fluorescence quenching.

There is no model system yet described analogous to those studied in the case of indole by Leonard and collaborators (20). The synthesis of such compounds in which a phenol group is linked to a nucleic acid base by a trimethylene bridge would certainly be of great interest. In the case of puromycin which contains an 0-methylated phenyl group linked to N, N-dimethylaminopurine the fluorescence of the former ring is quenched although it is not again possible to differentiate between energy transfer and stacking interactions as possible origins of the phenomenon.

In complexes formed between nucleic acids and tyrosine-containing oligopeptides (such as Lys-Tyr-Lys) the fluorescence of tyrosine is quenched (23) (table 2). In single-stranded polynucleotides or tRNA complexes, proton magnetic resonance spectra clearly show that the tyrosyl ring is stacked with the bases (32). This stacking interaction appears responsible for fluorescence quenching. However in double-stranded DNA complexes stacking interactions do not take place as shown by PMR studies. Nevertheless fluorescence is quenched as efficiently as in single-stranded polynucleotide complexes (table 2, unpublished results). Therefore it is clear that in those oligopeptide-DNA complexes that we have investigated, tyrosine fluorescence quenching is not due to stacking interactions.

Table 2
Apparent fluorescence quantum yields of tyrosine-containing peptides bound to poly(A) and DNA relative to that of the free peptide (Mayer R., Toulmé F. and Hélène C., unpublished results)

	poly(A)	DNA
Lys-Tyr-Lys	0.215	0.30
Lys-Tyr(OMe)-Lys	0.13	0.33
Lys-Tyr-Lys NHEt	0.235	0.24
Ac Lys-Tyr-Lys NHEt	0.255	0.23
Lys-Ala-Tyr-Ala-Lys NHEt	0.18	0.18

b) Hydrogen bonding interactions

The hydroxyl group of tyrosine is a good candidate for hydrogen bonding interactions with nucleic acid constituents. In chloroform solution p-cresol (the side chain of tyrosine) forms hydrogen bonds with nucleic acid bases (33). The ground state pK of the tyrosine hydroxyl group is around 10. But the pK of the first excited singlet state is much lower. Using the Förster cycle we have estimated a pK value of about 5 for this excited state. Excitation of tyrosine at pH 7 in aqueous medium does not lead to deprotonation because the excited state protonation equilibrium cannot be reached during the singlet lifetime. However if tyrosine is excited in the presence of good proton acceptors it may lose a proton in the excited state. Since the fluorescence quantum yield of the tyrosinate ion is very low, proton loss in the first excited singlet state will lead to fluorescence quenching. This hypothesis has already been put forward to explain the quenching of tyrosine fluorescence by phosphate dianions which occurs according to a diffusion-controlled process in aqueous solution (34). Phosphate monoanions are much less efficient quenchers of tyrosine fluorescence. Dimethylphosphate was used as a model compound to test the possible role of phosphate groups of nucleic acids in tyrosine fluorescence quenching. This compound did not quench tyrosine fluorescence but, on the contrary, led to a slight increase in quantum yield. It can thus be deduced that phosphate groups of nucleic acids are not expected to be involved in quenching the fluorescence of tyrosyl residues of proteins.

The fluorescence of tyrosine is quenched in the complexes formed by DNA with oligopeptides such as Lys-Tyr-Lys or Lys-Tyr (NH_2). As already reported above this is not due to stacking with the bases since stacking is not observed in these complexes. In order to test whether hydrogen bonding of the tyrosyl hydroxyl group could be important in the mechanism of fluorescence quenching, we have studied complex formation with the peptide Lys-Tyr(OMe)-Lys where the hydroxyl group of tyrosine is replaced by a methoxy ($-OCH_3$) group. The fluorescence of this peptide is quenched in DNA complexes with approximately the same efficiency as that of the non-methylated peptide (table 2). Proton magnetic resonance experiments revealed only very small upfield shifts of the methoxyphenyl protons thereby indicating that stacking of the aromatic ring with bases cannot account for the observed extent of fluorescence quenching. Therefore it can be concluded that neither stacking nor hydrogen bonding are responsible for tyrosine fluorescence quenching in the DNA-peptide complexes that we have investigated.

c) Energy transfer

Critical Förster distances for singlet energy transfer from tyrosine to nucleic acid bases have recently been calculated (30). They range from 10 to 17 Å at room temperature. Thus energy transfer appears as a likely source of tyrosine fluorescence quenching in protein-nucleic acid complexes if the tyrosyl residue is located close enough to the absorbing bases (10-20 Å). The transfer efficiency will of course depend not only on the distance but also on the relative orientation of the transition moments of tyrosine and of the base or base pairs.

Methylation of the hydroxyl group of tyrosine does not alter markedly the fluorescence spectrum of this molecule. Thus the efficiency of energy transfer is not expected to depend very much on whether the hydroxyl group is free or methylated. The nearly identical results that we have obtained with the two peptides Lys-Tyr-Lys and Lys-Tyr(OMe)-Lys (see above) could thus be explained as resulting from a similar structure of the complexes with DNA leading to similar quenching of the aromatic amino acid fluorescence.

Fluorescence quenching in the oligopeptide-DNA complexes could also be due to an interaction of the tyrosyl ring with some other group of the peptide itself as a result of a conformational change of the peptide upon binding to DNA. If this conformational change is the same for Lys-Tyr-Lys and Lys-Tyr(OMe)-Lys, then similar fluorescence quenching would be expected in both cases. However we have recently investigated the fluorescence behavior of several peptides with different amino acid sequences. For example Lys-Tyr-Lys and Lys-Ala-Tyr-Ala-Lys give very similar results independently of whether the terminal carboxyl and α -amino groups are substituted or not (by ethylamide or acetyl groups, respectively). Their fluorescence is quenched to a similar extent upon binding to DNA (table 2). It thus appears that a conformational change of the peptide bringing the tyrosyl group in close proximity to a quenching group inside the peptide itself is not very likely to explain tyrosine fluorescence quenching in peptide-DNA complexes.

4. Phenylalanine

Very few results have been published concerning phenylalanine fluorescence in the presence of nucleic acid bases. Due to its low extinction coefficient and its absorption at short wavelengths, it is difficult to measure phenylalanine fluorescence in the presence of nucleic acid constituents.

In frozen aqueous solutions at neutral pH, the fluorescence of purine and pyrimidine nucleosides is not affected by phenylalanine even at a 1:1 ratio. When cytidine is protonated at acidic pH a red shift of its fluorescence is observed in the presence of phenylalanine. In these aggregates the fluorescence of phenylalanine is so weak that it cannot be detected in the presence of nucleosides (18).

In the complexes formed at room temperature between polynucleotides or DNA and oligopeptides such as Lys-Phe-Lys, the phenylalanyl fluorescence is quenched (figure 1). The screening effect of the nucleic acid is of course very high at the excitation wavelength (260 nm). This reduces the apparent fluorescence intensity of the peptide by a factor as high as 100 (this of course depends on the nucleic acid absorbance). However this screening effect can be easily taken into account since Lys-Phe-Lys - nucleic acid complexes can be dissociated at high ionic strength and the fluorescence of the free peptide determined in the presence of the nucleic acid.

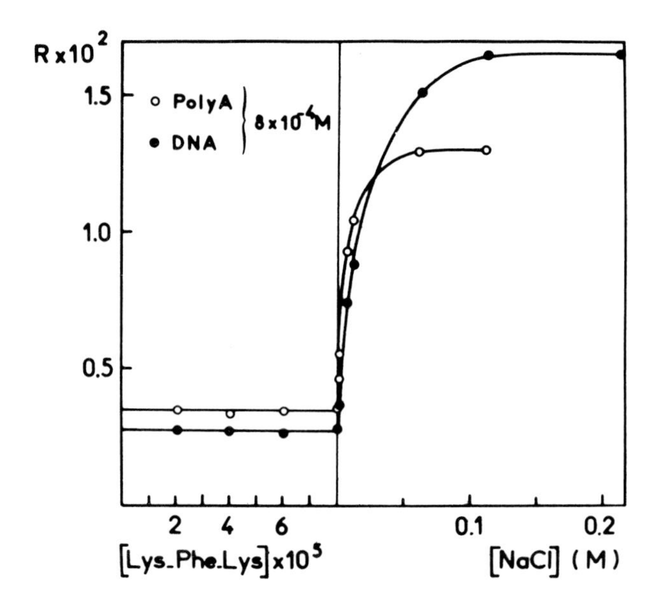

Figure 1 : Quenching of phenylalanine fluorescence in Lys-Phe-**L**ys complexes with poly(A) and DNA. The tripeptide is added to 8×10^{-4} M solutions of poly(A) or DNA in a pH 7 buffer containing 1 mM NaCl and 1 mM Na cacodylate at 2°C. Then NaCl is added to dissociate the complexes ; this leads to an increase in fluorescence intensity. R is the ratio of fluorescence intensities in the presence and absence of the nucleic acid. In the absence of screening effect R should be equal to 1 after NaCl-induced dissociation of the complexes. The observed difference at high NaCl concentration represents the high screening effect of the nucleic acid at the excitation wavelength (260 nm). Fluorescence was measured at 280 nm.

As already observed with tyrosine, proton magnetic resonance studies provide strong evidence for stacking of the phenylalanyl ring with bases in single-stranded polynucleotides but not in double-stranded DNA. Since the quenching of phenylalanyl fluorescence is even more efficient with DNA than with poly A we can conclude that stacking with bases is not responsible for fluorescence quenching in DNA complexes.

Critical distances for singlet energy transfer from phenylalanine to nucleic acid bases have not been calculated yet. The wavelength range covered by phenylalanine fluorescence corresponds to a good overlap with base absorption. This means that the transfer probability from Phe to bases is expected to be very efficient and to lead to fluorescence quenching. One must also be cautious about the reabsorption of Phe fluorescence by nucleic acids which could contribute to an apparent fluorescence quenching. In mixtures of DNA with Lys-Phe-Lys the screening effect of DNA as well as the reabsorption phenomena are taken into account when the complex is dissociated at high ionic strength.

In conclusion it seems that stacking of the Phe ring in single-stranded polynucleotides and energy transfer to base pairs in double-stranded DNA are likely mechanisms explaining Phe fluorescence quenching in Lys-Phe-Lys - nucleic acid complexes.

Complexes

The above discussion shows that several mechanisms can be responsible for the quenching of aromatic amino acid fluorescence in nucleic acid - protein complexes. A conformational change in the protein due to complex formation must always be considered as a possible source of quenching. However the generality of the quenching phenomenon when nucleic acids are the ligands (see table 1) points out to the involvement of more specific mechanisms. Among them direct interactions between the aromatic amino acid and the nucleic acid bases are most interesting. From the study of model systems we have shown that stacking of these two types of molecules will always lead to fluorescence quenching. Interactions with phosphate groups have also to be considered but do not appear to be likely causes of aromatic amino acid fluorescence quenching. Although there is no experimental evidence for quenching induced by hydrogen bonding of tyrosine to nucleic acid bases, this possibility must be contemplated in protein-nucleic acid complexes. In any case a close proximity of tyrosine or phenylalanine with bases even without specific interaction will be expected to lead to fluorescence quenching due to singlet energy transfer from the aromatic amino acid to the bases.

Determining the exact cause of fluorescence quenching in protein-nucleic acid complexes will represent a real challenge. If the crystal structure of the complex has been solved, then the mechanism of fluorescence quenching can be established. For example in the complex formed by staphylococcus nuclease with the inhibitor pTp (35), the fluorescence of Tyr 113 which lies parallel to the thymine ring at a distance of 4.5 Å is expected to be efficiently quenched.

In most cases however the crystal structure is not known. The explanation of fluorescence quenching then requires the conjugation of different physical methods such as nuclear magnetic resonance and circular dichroïsm together with fluorescence spectroscopy to obtain information on the environment of the aromatic amino acid, on its interaction (stacking, hydrogen bonding ...) with the nucleic acid ligand and on the environmental changes induced by complex formation.

(1) Longworth J. W. (1971) in "Excited states of proteins and nucleic acids" Steiner R. F. and Weinryb I. Ed., Plenum Press, pp. 319-484
(2) Brand L. and Gohlke J. R. (1972) Ann. Rev. Biochem., 41, 843-868
(3) Hélène C., Brun F. and Yaniv M. (1969) Biochem. Biophys. Res. Comm., 37, 393-398 ; (1971) J. Mol. Biol., 58, 349-365
(4) Rigler R., Cronvall E., Hirsch R., Pachmann U. and Zachau H. G., FEBS Letters (1970) 11, 320-323
(5) Engel G., Heider H., Maelicke A., Haar F. von der and Cramer F., (1972) Eur. J. Biochem., 29, 257-262
(6) Farelly J. G., Longworth J. W. and Stulberg M. P. (1971) J. Biol. Chem., 246, 1266-1270
(7) Bartman P., Hanke T. and Holler E. (1975) J. Biol. Chem., 250, 7668-7674

(8) Lapointe J. and Söll D. (1972) J. Biol. Chem., 247, 4975-4981
(9) Parfait R. (1973) Eur. J. Biochem., 38, 572-580
(10) Cuatrecasas P., Edelhoch H. and Anfinsen C. B. (1967) Proc. Nat. Acad. Sci. US, 58, 2043-2050
(11) Pongs O. (1970) Biochemistry, 9, 2316-2321
(12) Kelly R. C. and Von Hippel P. (1976) J. Biol. Chem., 251, 7229-7239
(13) Hélène C., Toulmé F., Charlier M. and Yaniv M. (1976) Biochem. Biophys. Res. Comm., 71, 91-98
(14) Pretorius H. T., Klein M. and Dry L. A. (1975) J. Biol. Chem., 250, 9262-9269
(15) Molineux I. J., Pauli A. and Gefter M. L. (1975) Nucl. Ac. Res., 2, 1821-1837
(16) Montenay-Garestier T. and Hélène C. (1968) Nature, 217, 844-845 ;
(17) Montenay-Garestier T. and Hélène C. (1971) Biochemistry, 10, 300-306
(18) Montenay-Garestier T. and Hélène C. (1973) J. Chim. Phys., 70, 1385-1390
(19) Hui Bon Hoa G. and Douzou P. (1970) J. Chim. Phys., Suppl. 197-203
(20) Mutai K., Gruber B. A. and Leonard N. J. (1975) J. Am. Chem. Soc., 97, 4095-4104
(21) Hélène C. and Brun F. (1971) Biochemistry, 10, 3802-3809
(22) Hélène C. and Dimicoli J. L. (1972) FEBS Letters, 26, 6-10
(23) Brun F., Toulmé J. J. and Hélène C. (1975) Biochemistry, 14, 558-563
(24) Toulmé J. J., Charlier M. and Hélène C. (1974) Proc. Nat. Acad. Sci. U. S. A., 71, 3185-3188
(25) Toulmé J. J. and Hélène C. (1977) J. Biol. Chem., 252, 244-249
(26) Dimicoli J. L. and Hélène C. (1974) Biochemistry, 13, 714-723
(27) Steiner R. F. and Kirby E. P. (1969) J. Phys. Chem., 73, 4130-4135
(28) Weber G. (1961) in "Light and Life", W. McElroy and G. Glass, Ed., Johns Hopkins Press, Baltimore, pp. 82-107
(29) Bushueva T. L., Busel E. P. and Burstein E. A. (1975) Studia Biophysica, 51, 173-182
(30) Montenay-Garestier T. (1975) Photochem. Photobiol., 22, 3-6
(31) Hélène C., Montenay-Garestier T. and Dimicoli J. L. (1971) Biochim. Biophys. Acta, 254, 349-365

(32) Dimicoli J. L. and Hélène C. (1974) Biochemistry, 13, 724-730
(33) Sellini H., Maurizot J. C., Dimicoli J. L. and Hélène, C. (1973) FEBS Letters, 30, 219-224
(34) Feitelson J. (1964) J. Phys. Chem., 68, 391-397
(35) Arnone A., Bier C. J., Cotton F. A., Day V. W., Hazen E. E., Richardson D. C., Richardson J. S. and Yonath A., (1971) J. Biol. Chem., 246, 2302-2316.

SPECIFICITY OF PHOTOCHEMICAL CROSS-LINKING IN PROTEIN-NUCLEIC ACID COMPLEXES

Joseph Sperling and Abraham Havron

Department of Organic Chemistry,
The Weizmann Institute of Science,
Rehovot, Israel

The active study of protein-nucleic acid interactions in recent years, has emphasized the critical role these interactions play in life processes and their control. The high affinities observed in protein-nucleic acid complexes, can be accounted for by assuming the occurrence of simultaneous interactions of a number of functional groups on the two partners. However, the molecular mechanism upon which these specific interactions are based is not yet understood in detail.

The tendency of proteins and nucleic acids to form stable covalent complexes as a result of ultraviolet irradiation (Smith, 1976) has increasingly been used as a probe for studying the structure of native nucleoprotein complexes (Schimmel et al., 1976). The proposed approach to the problem, thus, utilizes photochemistry in an attempt to "freeze" existing contact points in the nucleoprotein complexes, and thereby allowing the identification and chemical characterization of the interacting residues. The major advantage of this approach is that the photochemical cross-linking can be performed on naturally occurring protein-nucleic acid complexes, under optimal conditions where maximum binding and stability of the native structures occur. It is apparent, therefore, that the reliability of the photochemical approach depends on the ability of both purines and pyrimidines to form covalent adducts with a major number of amino acids - without particular preference toward specific ones. At the same time it requires that specific covalent bonds would be formed only between neighboring residues in the native structure. The compliance with these conditions would then mean that the photochemical reactions cross-link interacting regions on the macromolecules, or residues which are in close proximity in the native complex.

B. Pullman and N. Goldblum (eds.), Excited States in Organic Chemistry and Biochemistry, 79-84.

As for the first condition, it has been observed that a large variety of amino acids forms photo-adducts with uracil and thymine derivatives (Smith, 1969; Schott and Shetlar, 1974; Shetlar et al., 1975). The purines, although considered to be less reactive than the pyrimidines in photochemical reactions, also take part in cross-linking to proteins. This has been shown recently in the histone H4 - ATP system where a high yield of a cross-linked product was obtained upon UV irradiation of the complex (Sperling, 1976). In our work we address ourselves to the problem of specificity. We approached this problem by choosing a nucleoprotein complex of known structure, and asked whether the photochemical cross-linking occurred at the contact point between the partners which was known from x-ray studies.

The complexes of RNase A with each of its competitive inhibitors, uridine 2'(3'),5'-diphosphate (pUp) and cytidine 2'(3'),5'-diphosphate (pCp), were irradiated with light of λ=254 nm, or with light of λ>300 nm in the presence of a ketonic photosensitizer (acetone).

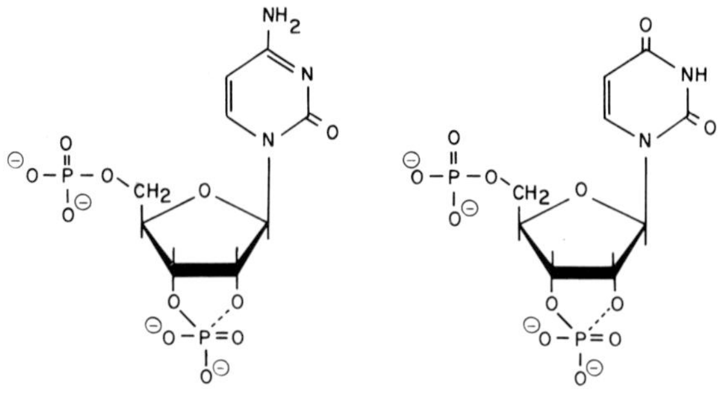

pCp pUp

We have shown previously (Sperling and Havron, 1976), by three different criteria, that the cross-linking was specific:
a) the denatured enzyme failed to cross-link with the inhibitors, b) the extent of covalent binding of pUp could be reduced by the addition of increasing amounts of another competitive inhibitor (3'-UMP), and c) a single tryptic peptide (Asn-67 - Arg-85) of RNase became covalenty linked to both pUp and pCp even in the presence of a large excess of either inhibitors. Here we wish to report the identification of the amino acid residues of RNase which participated in the covalent binding of the enzyme to pUp.

The complex of RNase and ^{14}C-pUp was irradiated with light of λ>300 nm in the presence of acetone as already described (Sperling and Havron, 1976). Monomeric RNase containing one covalently bound moiety of pUp (RNase-pUp) was separated from intact RNase and pUp by ion-exchange chromatography on Dowex-2 using a gradient of NH_4HCO_3. The modified enzyme, having more negative charges than native RNase, was eluted from the resin at 0.2 M salt, whereas intact RNase was eluted with the starting buffer (0.02 M NH_4HCO_3). RNase-pUp was completely inactive, as could be expected for a cross-linked enzyme containing one equivalent of covalently bound inhibitor at the binding site.

Purified RNase-pUp was digested exhaustively with trypsin and the resulting peptides were fractionated by preperative two dimensional separation on paper as described earlier (Sperling and Havron, 1976). A single radioactive peptide whose amino acid composition was consistent with the sequence Asn-67 - Arg-85 of RNase was thus obtained. This peptide, which contained one ^{14}C-pUp molecule per peptide chain was partially digested by thermolysin. The resulting peptides were purified on paper and their amino acid composition was determined after acid hydrolysis as shown in Table I. High voltage paper electrophoresis at pH 1.9 separated the peptides into three radioactive and ninhydrin positive bands. Chromatography in butanol:acetic acid:water: pyridine mixture (15:3:10:12 v/v) in the second dimension separated these bands as follows: the slowest moving one revealed a single radioactive and ninhydrin positive band (A). The next band was separated into three ninhydrin positive bands (B,C,D) of which only C was radioactive. The third revealed two ninhydrin positive bands (E,F) only E being radioactive. The peptide bonds between residues Ser-77 - Thr-78 and Ser-80 - Ile-81 are the most susceptible bonds to hydrolysis by thermolysin (Matsubara, 1970). Partial cleavage at these points should yield six peptides including parent peptide A.

TABLE I: AMINO ACID COMPOSITION OF PEPTIDES OBTAINED BY PARTIAL THERMOLYSIN DIGEST OF TRYPTIC PEPTIDE ASN-67-ARG-85 OF RNASE CONTAINING ONE COVALENTLY BOUND ^{14}C-PUP PER MOLECULE OF ENZYME

PEPTIDE	AMINO ACID COMPOSITION[A] 67			70					75					80					85
A	Asn 3/3	Gly 1.04	Gln 1.8/2	Thr 2/3	Asn 3/3	Cys 1.8/2	Tyr 1.5/2	Gln 1.8/2	Ser 2/2	Tyr 1.5/2	Ser 2/3	Thr 2/3	Met 0.9	Ser 2/3	Ile 0.75	Thr 2/3	Asp 3/3	Cys 1.8/2	Arg 0.9
B	Asn 2/2	Gly 1.04	Gln 2.4/2	Thr 0.84	Asn 2/2	Cys 0.9	Tyr 1.5/2	Gln 2.4/2	Ser 2/2	Tyr 1.5/2	Ser 2/2								
C												Thr 2/2	Met 1	Ser -	Ile -	Thr 2/2	Asp 1	Cys 0.93	Arg 0.81
D	Asn 2/2	Gly 0.83	Gln 1.6/2	Thr 1.6/2	Asn 2/2	Cys 0.9	Tyr 1.6/2	Gln 1.6/2	Ser 2/2	Tyr 1.6/2	Ser 2/2	Thr 1.6/2	Met 1.09	Ser 2/2					
E															Ile 1.1	Thr 0.5	Asp 1	Cys 1.1	Arg 0.98
F												Thr 0.83	Met 1	Ser 1.18					

[A]BASED ON THE THEORETICAL NUMBER OF ASPARTIC ACID RESIDUES (OR METHIONINE SULFONE RESIDUE IN PEPTIDE F), THE THEORETICAL NUMBER OF RESIDUES WHICH APPEAR MORE THAN ONCE IN A PEPTIDE CHAIN ARE GIVEN BY THE DENOMINATOR. CYS AND MET ARE IN THE OXIDIZED FORM: CYSTEIC ACID AND METHIONINE SULFONE, RESPECTIVELY.

Peptides B and D have been identified as Asn-67 - Ser-77, and Asn-67 - Ser-80, respectively. Since they are not radioactive the possibility that the photochemical cross-linking with pUp modifies any residue in the sequence Asn-67 through Ser-77 can be excluded. Peptides C and E are the C-terminal complements of B and D, respectively, and are both radioactive. The amino acid composition of peptide C is consistent with the sequence Thr-78 to Arg-85 which is deficient in Ser-80 and Ile-81. This peptide, however, contained only one equivalent of radioactive inhibitor. This can be explained by assuming that primarily only one of the missing residues cross-linked with excited pUp; the second one lost its identity in a secondary reaction either with reactive intermediates (free radicals) or by cross-linking to the same molecule of pUp. This assumption is supported by the fact that no radioactive peptides deficient either in Ser-80 or Ile-81 were obtained. Furthermore, based on the known reactivity of uracil derivatives in forming,under irradiation, addition products with alcohols, which can be considered analogous to the side chains of serine or threonine (Elad, 1976), it can be speculated that Ser-80 is the initial target for the excited pUp, whereas its next neighbor Ile-81 is subsequently modified. The amino acid composition of peptide E is consistent with sequence Ile-81 - Arg-85. It contains half the radioactivity expected for a 1:1 addition product with ^{14}C-pUp, while half of Thr-82 is missing. This result is difficult to account for without a detailed study of the photochemical mechanism of the cross-linking reactions and the chemical structure of the addition products. However, the published work on the photochemical reactions of uracil derivatives with alcohols indicates that two types of addition products across the 5,6-double bond of the pyrimidine ring can be obtained (Elad, 1976): I) C-C bond formation between C-6 of uracil and the α-carbon of the alcohol; II) C-O bond formation between C-6 and the hydroxylic oxygen of the alcohol. The latter ether type photoproduct is sensitive to acid and is hydrolyzed to give the unmodified starting materials. It can thus be speculated that Thr-82 in RNase underwent partially a type II modification, and modified peptide E lost part of its radioactivity during work-up with concomitant restoration of Thr-82.

Amino acid residues Ser-80, Ile-81, and Thr-82 which were modified by pUp, are part of peptide 77-82 which constitutes the bottom of the binding site on RNase for the pyrimidine ring of nucleotides (Richards and Wyckoff, 1973). The specific covalent binding of these consecutive residues with the inhibitor means that the photochemical cross-linking occurs specifically between a region on the protein which is in close association with the pyrimidine ring in the native structure of the enzyme-inhibitor complex.

As was indicated before the cross-linking could be induced

either by direct exitation of both the enzyme and the inhibitor by light of λ=254 nm, or through photosensitization with acetone and light of λ>300 nm. Similar results were obtained in both cases, however, the analysis of the photosensitized reaction was less ambiguous since it was not accompanied with protein - protein cross-linking and destruction of light-sensitive amino acids, which complicate the analyses of the cross-linked products. In the photosensitized reaction most of the incident light is absorbed by acetone. The ketone, at its triplet state, is able to transfer the exitation energy to the pyrimidine rather efficiently (Elad et al., 1971). It is feasible, therefore, that the excited nucleotide at the enzyme's binding site cross-links with neighboring amino acid residues to form the specific adducts. An alternative pathway involves energy transfer to amino acid residues with subsequent generation of free radicals (Sperling and Elad, 1971) which may eventually cross-link to the pyrimidine.

The observed specificity of the cross-linking favours the possibility that the triplet state of pUp is an intermediate species in both the direct and the photosensitized reactions (Fisher et al., 1974; Varghese, 1974). The alternative mechanism can be ruled out since it may lead to non-specific photoproducts. Concerning the photosensitized reaction, it should be noted that the binding site of the enzyme is accessible to acetone which is present in a concentration of 1 M. This relatively high concentration of acetone (as compared with 1-10 mM of enzyme) does not affect the structure of the enzyme as reflected by the total retention of its nucleolytic activity in the presence of acetone (Sperling and Havron, 1976).

It can be concluded that the specificity of the photochemical cross-linking is achieved by the excitation of a molecule of pUp at the binding site of the enzyme and its subsequent reaction with neighboring amino acid side chains. This reaction takes place in a faster rate than the dissociation of the enzyme-inhibitor complex. Triplet pUp molecules which are not bound at the enzyme's binding site prefer either to revert to ground state, or to form other products (e.g. dimers).

References

Elad, D., in Aging, Carcinogenesis, and Radiation Biology, Smith, K.C., Ed., New York, Plenum Press, p 243.

Elad, D., Rosenthal, I., and Sasson, S. (1971), *J. Chem. Soc. (C)*, 2053.

Fisher, G.J., Varghese, A.J., and Johns, H.E. (1974), *Photochem. Photobiol.*, *20*, 109.

Matsubara, H. (1970), *Methods Enzymol.*, *19*, 648.

Richards, F.M. and Wyckoff, H.W. (1973), in Atlas of Molecular Structures in Biology, Phillips, D.C., and Richards, F.M.,

Ed., Clarendon Press, Oxford, P. 1.
Schimmel, P.R., Budzik, G.P., Lam, S.S.M., and Schoemaker, H.J.P. (1976), in Aging, Carcinogenesis, and Radiation Biology, Smith, K.C., Ed., New York, Plenum Press, P. 123.
Schott, H.N. and Shetlar, M.D. (1974), *Biochem. Biophys. Res. Commun.*, *59*, 1112.
Shetlar, M.D., Schott, H.N., Martinson, H.G., and Lin, E.T. (1975), *Biochem. Biophys. Res. Commun.*, *66*, 88.
Smith, K.C. (1969), *Biochem. Biophys. Res. Commun.*, *34*, 354.
Smith, K.C. (1976), in Aging, Carcinogenesis, and Radiation Biology, Smith, K.C., Ed., New York, Plenum Press, P. 67.
Sperling, J. and Elad, D. (1971), *J. Amer. Chem. Soc.*, *93*, 967.
Sperling, J. (1976), *Photochem. Photobiol.*, *23*, 323.
Sperling, J. and Havron, A. (1976), *Biochemistry*, *15*, 1489.
Varghese, A.J. (1974), *Biochim. Biophys. Acta*, *374*, 109.

EXCITATION AND IONIZATION OF 5-HALOURACILS: ESR AND ENDOR OF SINGLE CRYSTALS

Jürgen Hüttermann

Institut für Biophysik und physikalische Biochemie, Universität, D-84 Regensburg Germany.

The response of 5-halogen substituted analogs of the nucleic acid base uracil to excitation by UV-light or ionizing radiation is a topic of interest in radiation biology since nearly two decades. These components and their nucleoside(-tide) derivatives, (I), can be incorporated into DNA of actively proliferating cells in place of the corresponding thymine residues. Irradiation of cells containing the halogenated analogs produces an up to four-fold increase in lethal, biological damage. The degree of radio-sensitization by 5-halouracils depends, among others on the fraction of thymine residues replaced and on the

(I)

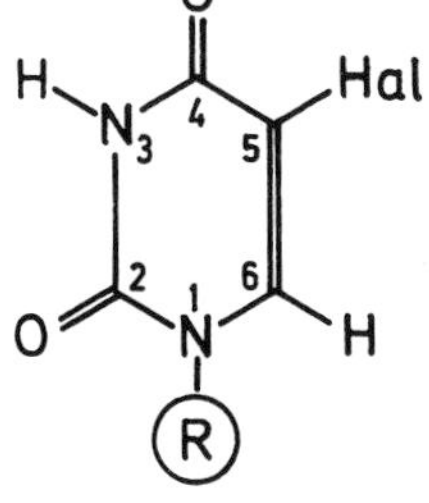

halogen substituent. 5-Fluorouracil when incorporated instead of uracil in certain, single-stranded DNAs evokes only a small increase in damage. The other 5-halouracils, replacing thymine, yield a sensitization which increases in the order Cl, Br, and I as substi-

B. Pullman and N. Goldblum (eds.), Excited States in Organic Chemistry and Biochemistry, 85-98.

tuents. The impact of these findings on the role of DNA as target of primary importance in radiation-biology and, moreover, on the radiotherapy of neoplastic diseases has given impetus to numerous investigations of the various aspects of 5-halouracil-sensitization. Review articles about UV-excitation (1) and bio-medical implications of X-irradiation of 5-halouracils (2) have become available recently.

The differences in the primary response to ionizing radiation between 5-halouracils and thymine or uracil are not completely understood. Extensive studies have been performed in dilute aqueous solutions on the structure and yields of short-lived intermediates produced by reaction with radicals from water radiolysis (3). The hydroxy radical, $OH^{\bullet}$, attacks both the 5- and 6-carbon site of the unsaturated 5,6-bond. Addition to the carbon adjacent to the halogen-substituent leads to rapid oxidative dehalogenation whereas reaction at the 6-carbon results in a more stable intermediate. The relative contribution of both pathways to the overall reaction depends on the halogen substituent. For F, Cl, and Br about 20, 35, and 35% of the molecules do not undergo halogen elimination.

Addition of the solvated electron, e_{aq}, to the 5-halouracil base leads to formation of base anions. These species either immediately eliminate halide ions under formation of a highly unstable uracil-5-yl radical or protonate at the 6-carbon yielding a neutral, more stable intermediate. Again, the halogen substituent determines the fraction of both reactions, the halide ion production being quantitative for Br as is protonation for F. At neutral pH, formation of Cl^- and protonation take place at nearly equal rates in the chloro derivative.

Based on these findings, the main difference between 5-halouracils and thymine in the 'indirect' system of aqueous solutions is halogen elimination in the former components which may be a dominant process. As a consequence, dehalogenation together with subsequent intermediates formed in this reaction have been proposed to be the initial links in the chain of processes leading to an increased biological damage (4). Little attention, however, has been given so far to the response of 5-halouracils to direct excitation and ionization which certainly is also involved in the radiosensitizing reactions. These effects can be studied in the solid phase. We have used electron-spin-

(ESR) and electron-nuclear-double-resonance (ENDOR) spectroscopy of single crystals to elucidate the chemical structure of free radical intermediates formed in several 5-halouracil bases and nucleoside upon X-irradiation between 4.2K and room temperature. In addition, we have analyzed some of the radical reactions taking place in this temperature range by ESR.

A. RADICAL STRUCTURES

The ESR-spectra of all 5-halouracils irradiated at room temperature reveal the presence of several different radicals. An example is given in Fig. 1 for the nucleoside 5-iododeoxyuridine which exhibits two distinct groups of lines. One is a multiline pattern shifted down-field whereas the other group mainly consists of a doublet centered around a g-factor close to that expected for organic radicals containing light atoms. The large g-factor shift of the low-field group can be attributed to a radical in which the unpaired electron interacts with an iodine nucleus. In the particular spectrum shown, the magnetic field is directed along the carbon-iodine bond providing for an exceptionally simple pattern which is assigned to an α-iodo radical fragment R_2CI-CH_2R'. The coupling of the two

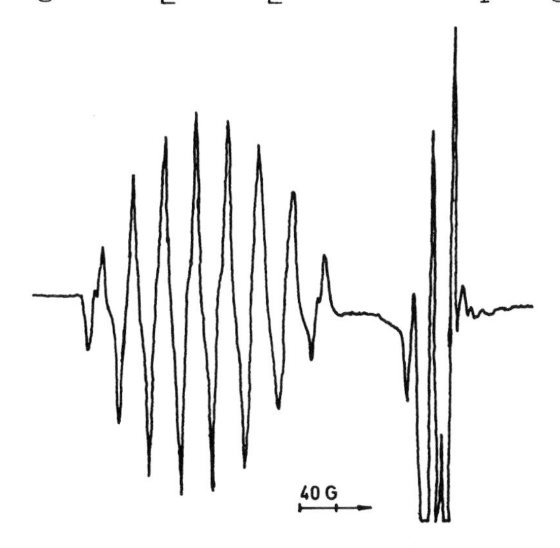

Fig. 1: ESR-spectrum of IUdR-single crystal irradiated at 300K and measured at 35GHz. Magnetic field is oriented along C-I bond.

β-protons is nearly equivalent and equal to the α-iodine interaction giving rise to the octet line group. Similar α-halogenradicals of the general structure (II) are found in all but one of the 5-halouracil derivatives investigated. These include the bases 5-fluoro- (FU) and 5-chlorouracil (ClU), the deoxyribosides 5-chloro- (ClUdR), 5-bromo- (BUdR) and 5-iodo-deoxyuridine (IUdR) and one riboside, 5-chlorouridine (ClUR). The base 5-bromouracil (BU) is an exception to this scheme. Its spectra indicate a radical involving bromine hyperfine interaction but cannot be attributed to radicals of type (II).

(II)

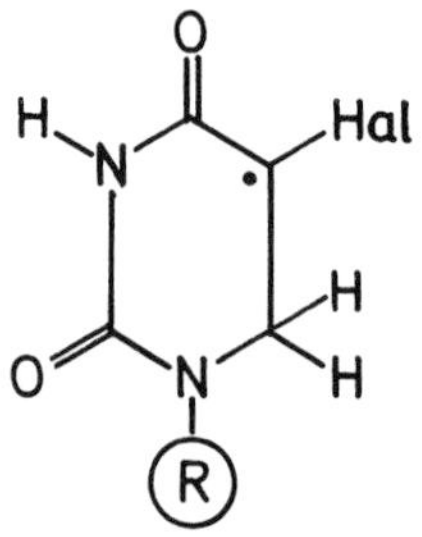

The determination of α-halogen hyperfine coupling parameters is complicated and has remained scarce in the ESR-literature. Except for fluorine, all halogens have a nuclear spin $I > 1/2$ and consequently possess large nuclear electric quadrupole moments. These give rise to highly complex spectra caused by a breakdown of the usual magnetic dipole transition rules by quadrupolar interaction. We have utilized numerical simulation of spectra (5) using a paramerized α-halogen spin-hamiltonian:

$$\mathcal{H} = \beta H g S + S A_{hal} I_{hal} + I_{hal} Q_{hal} I_{hal} + g_N \beta_N H I_{hal}$$

The spectra based on this hamiltonian were calculated and fitted, by trial and error, to the experimental spectra in the canonical orientations x, y, and z. These directions are shown instructure (III). As a final test for the validity of the parameters gained this way, spectra were simulated and compared to experimental spectra obtained upon rotation of the magnetic field in an appropriate crystal plane. The principal values of the α-halogen interaction tensors determined by this procedure are listed in Table 1.

(III)

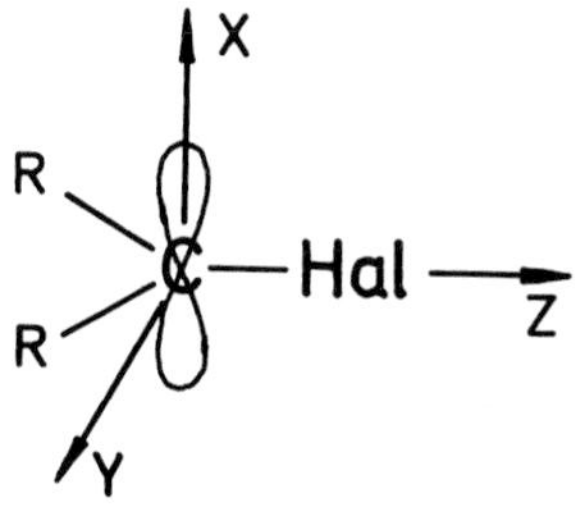

The complex ESR-spectra of α-halogen radicals usually prevent a precise determination of additional interactions like those of the β-protons in (II) except for the z-direction (cf. Fig.1). We have used ENDOR-spectroscopy to evaluate these and other proton interactions. The two β-proton couplings are found to be nearly equivalent with values around 38 G in the deoxyribosides ClUdR, BUdR and IUdR indicating that the pyrimidine ring remains approximately planar upon radical formation. Ring puckering and corresponding inequivalence of the β-protons is observed in all 5-halouracil bases and in the riboside ClUR. Weak proton interactions not resolved in ESR-spectra have been analyzed in ClUdR and BUdR. In both components, a coupling with the proton bonded to the 3-nitrogen can be determined which results from a fraction of about 0.06 of the spin-density delocalized to that position.

Turning to the radicals other than the α-halogen type, a more diverse scheme of structures is obtained. Still, doublet line groups centered around the free spin g-factor as observed in Fig.1 seem to be present in all 5-halouracil derivatives but are frequently masked by lines of other radicals. For BUdR and IUdR, a radical originating from hydrogen addition to the 4-carbonyl oxygen, structure (IV), has been proposed

(IV)

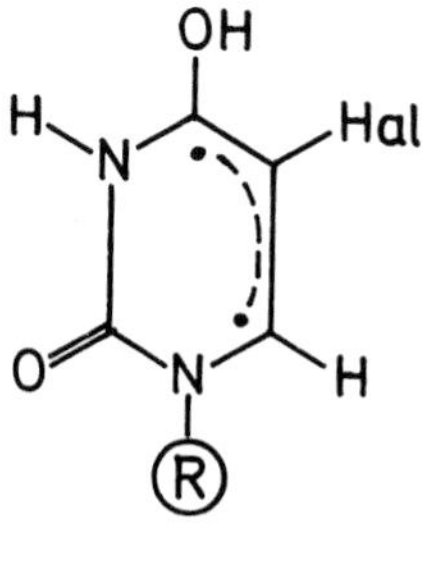

Table 1. α-Halogen interactions in radicals (II) of 5-halouracil derivatives

Substance	Nucleus	Hyperfine coupling [G]	e^2qQ [MHz]	g-tensor	Ref.
FU	^{19}F	x: + 178.0 y: − 18.0 z: − 9.0	---	2.0029 2.0052 2.0066	(7)
ClU	^{35}Cl	x: + 17.1 y: − 4.5 z: − 6.2	72.0	2.0018 2.0083 2.0072	(8)
ClUR	^{35}Cl	x: + 18.9 y: − 5.2 z: − 5.8	72.0	2.0011 2.0075 2.0057	(8)
ClUdR	^{35}Cl	x: + 16.8 y: − 3.9 z: − 4.8	72.0	2.0021 2.0086 2.0068	(15)
BUdR	^{81}Br	x: + 100.0 y: − 20.0 z: − 32.0	330.0	2.0003 2.0240 2.0215	(16)
IUdR	^{127}I	x: + 90.0 y: − 50.0 z: − 40.0	1270.0	1.9870 2.0390 2.0500	(17)

some time ago to account for the doublet on the basis of ESR-measurements (6). Recently, more detailed ESR-ENDOR studies have lead to an unequivocal assignment of (IV) in FU and ClUR. Similar doublets have also been observed in the bases ClU and BU but cannot be analyzed owing to strong overlap with other radical patterns (7,8,6).

The origin of the doublet line group of (IV) is a spin-density at the 6-carbon which interacts with the proton in α-position. The degree of delocalization from the primary radical site C_4 to the 6-position varies with the 5-halogen substituent. In FU, the spin-density remaining at C_4 is sufficient to resolve the OH-proton coupling in ESR-spectra whereas in ClUR this interaction can be detected only by ENDOR. A small fraction ($\sim$ 0.1) of negative spin-density is located on the 5-carbon in both components giving rise to a small but resolved halogen interaction. This finding is important since it reveals that the carbon-halogen bond remains intact upon formation of radicals (IV).

Further radicals at room temperature are found in some 5-halouracil bases. ClU and BU stabilze species of the structure (V) in which the unpaired electron interacts with a nitrogen and a halogen nucleus (8,6). The radical is formed from loss of a hydrogen from the 1-nitrogen position. The assignment of (V) in BU must presently be regarded as tentative but is firm in ClU. The radical is not observed in FU at 300K whereas it is stabilized at 77K in a co-crystal with 1-methylcytosine (9). Another species, the base cation is formed in ClU under conditions of very low irradiation doses. This radical is unstable at 300K.

(V)

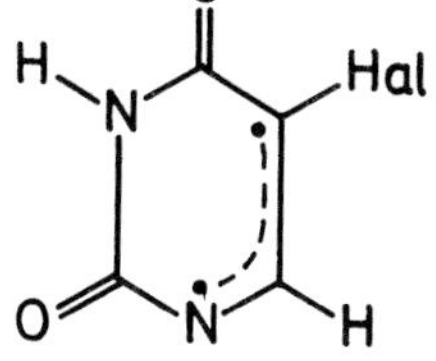

ESR-measurements below room temperature have been performed so fa only with nucleoside derivatives. A common feature of the 5-halodeoxyuridines ClUdR, BUdR and IUdR irradiated at 77K or below is the occurrence of an alcoxy radical RCH_2O at the 5'-position of the deoxyribose moiety, structure (VI). This species is characterized by a large g-factor anisotropy induced by the oxygen spin-orbit coupling and by two inequivalent β-proton interactions of unusually large mag-

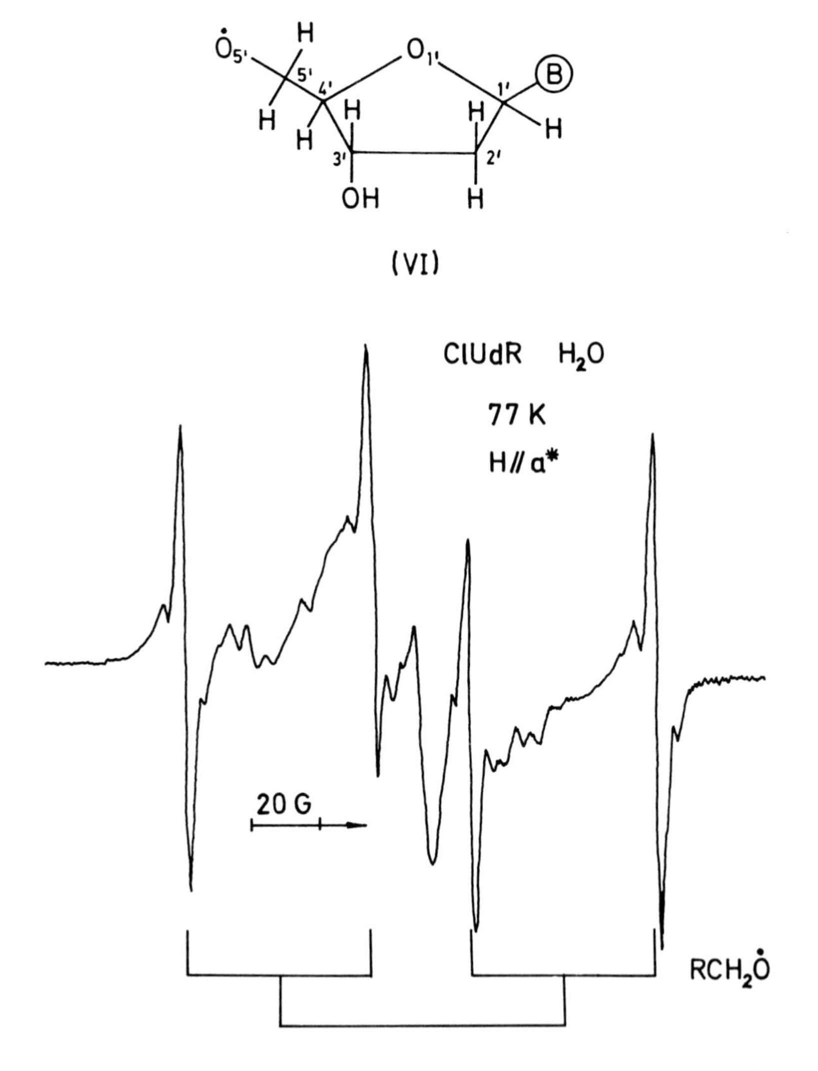

Fig.2: ESR-spectrum of RCH_2O radical (VI) in ClUdR single crystal at 77K.

nitude (10). A typical spectrum of radical (VI) is presented in Fig.2. Radicals of this type have been detected at low temperatures in several non-halogen substituted nucleosides(-tides) and are considered a primary oxidation event (10,11,12). In order to be stabilized, the alcoxy radical seems to require certain crystal field conditions which, in the components studied are provided for by a specific hydrogen-bonding scheme around the primary alcohol group. This network is missing in the riboside ClUR crystals in which, consequently, the RCH_2O species (VI) is not observed.

Another radical common to the 5-halodeoxyuridines ClUdR, BUdR and, at present not well documented, IUdR is structure (VII) which results from hydrogen loss at the 5'-carbon position (13). The unpaired electron

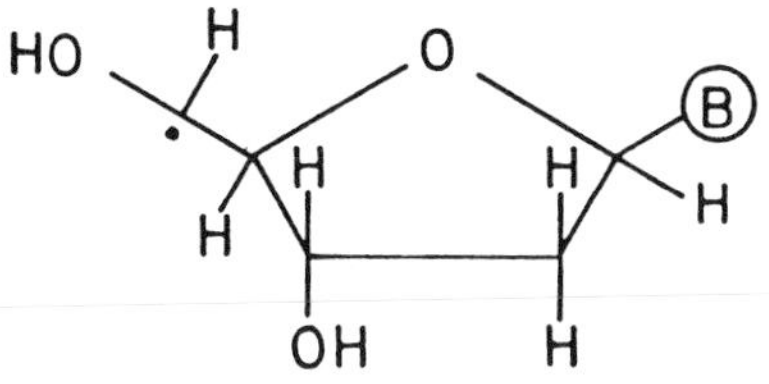

(VII)

in (VII) interacts with the α-proton remaining bonded to $C_{5'}$, the proton in β-position at the 4-carbon, and the terminal hydroxy-proton. Again, for reasons stated in the subsequent section, this radical is absent in the riboside ClUR. This compound instead stabilizes anion radicals at 4.2K whereas at 77K the same radicals are observed which appear in the room temperature spectra, (II) and (IV). Both these radical species are also present in ClUdR, BUdR, and IUdR crystals when irradiated between 20K and 77K. Measurements at 4.2K have still to be performed in these components.

B. RADICAL REACTIONS

To gain a more complete picture of the radiation-damage processes in 5-halouracils in the direct system of single crystals we have analyzed the reactions taking place between several of the radicals described above. A reaction sequence occurring in the deoxynucleosides ClUdR and BUdR which involves the two deoxyribose located species (VI) and (VII) and the α-halogen radical (II) has been studied in detail by kinetic measurements between 90K and room temperature. The reaction starts with the decay of the alcoxy radical (VI) between about 90 and 120K. The radical converts into

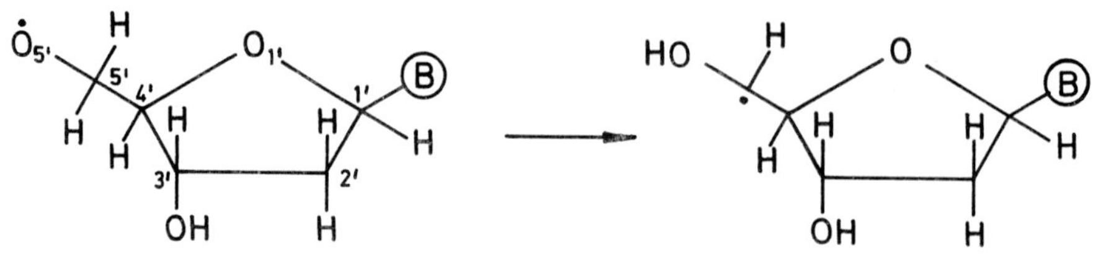

(VI) (VII)

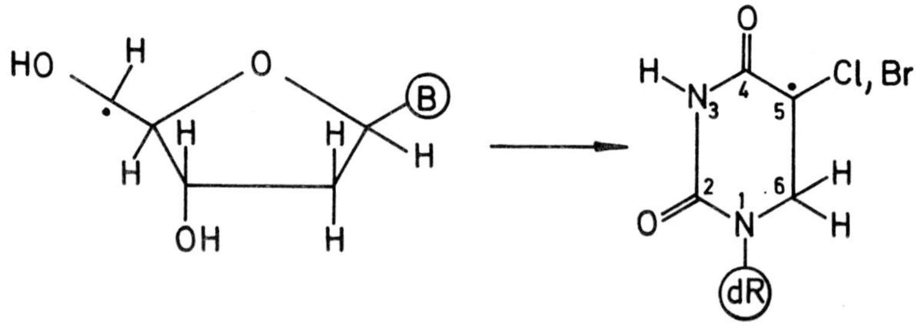

(VII) (II)

structure (VII) probably by abstracting a hydrogen from a neighbouring molecule at the 5'-carbon site. The apparent activation energy for this process is 5±1 kcal/mol for ClUdR and 7±1 kcal/mol for BUdR. The hydroxyalcyl species (VII) decays between about 200K and 240K under concurrent formation of the α-halogen radical (II). Activation energies for this reaction cannot be given since kinetic irregularities brought about by 'phase transitions' in both crystal systems occur in this temperature range. The decay-formation process (VII) → (II) involves transfer of a hydrogen from a non-exchangeable site of the deoxyribose to the 5,6-double bond of the pyrimidine ring (14).

The reaction sequence (VI) → (VII) → (II) is one pathway of the formation of the α-halogen radicals. It apparently is operative only in crystals stabilizing the alcoxy species (VI) and in which the geometrical arrangement of the molecules provides for the reconstitution of the alcohol group by hydrogen-abstraction from a near-by site. Another pathway which has been assessed indirectly by using deuterated crystals is protonation of the 5-halouracil base anion at the 6-carbon site. This latter reaction seems to be the only source of α-halogen radicals (II) in crystals like

ClUR which do not stabilze the alcoxy species and, consequently exhibit only low concentrations of radicals (II) at room temperature.

Arguments concerning the origin of the other radicals in 5-halouracils at 300K can be only of speculative nature at present. Comparison with radiation-chemical findings suggest that radical (IV) may also be due to a protonation reaction of the base anion. If so, the parameters distributing protonation between the 4-oxygen and the 6-carbon site resulting in radicals (IV) and (II), respectively, are not yet understood. It appears reasonable to assume involvement of the electronegativity of the 5-halogen-substituent in this process.

Finally, the radical resulting from hydrogen loss at the 1-nitrogen, structure (V), is frequently observed in other nucleic acid bases and is considered to be formed by de-protonation of the pyrimidine base cation. This mechanism could also operate in the 5-halouracil bases. The site of oxidation, however, is shifted to the primary alcohol groups of the deoxyribose moiety in 5-halodeoxyuridines.

C. CONCLUSIONS

The primary processes following excitation and ionization by X-irradiation of 5-halouracil derivatives as studied on free radicals in single crystals exhibit striking differences with the reactions observec in dilute aqueous solutions. Most prominently, halogen elimination which is the dominant process in the aqueous phase appears to be quenched completely in the solid state. All radicals observed at the 5-halouracil base in crystals reveal an intact carbon-halogen bond. Moreover, some of these species like the α-halogen radicals (II) are formed by mechanisms similar to those which, in aqueous solutions lead to dehalogenation.

It is yet premature to relate the present knowledge about free radical structures and reactions in solid derivatives of 5-halouracils to the radio-sensitizing properties of these components when incorporated into nucleic acids. In view of the findings presented here, the model based on halogen-elimination which has been derived from aqueous solution studies should, however, be treated with some reservations regarding its ubiquitous applicability.

D. ACKNOWLEDGEMENT

The contributions of Drs.W.A.Bernhard, E.Haindl, A.Müller and M.C.R.Symons and of W.Neumüller, H.Oloff and G.Schmidt to various experimental and theoretical parts of this work are gratefully acknowledged. Financial support was obtained by a grant with the Deutsche Forschungsgemeinschaft.

E. REFERENCES

(1) F.Hutchinson, Quart.Rev.Biophys.6, 201 (1973).

(2) W.Szybalski, Cancer Chemotherapy Reports Part 1, 58, 539 (1974).

(3) P.Neta, J.Phys.Chem. 76, 2399 (1972).

(4) J.D.Zimbrick, J.F.Ward, and L.S.Myers,Jr., Int. J.Radiat.Biol. 16, 505 (1969).

(5) Program MAGNSPEC3. QCPE 150 (Indiana University, Chemistry Dept.).

(6) J.Hüttermann and A.Müller, Int.J.Radiat.Biol. 4, 297 (1969).

(7) W.Neumüller, Thesis (University of Regensburg), (1976).

(8) H.Oloff and J.Hüttermann, J.Magn.Res., in press (1977).

(9) R.A.Farley and W.A.Bernhard, Radiat.Res. 61, 47 (1975).

(10) W.A.Bernhard, D.M.Close, J.Hüttermann, and H. Zehner, J.Chem.Phys., in press (1977).

(11) H.C.Box and E.E.Budzinski, J.Chem.Phys. 62, 197 (1975).

(12) R.Bergene and R.A.Vaughan, Int.J.Radiat.Biol. 29 145 (1976).

(13) J.Hüttermann, W.A.Bernhard, E.Haindl, and G. Schmidt, Int.J.Radiat.Biol., submitted for publication.

(14) J.Hüttermann, W.A.Bernhard, E.Haindl, and G. Schmidt, J.Phys.Chem., in press (1977).

(15) J.Hüttermann, W.A.Bernhard, E.Haindl, and G. Schmidt, Mol.Phys. 32, 1111 (1976).

(16) E.Haindl and J.Hüttermann, J.Magn.Res., submitted for publication.

(17) J.Hüttermann, G.W.Neilson, and M.C.R.Symons, Mol.Phys. 32, 269 (1976).

DISCUSSION

SEVILLA :

Since reductive dehalogenation of halouracils has been shown to occur in aqueous solutions at room temperature as well as in aqueous glasses at 77°K, how can you lationalize the back of observation of the dehalogenated species in γ-irratiated single crystals ?

HUTTERMAN :

The main difference between the single crystals and other systems is absence of water which provides for solvation of the halid ion. Therefore, recombination of the halide, if eliminated, becomes possible in the crystals. Also, the embedding of halouracils in the crystals may be such as to prevent permanent halide elimination by not allowing further reactions or the uracil-yl radical formed in halide elimination. In the crystals, the molecular packing is usually such that an empty space is created around the halogen. Hydrogen donnors which are known to increase the halide ion yield, such as the protons of the deoxyribose are several Angströms away from the halogen in the crystals. We do not know at present any details of the mechanisms preventing de-halogenation in the crystals; it may be due to the combination of the two effects pointed out.

SEVILLA :

How, in your opinion, can the results you showed for photoionization of the nucleic acid bases, be tied into a scheme applicable for ionizing radiation? Further, more specifically, do you think the site of oxidation in nucleosides (-tides) upon ionizing radiation should be located on the bases ?

HUTTERMAN :

First it is clear that ionizing radiation will produce -ion radicals in the DNA bases. Their stability and subsequent reactions will depend on the environment in which the ion is produced. The matrices we use will tend to stabilize the initial ion radicals produced perhaps more than in other matrices such as single crystals. As to your second question, the initial site of oxidation in nucleoside or nucleotides of course must include the DNA base itself. It is therefore curious that in γ-irradiated single crystals of these molecules only the π -anions have been observed. The lack of observation of the π-cations could be due to their rapid reaction or to some difficulty in their observation. I prefer the latter explanation at this time.

OPTICAL STUDIES ON T4 GENE PRODUCT 32 PROTEIN DNA INTERACTION

Marcos F. Maestre, Jan Greve and Junko Hosoda

University of California, Space Sciences Laboratory
Berkeley, California.

Introduction

Bacteriophage T4 gene 32 coded protein (gp32) was isolated by Alberts and coworkers (1,2,3). It binds tightly, cooperatively and preferentially to single-stranded DNA. It facilitates both the denaturation and the renaturation of DNA. Genetic studies have demonstrated that this protein is essential for genetic recombination (4), DNA replication (5), repair of radiation damaged DNA (6), and protection of replicating T4 chromosome from degradative activities (7,8). In vitro, gp32 enhances the rate at which T4 DNA polymerase utilizes single-stranded DNA tempertates by 5 to 10-fold (9). Gp32 is also one of the essential components of the reconstructed DNA replicating apparatus (11).

It has been suggested that gp32 binds to the phosphate group(s) of polynucleotides or DNA rather than to bases so that bases in the complexes are in exposed positions (3,9,10). Sedimentation behavior of gp32 fd DNA complex indicated that the complex has far extended (or rigid) structure than free fd DNA (11). Delius et al. (12) have shown that the counter length of gp32 fd DNA complex spread with cytochrome c and 30% formamide on a 10% formamide hypophase was approximately 3.0 micron (about 4.6 A per nucleotide in contrast to the length of naked fd DNA spread in the same way being 1.9 micron (2.9 A/nucleotide). If one assumes that the base distance in the gp32 complex is the same in solution as in cytochrome c membrane, it can be concluded that the base distance in the complex is 35% longer than that in (B-form of) double-stranded DNA and over 50% longer than that in the single-stranded DNA. Destabilization of double-helical form of polynucleotide, as manifested as lowering of the melting tempera-

B. Pullman and N. Goldblum (eds.), Excited States in Organic Chemistry and Biochemistry, 99-111.

ture, by gp32 or any one of other DNA-melting proteins is explained thermodinamically; the protein which has higher affinity to single-stranded form than to double-helical form of polynucleotide shifting equilibrium toward (protein-bound) single strands. Jensen et al. (13) demonstrated that poly [d(A-T)·d(A-T)] melting by gp32 followed closely the thermodynamically predicted patterns. However, Alberts and Frey noted that T4 DNA was not denatured by gp32 at a variety of ionic conditions at temperatures up to 37°. Jensen et al. (13) examined several other naturally occuring DNAs and concluded that gp32 could not denature these naturally occuring DNAs because their destabilization by gp32 is kineticalky blocked.

Moise and Hosoda (14) showed that by limited proteolysis of gp32 a subunit named gp32*-I can be obtained. It has a molecular weight of 27,000 and is formed by removal of approximately 8,000-dalton peptide(s) (A-peptide) from COOH-terminal of gp32 (14). It appears that gp32*-I is a far stronger melting protein than gp32 because it melted T4 DNA almost completely under a condition original gp32 could not (15). A gp32*-I produced by chymotrypsin digestion of gp32 lowered Tm of T4 DNA by 50-60° (14). This coincidental removal of the COOH-peptide from gp32 and the apparent elimination of the kinetic block in T4 DNA-melting seems not only of great interest for gp32 induced DNA melting mechanism, but also may have profound meaning for the role of gp32 in vivo. Moise and Hosoda (14) proposed that A-peptide is a regulatory peptide which controls the dual function of gp32 melting protein and renaturing protein. In a regular conformation where A-peptide is free, it acts as an inhibitor of melting activity thus gp32 behaves as protector of single strands and promoter of renaturation but not as melting protein. This probably is very important to prevent uncontrolled melting of T4 chromosome. Only at a certain position within DNA-replication apparatus, A-peptide is displaced from its regular position (or folded at a different position) through interaction with other protein component(s) (activation of gp32). In this active form, gp32 behaves as melting protein contributing to the replicating fork movement. This model implies that the interaction between the activated form of gp32 and T4 DNA is more comparable to the interaction between gp32*-I and DNA in vitro than to the interaction between gp32 and DNA.

Techniques

Circular Dichroism: It is a well-known fact that circular dichroism (CD) is very sensitive to secondary structure of polynucleotides. Therefore, we decided to study the complexes formed between gp32 and poly (dA), poly (dT), poly (dA·dT) and poly [d(A-T)·d(A-T)] and the interaction of gp32*-I with T4 DNA and poly [d(A-T)·d(A-T)] by means of C.D. spectroscopy conclude that gp32 keeps the single-stranded polynucleotide in a C-DNA type conformation. It will also be shown that the conformation of poly [d(A-T)] in the gp32*-I complex is closer to a C-DNA conformation than that of poly [d(A-T)] in the complex with gp32. Moreover, it is argued that the single-stranded T4 DNA molecule in the complex with gp32*-I is also in a C-DNA conformation.

Fluorescence Detected Circular Dichroism (FDCD)

In an FDCD experiment, a solution is excited with equal intensities of left and right circularly polarized light, and the resultant fluorescence of the sample is detected. This is in contract to a CD experiment in which differences in absorption are measured. The method combines the specificity of fluorescence with the conformation dependence of CD. One obvious use for this technique is in obtaining information about the conformation of that portion of a macromolecule immediately surrounding a fluorescent chromophore.

The signal which one obtains from an FDCD instrument is related to other quantities by

$$(1) \quad \Theta_F = -28.64\left[\Delta E_F/2E_F - \Delta A/2A + \frac{2.303\ \Delta A}{2(10^A - 1)} \right]$$

Θ_F = signal (in degrees) read from the chart paper

ΔE_F = molar CD of fluorophore

E_F = molar absorptivity of fluorophore

A = total absorption of sample

ΔA = total CD sample

The total CD enters into the above equation since the relative intensities of left and right circularly polarized light reaching a sample are dependent upon the optical activity of the medium in front of the fluorophore (16). An FDCD measurement of the gp32 interaction with nucleic acids then involves three spectra. The FDCD spectrum, the regular CD spectrum and the absorbance spectrum of the complex. All this data will have to be processed together with the molar absorbitivity of the fluorophore to obtain the true CD (fluorophore). (Due to experimental difficulties for the changes in the fluorphore in gp32, in this case tryptophan, the data will be presented as a dimensionless ratio $\Delta E_F/E_F \times 10^3$.)

Results and Interpretations

The data is presented in two ways: a) as regular CD or FDCD spectra; b) as difference CD or FDCD spectra comparing two states i.e. complexed vs. uncomplexed gp32-DNA or native vs. heat denatured DNA. The rational for (b) is that obscuring optical activities that remain constant during complexation are eliminated thereby enabling comparisons with theoretical computations of CD that involve interstrand interactions _vs_. intrastrand interactions.

Figs. 1-3 show the CD of poly d(A-T)·(poly d(T-A) and poly dA·poly dT as a function of temperature and complexation with gp32 and gp32*-I. Melting occurs approximately 40°C lower in temperature for the gp32-d(AT) complex as compared to simple heating. Similar behavior occurs for poly(dA)·poly(dT). Two main characteristics of the CD alterations are that changes occur in regular heat denaturation in the 260-300 nm wavelength region of the spectrum, the so-called premelting behavior (17). However, in the gp32 polymer complex that 260-300 nm remains essentially unchanged as a function of temperature. The general change upon complexation is a depression of the CD signal in that region toward negative values. Similar behavior is seen in the gp32 single-stranded poly(dA) or poly(dT) complexation in Fig. 4. Thus, we though that if CD spectra could be compared as a difference CD between differing states of the gp32-DNA complex it would be simpler to compar with theoretical difference CD curves.

Fig. 5 shows the essential differences between the polymers in their melted states, heat denaturation _vs_. protein denaturation. Most strand separation occurs by 25°C in the gp32-polymer complex (0.01M KCl, 2mM tris pH 7.2, 0.1 mM EDTA). It is evident from the large magnitude of the CD signals at 245 nm that there is still considerable optical interaction among the nucleic acid bases. On regular melting there is a considerable reduction in the negative 245 nm trough (Fig. 3). When differences are taken (Figs. 6-7) and compared to the computed differences (18) we can see that for both types of polymers the interaction among the strands complexed to gp32 gives a closer fit to the assumed model. In detail the model assumes that both strands of the polymers have the same geometry ("B" family geometry) and that the changes that occur upon melting are solely due to strand separation, i.e. the _inter-strand_ interactions go to zero leaving only _intra-strand_ optical interactions. Greve _et al._ (18) used the computations of Cech (19) to produce the difference curves in Figs. 6 and 7. As of yet, it is not possible to calculate the secondary structure of a polynucleotide exactly from its CD spectrum and any discussion of such will be of qualitative nature. The CD of a single-stranded polynucleotide is the sum of the intrinsic CD of the bases in the helix and the CD due to interactions between optical transitions on the bases. As the CD of mononucleotides is small the main CD contribution arises from the latter effect. These interactions can be described as dipole-dipole interactions (20) and are proportional to the inverse

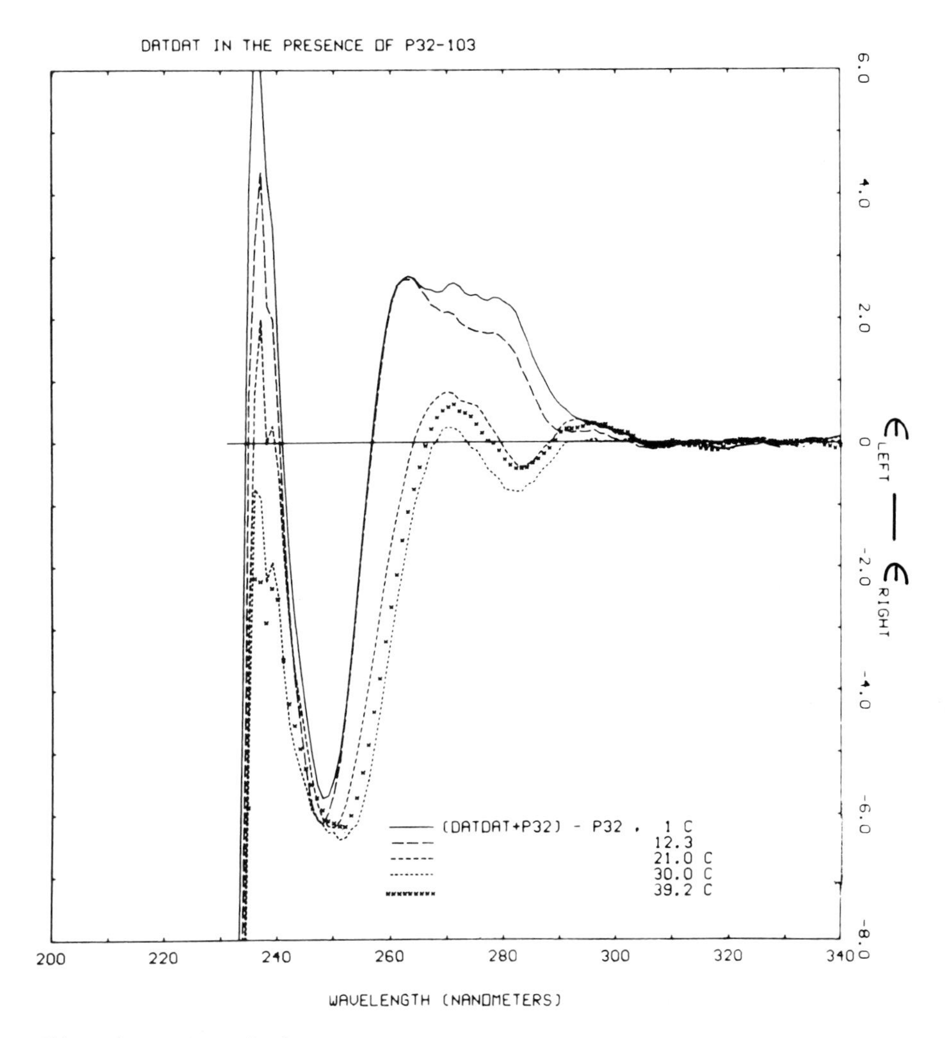

Fig. 1 Poly d(AT)·poly d(TA) and gp32 interaction in 10 mM KCl 2 mM Tris (pH 7.8) 0.1 mM EDTA.

third power of the distance between the bases. Therefore, it is expected that an increase in this distance reduces the CD considerably. In particular, if upon complexation the DNA strand stretches out 35% as suggested by Delius (12), the CD has to decrease enormously. From the measurements on single-stranded polymers presented above, it is clear that such a decrease is not found in general. The negative CD band stays either the same [poly(dT)] or becomes larger [poly(dA)]. Only a reduction in the positive CD band is found which is, in our opinion, due to another effect (see below). The same effects are found upon

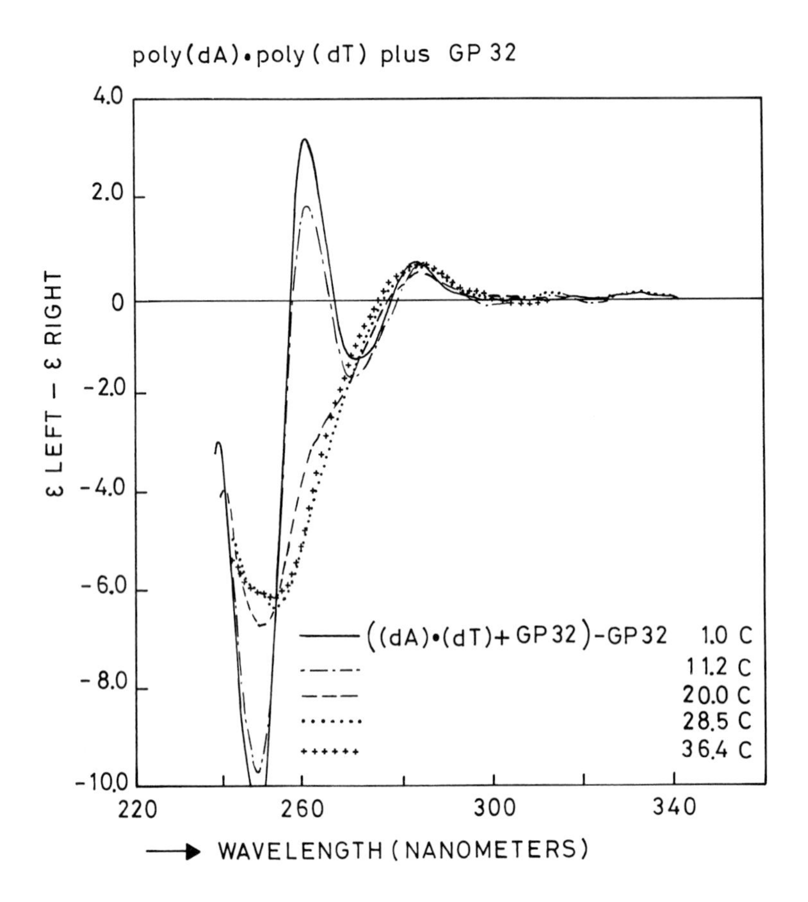

Fig. 2 Poly d(A)·poly d(T) and gp32 interaction. Buffer: 10 mM KCl 2 mM Tris (pH 7.8), 0.1 mM EDTA.

comparison of the CD spectra of complexed and single-stranded (heat denatured double-stranded) poly d(A-T). Although a uniform stretching of the helix must be ruled out for these reasons, a non-uniform stretching might explain the optical measurements. One can imagine that upon binding the sugar phosphate backbone is locally stretched between two bases, whereas little or no stretching occurs on both sides of this deformed place. Such a mechanism would agree with Jensen's (13) result that the basic binding unit is a dinucleotise monophosphate. Moreover, the fact that the CD is not considerably smaller than that of the free

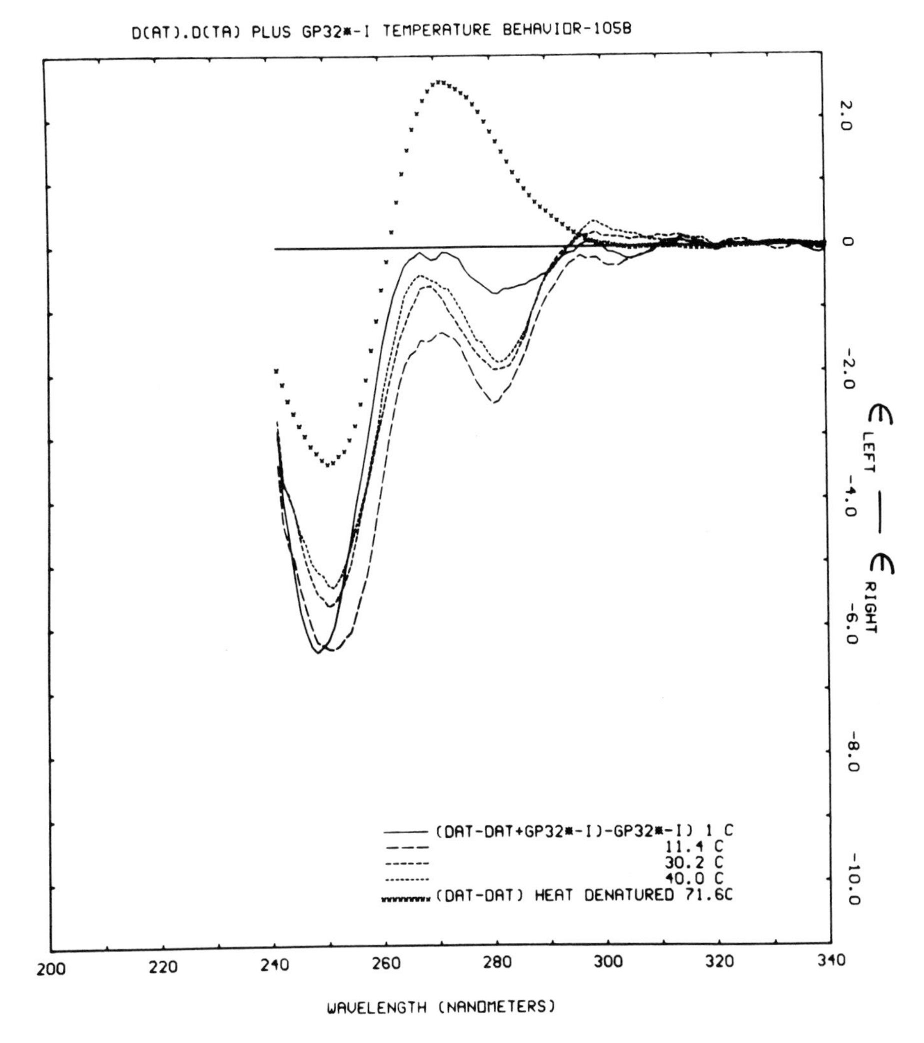

Fig. 3 Poly d(AT) poly d(TA) and gp32*-I interaction in buffer 10 mM KCl, 0.1 mM EDTA, 2 mM Tris, pH 7.8, 10 mM MgSO4. Notice that the alteration of CD spectrum in the 260-30 nm region is much larger than that of gp32 (Fig. 1).

polynucleotide would be explained since Cantor _et al_. (21) showed that di- and tri-nucleosides do have a large CD signal.

To determine the configuration of the stacked bases relative to helix axis, we have to know the CD spectra of single-stranded polynucleotides in e.g. A, B and C conformation. These spectra are unknown yet. However, the CD spectrum of double-stranded polynucleotides above 270 nm is mainly due to intra strand interactions (22). Therefore, we can determine the structure of the

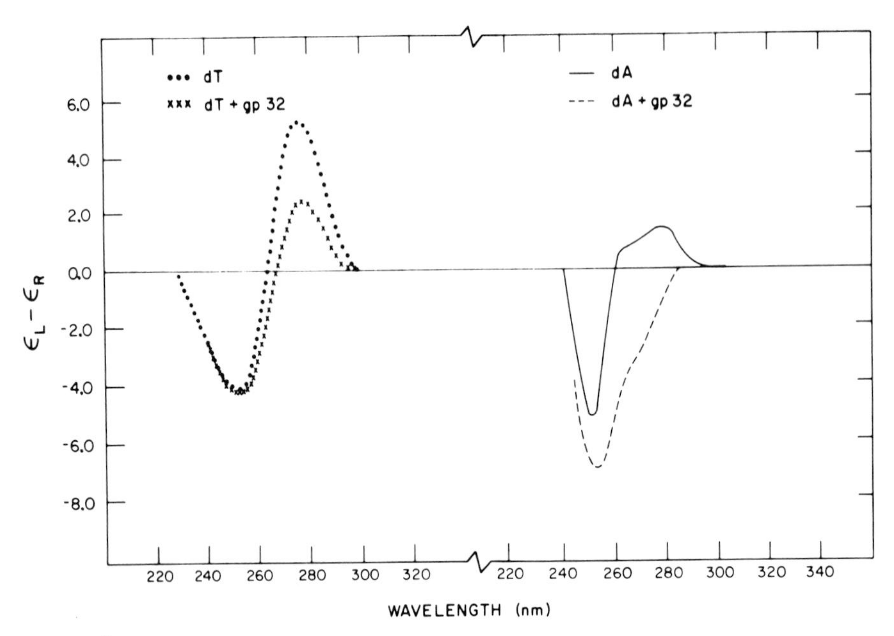

Fig. 4 CD of gp32-single-stranded polymers vs. CD of polymers. Buffers as in Figs. 1-3.

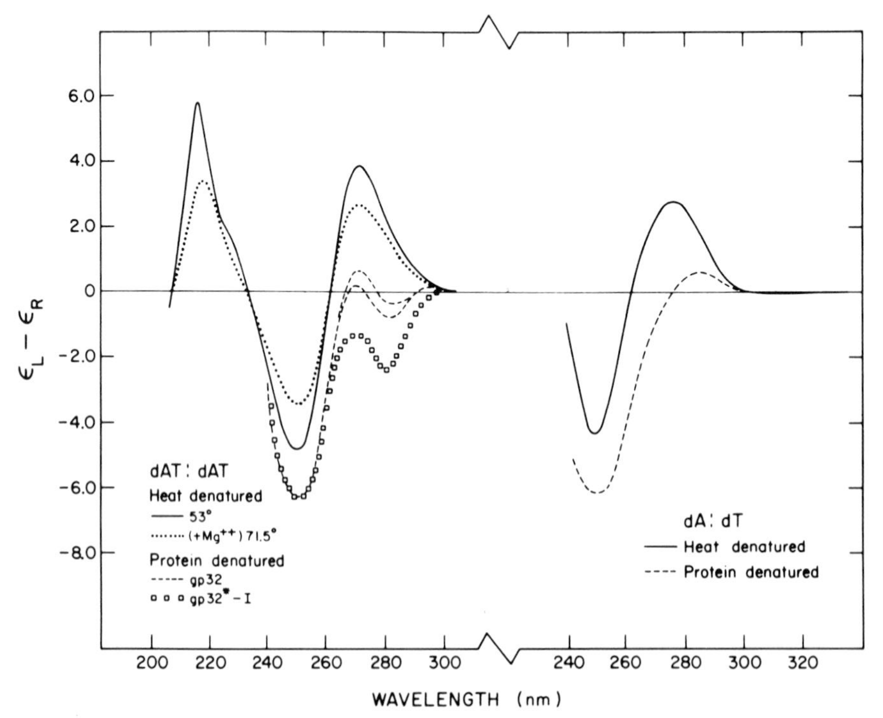

Fig. 5 Comparison of CD of polymers denatured by heat vs. protein denaturated. 71.5°C curve has Mg added.

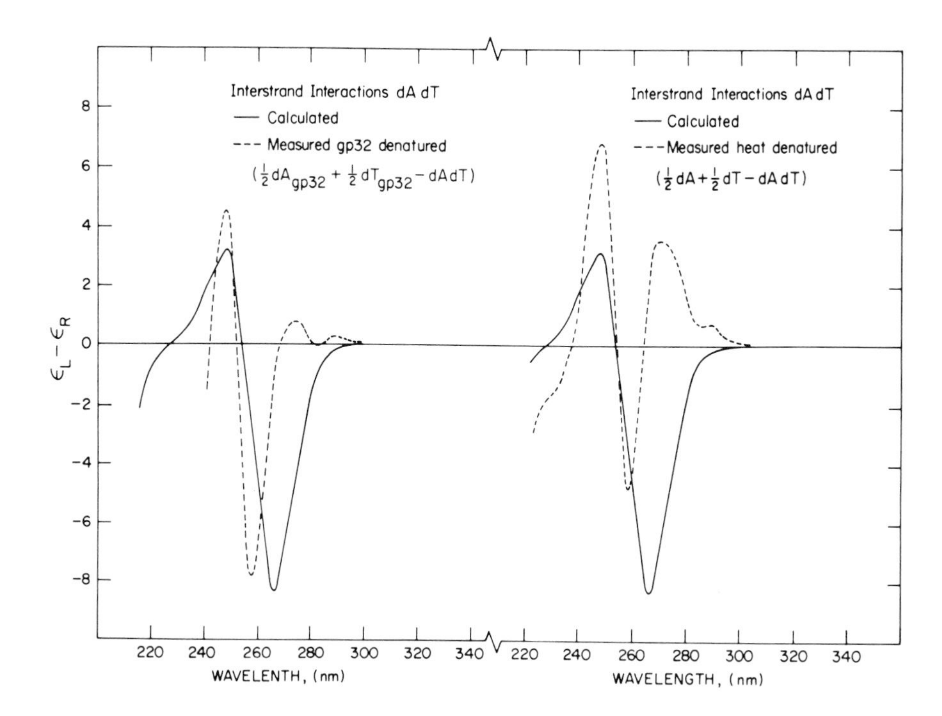

Fig. 6 Difference poly(dA) poly(dT) CD spectra (single-stranded -double stranded) compared with theoretical computation.

single-stranded polymers by comparing the CD in the wavelength region above 270 nm with that of the double-stranded polymers.

For poly (dA•dT) at low N/P ratio where saturation is assured, little or no change is found in this wavelength region upon complexation. We therefore propose that upon binding the protein keeps the bases stacked in the same conformation as they have in the double-stranded polymer at 1°C. Since under these conditions poly (dA•dT) has a structure which is close to a double-stranded C-DNA conformation, we propose that the single strands are kept in a single strand C-DNA conformation (N.B. With single strand C-DNA conformation we mean that the conformation is the same as the single strand has in the double helical molecule in the C configuration). For the poly [d(A-T)•d(A-T)]-g P32 denaturation the situation is more complex. Clearly large changes are found in the CD spectrum above 270 nm. Therefore a change in conformation must occur upon protein binding. Before

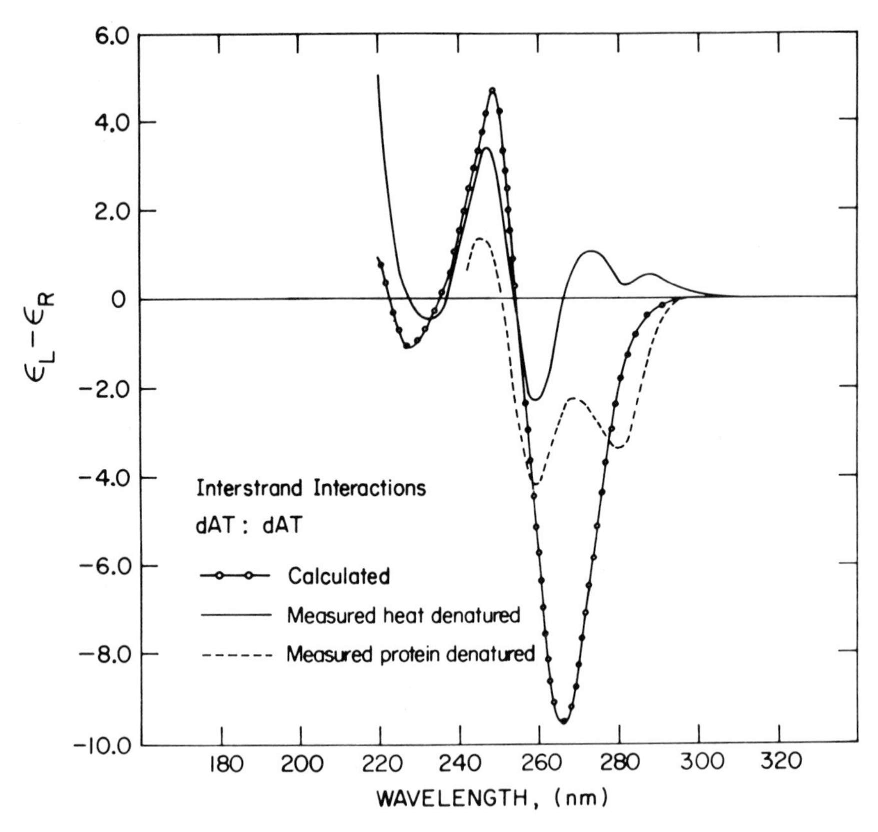

Fig. 7 Difference poly d(AT)poly(dTA) (single double-stranded) compared with theoretical computation.

binding poly [d(A-T)·d(A-T)] is in a structure which is in between a B and a C DNA conformation (Greve, Maestre and Levin, 1976)(18). Upon binding of P32 the nature of the CD spectrum above 270 nm is changed more towards that as found for double-stranded polymer-films at low relative humidity where the poly [d(A-T)·d(A-T)] is in the C-conformation (M.F. Maestre, unpublished results).

Therefore we conclude that the protein keeps the poly d(A-T) strand also in a conformation which is close to a C-conformation. This conformation is closer to a C-conformation than that of the double-stranded polymer in solution at 1°.

Comparison with gp32*I complexation. From Fig. 3 and Fig.8 follows that gp32*-I has a slightly different action than gp32. Firstly, gp32*-I shifts the melting temperature of T4 DNA by about 60°C so that under physiological conditions the T4 DNA is strand separated. Secondly in the interaction with poly [d(A-T)·d(A-T)] complexes are already formed at 1°C and complete strand

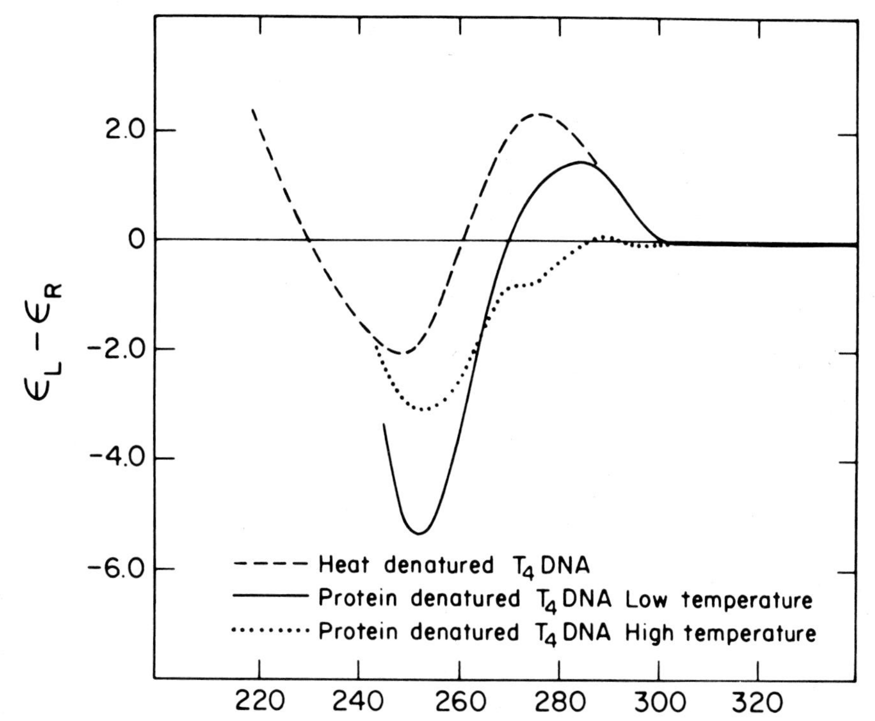

Fig. 8 Wavelength (nm). Heat denaturation vs. gp32*-I protein denaturation of T4 DNA.

separation occurs at about 10°C. This is about 15 degrees lower than in the case with gp32. Thirdly the interaction with gp32*-I seems to result in a single-stranded poly [d(A-T)] structure which is slightly different (closer to a C-DNA conformation).

The FDCD measurement on the gp32-polymer complexes as a function of temperature showed that the assymetry factor ΔE (fluorescer)/E(fluorescer) decreases as the interaction with the polymer increases. Whether this reflects a secondary structural change in the tryptophan moeties of the proteins or a polarization phenomenon due to orientation of the protein with respect to the DNA molecule cannot be determined at the present time. Similar behavior is seen in the gp32 interactions with single stranded poly d(A) and poly d(T). However, there is an increase of dissymetry magnitude on the complexed gp32 protein over the free protein.

Acknowledgements

This work was supported by USPH grant AIO 8427-08, by NASA grant NGR 05-003-460 and by a grant of the Netherlands Organization for the Advancement of Pure Research (ZWO).

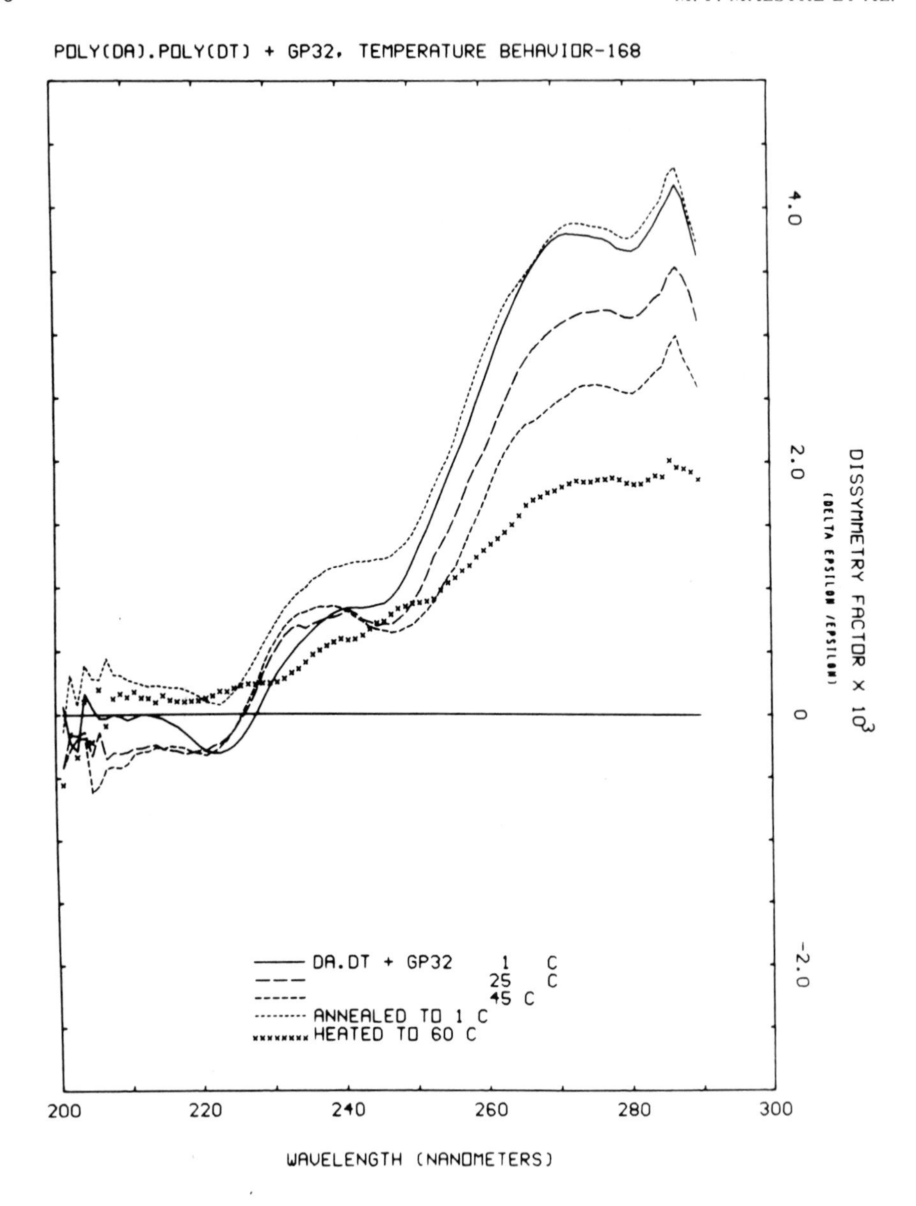

Fig. 9 Dysymmetry factor DE/E for the tryptophan fluorophore of gp32 as it interacts with poly d(A)·poly (dT) as a function of temperature.

References

1. Alberts, B.M., Amodo, F.J., Jenkins, M., Butmann, E.D. and Ferris, F.L. Cold Spring Harbor Symp. Quant. Biol. 33:289-305 (1968).
2. Alberts, B.M. Fed. Proc. Fed. Amer. Soc. Exp. Biol. 29:1154-

1163 (1970).
3. Alberts, B.M. and Frey, L. Nature 227:1313-1318 (1970).
4. Berger, H., Warren, A.J., Fry, K.E. J. Virol. 3:171-175 (1969).
5. Epstein, R.H., Bolle, A., Steinberg, C.M., Kellenberger, E., Boy de la Tour, E., Chevalley, R., Edgar, R.S., Susman, M., Denhardt, G.H. and Lielaudi, A. Cold Spring Harbor Symp. Quant. Biol. 28:375-394 (1963).
6. Wu, J.R. and Yeh, Y.C. J. Virol. 12:758-765 (1973).
7. Curtis, M.J. and Alberts, B. J. Molec. Biol. 102:793-798(1976).
8. Mosig, G. and Bock, S. J. Virol. 17:756-761 (1976).
9. Huberman, J., Kornberg, A., and Alberts, B.M. J. Mol. Biol. 62:39-52 (1971).
10. Jensen, D.E., Kelly, R.C., von Hippel, P.H. J. Biol. Chem. 251:7240-7250 (1976).
11. Alberts, B.M., Morris, F.C., Mace, D., Sinha, N., Bittner, M., Moran, L. Proc. 1975 ICN-UCLA Symp. Molec. Cell Biol., W.A. Benjamin, Inc., Menlo Park, California.
12. Delius, H., Mandell, N.J., Alberts, B. J. Mol. Biol. 67: 341-350 (1972).
13. Jensen, D.E., Kelly, R.C., von Hippel, P.H. J. Biol. Chem. 251:7215 (1976).
14. Moise, H., Hosoda, J. Nature 259:455-458 (1976).
15. Hosoda, J., Takacs, B., Brack, C. FEBS Letters 47:338-342 (1974).
16. Turner, D.H., Tinoco, I. and Maestre, M.F. J. Amer. Chem. Soc. 96:4340 (1974).
17. Gennis, R.B. and Cantor, C.R. J. Mol. Biol. 65:381 (1972).
18. Greve, J., Maestre, M.F. and Levin, A. Biopolymers, in press 1977.
19. Cech, C.M. Thesis, U.C. Berkeley and Biopolymers 15:131-152 (1976).
20. De Voe, H. J. Chem. Phys. 43:3199 (1965).
21. Cantor, C.R., Warshaw, M.M., and Shapiro, H. Biopolymers 9: 1059 (1970).
22. Greve, J., Maestre, M.F. and Levin, A. Manuscript in preparation.

THE CHEMISTRY OF EXCITED STATES OF AROMATIC AMINO ACIDS AND PEPTIDES

E. HAYON

Tech. Center, General Foods Corp.,
White Plains, NY 10625

A brief review is presented on recent work dealing with the singlet excited states and triplet states of tyrosine, tryptophan and related systems in solution. The nature of the excited state precursors leading to the electron ejection from these aromatic amino acids is discussed. Experimental evidence in support of monophotonic *versus* biphotonic processes is presented. A correlation is presented between the quenching rate constants of the singlet excited state or the triplet state of an aromatic amino acid and the reactivity of the quencher towards the hydrated electron.

INTRODUCTION

The photo-induced modifications of proteins result from the direct absorption of excitation energy (light) or, indirectly, from sensitized reactions (e.g. energy transfer, photo-oxidation) and free radical reactions (redox or addition).

The direct absorption of excitation energy by proteins is due, mainly, to the presence of aromatic residues-phenylalanine, tyrosine and tryptophan. It is the nature of the excited states of these aromatic amino acids, as perturbed by their presence in a polypeptide chain and the structural influence of the protein, that determines the macromolecular luminescence of natural proteins and the chemistry that results from the optical excitation of the protein. This article will deal with the excited state chemistry of tyrosine, tryptophan and related systems. For a fairly recent review on phenylalanine, see ref. 1-2.

B. Pullman and N. Goldblum (eds.), Excited States in Organic Chemistry and Biochemistry, 113-122.

TYROSINE

Recent work has focused attention on (a) the excitation of phenol and tyrosine into higher singlet excited states and its influence on the fluorescence quantum yield[3-5], and (b) the nature of the excited state precursor(s) leading to electron ejection.[6-11]

As previously observed[12,13] for indole and tryptophan (see more below), it was found[3-5] that the fluorescence quantum yield, ϕ_F, of phenol and tyrosine was not independent of the excitation wavelength but exhibits a considerable drop with decreasing wavelength between 250 and 220nm. Above 250nm, ϕ_F is constant. For phenol, it was found[4] that the ϕ_F below 250nm is also markedly dependent on the solvent. The decrease in the fluorescence quantum yield is least in methanol, intermediate in water and greatest in cyclohexane. A temperature dependence on ϕ_F has also been reported[5] for phenol in water. The decrease in ϕ_F between 250 and 220nm increases with increase in temperature between 2° and 80°C by about a factor of 2.

Below about 250nm, the second (higher energy) absorption band of phenol and tyrosine starts absorbing, and it is excitation of this electronic band which gives rise to the decrease in ϕ_F with increase in excitation energy. The results suggest that a very fast process competing with the $S_2 \longrightarrow S_1$ internal conversion may be responsible for the lowering of the fluorescence quantum yield. Alternatively, a process from a high vibrational level of the S_1 excited state could also account for these observations. In the case of phenol, the dielectric proterties of the solvent do not appear to be of importance. Hydrogen bonding with the solvent and other molecular interactions may, however, be playing an important role.

The photoionization of phenol, p-cresol, tyrosine and tyrosyl peptides was shown,[6-8] using the fast-reaction technique of laser flash photolysis, to occur primarily from the triplet state via a biphotonic process. The following experimental results, in support of this mechanism, were obtained: (a) the electron ejection was produced within the duration (~15nsec) of the laser pulse; (b) no additional formation of e^-_{aq} or phenoxy radicals was observed at longer times, or simultaneous with the decay of the triplet state. The lifetime of the triplet varied from about 3-10μsec for different phenolic compounds; and (c) a strong dependence upon the light intensity (I) (slightly less than I^2) was found. Such a dependence upon I was observed for light flashes of about 7 μsec duration[6,7] as well as 15nsec duration.[8] Since the lifetime of the lowest singlet excited state is about 2-4nsec, it excludes the possibility that S_1 is involved in the biphotonic process. It does not, however, eliminate the possibility of some small contribution from the S_1 level, from a vibrationally excited S_1 and/or a higher singlet excited state, in the electron ejection mechanism.

It was suggested[9] that the photoionization mechanism depends on the wavelength of excitation: for excitation between 240-270nm it is monophotonic. Recent results[10-11] have confirmed the biphotonic mechanism from 265nm and higher wavelengths.

On optical excitation of the phenolate and tyrosinate ion in alkaline solution[7,8] a linear dependence upon I was found, and a monophotonic mechanism was postulated. The experimental data, though not the interpretation, of the results of Lachish et.al.[10] support this mechanism. Since in alkaline solution the lifetime of the triplet is too short-lived, less than 10nsec, and no fluorescence has been observed, it is not possible at this time to identify the precursor involved in the photoionization process.

QUENCHING REACTIONS

In the presence of oxygen, ^{3}Tyr-OH is rapidly quenched with $k_q=4.8\times10^9 M^{-1}s^{-1}$. The quenching reaction leads[6,8] to an electron transfer with the formation of an increased yield of phenoxy radicals and the superoxide radical O_2^-:

$$^3\text{Tyr-OH} + O_2 \longrightarrow \text{Tyr-O}\cdot + O_2^- + H^+$$

Similarly, ^{3}Tyr-OH is quenched by RSSR, RSH and RSR substrates[14,8] effectively via an electron transfer mechanism:

$$^3\text{Tyr-OH+RSSR} \longrightarrow \text{RSSR}^{\overline{\cdot}} + \text{P}\cdot$$

A correlation between the quenching rate constants of the triplet state of tyrosine and the rate constants of e^-_{aq} with the same organosulfur compounds has been reported,[8] see Figure 1.

Energy transfer from ^{3}Tyr-OH to tryptophan to populate the triplet level of Trp has been reported[8], with $k_q=6.0 \times 10^9 M^{-1}s^{-1}$.

$$^3\text{Tyr-OH} + \text{Trp} \longrightarrow \text{Tyr-OH} + {^3\text{Trp}}$$

TRYPTOPHAN

Indole exhibits two close-lying singlet states, S_1 and S_2, transitions to which from S_o lies in the 260-290nm region[15,16]. The influence of excitation wavelength, oxygen and solvents upon the fluorescence quantum yields of indole and tryptophan have recently been re-examined in detail.[5,12,13,17]

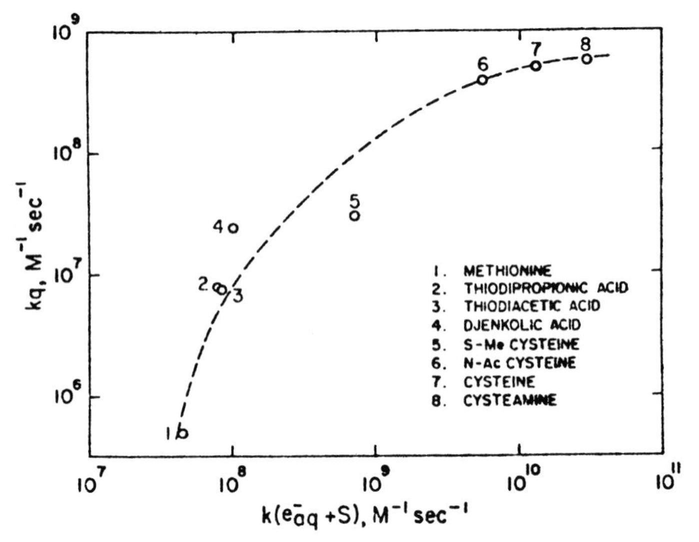

Figure 1. Correlation between kq of the triplet state of tyrosine(10^{-3}M, 25°) by RSH to RSR compounds and the rate constants of e^-_{aq} with the same organosulfur compounds.

The fluorescence quantum yield of both indole and tryptophan are reported[12] to be wavelength dependent indicating that CO_2^- and side chain absorption is not responsible for the effect. The ϕ_F decreases with increase in the excitation energy. While the pH and oxygen affect the absolute flourescence quantum yields, their presence influences the dependence of ϕ_F upon wavelength just the same. Similar wavelength dependence have been observed[12,5] in different solvents. In general, in aliphatic alcohols ϕ_F remains unchanged while in water, cyclohexane and n-hexane $\emptyset_F$ decreases with decrease in wavelength between 220-240nm. Below 220nm and above 240nm, $\emptyset_F$ is independent of wavelength.

Temperature has also been shown to affect $\emptyset_F$ of indole. The decrease in $\emptyset_F$ with decreasing excitation wavelength increases substantially with increase in temperature[5] in the range 25-65°C. Based on a comparison between H_2O, D_2O and aqueous 2-propanol it appears that neither the dielectric constant nor the dielectric relaxation time are the sole factors responsible for the temperature dependence of $\emptyset_F$ at different wavelengths.

It is apparent that the influence of the environment on the excitation wavelength dependence of the fluorescence yield of indole and tryptophan is due to one or more processes which are competing with the $S_1 \longrightarrow S_0$ internal conversion. The population of S_1 (i.e. $\emptyset_F$) can be affected by competing processes occuring at the S_2 level and/or from vibrationally excited levels of S_1. Since it has been shown[18] that the photoionization of indole and tryptophan (see more below) occurs from singlet excited states, including vibrationally excited states or the S_2 state, it has been suggested[5] that electron ejection may be formed from a CTTS state which can be populated from S_2 at a rate competitive with internal conversion to S_1.

The quenching of indole and tryptophan fluorescence has been studied for various quenchers (see e.g. refs. 19-21). A wide range of organic and inorganic substances have been shown to quench the fluorescence with rate constants ranging all the way to the diffusion-controlled limit (see Table I). It was suggested[20] that these quenchers are also reactive towards hydrated electrons. A direct correlation is demonstrated for the first time in Figure 2.

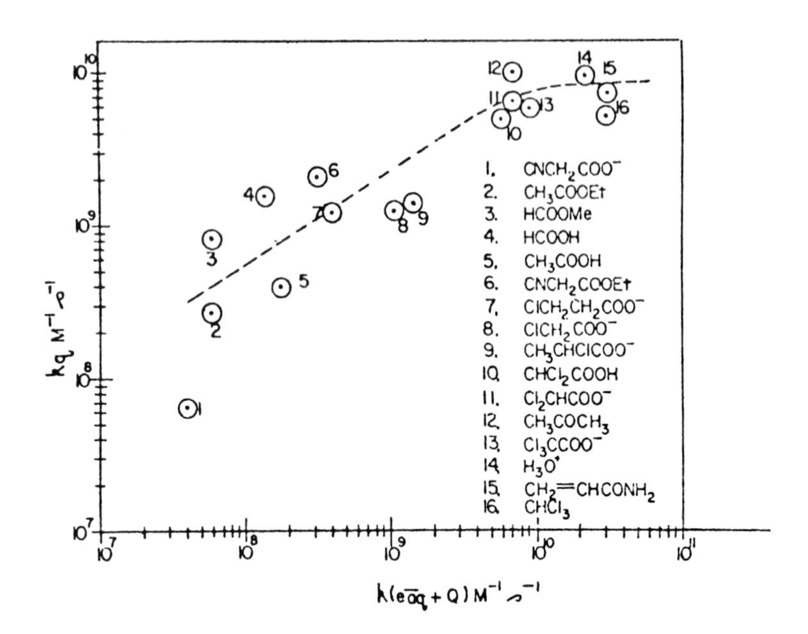

Figure 2. Correlation between kq of the singlet excited state of indole(23°)by various quenchers and the rate constants of e^-aq with the same quenchers.

The results are obtained from refs.20-24. It can be seen that both k_q and $k(e^-_{aq}+Q)$reached the diffusion-controlled limit. Such a correlation suggests a strong charge-transfer character of the S_1 states of indole and tryptophan since these states are efficiently quenched by substances with a high electron affinity.

It has not been shown whether a complex is formed between the singlet excited state and the quencher, and the dependence upon the S_1 energy level. Recently, the quenching of triplet states by inorganic ions has been shown[25] to depend largely on the proximity of the charge-transfer level of the triplet-anion complex to the triplet level of the molecule.

PROTOIONIZATION

The photo-induced electron ejection from indole tryptophan and related systems in water was shown[18] to occur mainly from singlet excited states. In neutral solution of tryptophan, optical excitation at 265nm using single pulses of 15 and 3.6nsec demonstrated a small dependence of the yield of photoionization upon the light indensity.Plots of ϕe^-_{aq} versus intensity on a log-log scale gave slopes of 1.2(for 15nsec pulse) and 1.5(for 3.6nsec pulse) instead of the expected slopes of 1 and 2 for a monophotonic and biphotonic process, respectively.These results were confirmed by Lachish et al[10] and Bryant et al[26].

In alkaline solution at pH 10.6 where τ_F=9.0nsec, as compared to ~3.0nsec at pH 6.8, it was found using a 3.6nsec pulse that ~10% of the total yield of e^-_{aq} was formed _after_ the end of the pulse.From the above results, the following conclusions[18] can be reached on the electron ejection mechanism using 265nm momochromatic excitation: (_a_) essentially all the electrons have the singlet excited state and not the triplet state as the precursor; (_b_) vibrationally singlet excited states may be involved, as well as S_2 states; (_c_) decreasing the excitation wavelength will lend to an increase in the population of the S_2 level and hence an increase in the photoionization yield.

TABLE I. QUENCHING RATES OF INDOLE FLUORESCENCE

QUENCHER	k_q, $M^{-1}s^{-1}$ a	$k(e^-_{aq}+Q)$, $M^{-1}s^{-1}$ b
HCOOH	$1.6x10^9$	$1.4x10^8$
$HCOO^-$	10^7	10^4
HCOOMe	$8.4x10^8$	$5.9x10^7$
CH_3COOH	$4.0x10^8$	$1.8x10^8$
CH_3COO^-	10^7	$1.2x10^6$
CH_3COOEt	$2.7x10^8$	$5.9x10^7$
$ClCH_2COOH$	$5.2x10^9$	$6.9x10^9$
$ClCH_2COO^-$	$1.3x10^9$	$1.2x10^9$
$CNCH_2COO^-$	$6.5x10^7$	$4.0x10^7$
$CNCH_2COOEt$	$2.1x10^9$	$3.2x10^8$
Cl_2CHCOO^-	$6.5x10^9$	$7.0x10^9$
$CH_3CHClCOO^-$	$1.4x10^9$	$1.4x10^9$
$ClCH_2CH_2COO^-$	$1.2x10^9$	$4.0x10^8$
Cl_3CCOO^-	$6.3x10^9$	$8.5x10^9$
CH_3COCH_3	$1x10^{10}$	$5.9x10^9$
H_3O^+	$1x10^{10}$	$2.3x10^{10}$
NH_2CONH_2	$1x10^7$	$2.7x10^5$
CH_2=$CHCONH_2$	$7.1x10^9$ c	$3.1x10^{10}$ d

[a] from ref.21, [b] from ref.22, [c] from ref.23, [d] from ref.24, [e] from ref.20.

Steen et al[5] have experimentally confirmed this point; (d) the yield of electron occurs primarily via a monophotonic process, though some contribution occurs via a biphotonic reaction. The latter becomes increasingly important as the τ_F increases, the pulse duration decreases and the light intensity at 265nm or below increases.

From the correlations between the quenching rate constants and the reactivity of the quencher towards hydrated electrons, Fig.1 and 2, it is suggested that such a correlation can be expected from the excited state precursor involved in the photoionization reaction. For tyrosine the correlation is observed for the quenching of the triplet state and, for tryptophan for the quenching of the singlet excited state. These results assume no ground state complex formation with the quencher.

The triplet-triplet absorption of indole and tryptophan has a maximum at 450nm[18,27]. It has not, as yet, been demonstrated whether photo-induced biphotonic processes occur from the triplet state.

Two similar transient absorptions with λ_{max} 450nm have been observed[18] for tryptophan and other derivatives in neutral water, with lifetimes of 43.5nsec and 14.3 μsec, and denoted as T_1 and T_2. The following observations and suggestions were offered: (a) T_1 is an excited state, possibly a triplet state since its absorption spectrum is similar to T_2. The latter has been shown to be a triplet since it can be populated by energy transfer from ^{3}Tyr, can populate anthracene to form ^{3}anthracene, and is quenched by O_2 and RSSR; (b) T_1 is observed only when an amino group is present (not observed in indole and indole-3-propionic acid) in its protonated NH_3^+ form. It is not observed on excitation of Trp at pH 9.5, N-Me-Trp, N-Acetyl-Trp and Gly-Trp-Gly; (c) T_1 is not a precursor of T_2 since the temperature dependence of T_1 and T_2 are quite different; (d) no transient absorption has been observed to arise from the decay of T_1. No clear understanding exists at present on the nature of T_1. It may be of interest to see if T_1 is dependent on the excitation wavelength.

TABLE II. QUANTUM YIELDS FOR THE PHOTOIONIZATION OF AROMATIC AMINO ACIDS AND PEPTIDES IN WATER AT 25° EXCITED AT 265nm.

PEPTIDE	pH	ϕ^-_{aq}	RATIO
Tyrosine	7.5	0.10	1.00
Tyrosylglycine	6.0	0.05	0.50
Glycyltyrosylglycine	6.0	0.04	0.40
Tryptophan	6.0	0.08	1.00
Tryptophanylglycine	5.2	0.05	0.63
Glycyltryptophan	5.0	0.03	0.38
Glycyltryptophanylgylcine	5.2	0.04	0.50

[a]The relative yields are considered to be of greater significance.

The triplet of tryptophan can be quenched effectively by disulfides via an electron transfer reaction, but with a low efficiency

$$^3Trp + RSSR \longrightarrow Trp^{\dot{+}} + RSSR^-$$

A comparison of the quantum yields for the photoionization of aromatic amino acids and peptides in water optically excited at 265nm, is shown in Table II. It can be seen for both tyrosine and tryptophan that the yield of electron ejection decreases for the peptide derivatives compared to the free aromatic amino acid.

REFERENCES

1. L.I.Grossweiner, Current Topics in Radiation Research Quaterly, 11, 141 (1976).
2. D.V.Bent and E. Hayon, J.Amer.Chem.Soc., 97,2606 (1975); L.J.Mittal, J.P.Mittal and E.Hayon, ibid., 95,6203 (1973).
3. J.Zechner, G.Kohler, G.Grabner and N.Getoff, Chem. Phys.Letters., 37, 297 (1976).
4. G.Kohler and N.Getoff, J.Chem.Soc., Faraday I, 72, 2101 (1976).
5. H.B.Steen, M.K.Bowman and L.Kevan, J.Phys.Chem., 80,482 (1976).

6. J.Feitelson and E.Hayon, J.Phys.Chem., 77, 10 (1973).
7. J.Feitelson, E.Hayon and A.Treinin, J.Amer.Chem. Soc., 95, 1025 (1973).
8. D.V.Bent and E.Hayon, J.Amer.Chem.Soc., 97, 2599 (1975).
9. M.P.Pileni, D.Lavalette and B.Muel, J.Amer.Chem. Soc., 97 2283 (1975).
10. U.Lachish, A.Shafferman and G.Stein, J.Chem.Phys. 64, 4205 (1976).
11. L.I.Grossweiner and J.F.Baugher, J.Phys.Chem., 81, 93 (1977).
12. I.Tatischeff and R.Klein, Photochem.Photobiol., 22 221(1975.
13. I.Tatischeff, R.Klein and M.Duquesne, Photochem. Photobiol., 24, 413 (1976).
14. J.Feitelson and E.Hayon, Photochem.Photobiol., 17 265 (1973).
15. L.J.Andrews and L.S.Forster, Photochem.Photobiol., 19, 353 (1974).
16. F.M.Spoinkel, D.Shillady and R.W.Strickland, J.Amer.Chem.Soc., 97, 6653 (1975).
17. H.B.Steen, J.Chem.Phys., 61 3997 (1974).
18. D.V.Bent and E.Hayon, J.Amer.Chem.Soc., 97, 2612 (1975).
19. J.Feitelson, Isr.J.Chem., 8, 241 (1970).
20. R.F.Steiner and E.P.Kerby, J.Phys.Chem., 83, 4130 (1969).
21. R.W.Ricci and J.M.Nesta, J.Phys.Chem., 80, 974 (1976).
22. M.Anbar, M.Bambenek and A.B.Ross, Natl.Stand.Ref. Data Ser., Natl, Bur.Stand., No.43 (1973).
23. M.R.Eftink and C.A.Ghiron, J.Phys.Chem., 80, 486 (1976).
24. V.Madhavan, N.N.Lichtin and E.Hayon, J.Amer.Chem. Soc., 97, 2989 (1975).
25. A.Treinin and E.Hayon, J.Amer.Chem.Soc., 98, 3884 (1976).
26. F.D.Bryant, R.Santus and L.I.Grossweiner, J.Phys. Chem., 79, 2711 (1975).
27. R.Santus and L.I.Grossweiner, Photochem.Photobiol., 15, 101 (1972).

CHIROPTICAL PROBES OF PROTEIN STRUCTURE

Thomas M. Hooker, Jr. and Warren J. Goux

Department of Chemistry, University of California,
Santa Barbara, California 93106, U.S.A.

Although x-ray diffraction studies of crystals is the definitive technique for the investigation of molecular structure in the solid state, it is certainly possible, even probable, that there are significant differences in the microscopic structure of protein molecules in the solid state and in solution. Even though it may be unlikely that significant deviations occur in the path of the polypeptide chain it is likely that the conformations of certain amino acid side chains, especially those on the surface, may vary significantly in the two states. Since it is the side chains that constitute the functional portions of enzyme molecules insofar as biological activity is concerned, this is a problem of considerable significance. The recent discovery that lysozyme can be crystallized in at least two different forms (1-3), and Vallee's (4) results which indicate a conformational change of a tyrosine residue when crystalline carboxypeptidase is dissolved may be taken as evidence supportive of this point of view.

Optical rotation and circular dichroism are among the most sensitive of physical methods which are commonly used to probe the conformation of molecules in solution. Until relatively recently, most investigations of the chiroptical properties of proteins have been concerned with the amide chromophores of the polypeptide backbone. Although such investigations are of considerable value insofar as the elucidation of secondary structure is concerned, it is difficult to derive information directly related to changes in local structure from them. This is so primarily because of the fact that exciton-type interactions among the amide chromophores occur in the ordered portions of the structure, so the electronic states involved are delocalized over large portions of the molecule.

B. Pullman and N. Goldblum (eds.), Excited States in Organic Chemistry and Biochemistry, 123-136.

On the other hand, most proteins contain only a relatively few side chain chromophores that absorb in the region above 220 nm. Thus, these chromophores might be utilized as built-in probes of local structure and environment. However, this will be possible only if it proves feasible to sort out the various intergroup interactions which give rise to the optical activity of these side chain chromophores. To do so requires the development of a tractable theory which can be used to predict the chiroptical properties of such chromophoric groups as a function of conformation and environment.

The development and utilization of such a theoretical formalism has been a major goal of our research for several years. The purpose of this report is to describe the theoretical chiroptical formalism which has been developed for this purpose in collaboration with Professor John Schellman of the University of Oregon, and to illustrate the application of the theory to a relatively complex molecule.

THEORETICAL FORMALISM

Almost ten years ago Bayley, Nielsen and Schellman (5) published a paper describing a general matrix formulation for the rotatory strength which incorporates all of the usual mechanisms of optical activity in a systematic fashion. They also demonstrated that the general formalism can be restricted in such a way that spectroscopic data derived from experimental investigations of simple molecules can be utilized to calculate the chiroptical properties of larger molecules containing multiple chromophores.

However, as Bayley, _et al_. pointed out in their original paper, the utilization of a basis set limited to only spectroscopically observable states can lead to potentially serious complications if the electronic transitions involved are not degenerat This problem arises as a result of a quantum mechanical transformation that is implicit in the rotatory strength, but which is not always valid if a limited basis set is employed. Schellman suggested, prior to the publication of his original paper, that it should be possible to avoid this difficulty if a theory based upon momentum moments instead of transition dipole moments were utilized (6).

It is possible to write expressions for the momentum moments associated with the electronic transitions of a molecule that are analogous to those described by Bayley, _et al_. for the dipole moment formalism. Thus, one can write for the linear and angular momentum, respectively:

$$\vec{P} = \bar{\bar{C}}^{-1}\vec{P}^{o}\bar{\bar{C}} \qquad\qquad \vec{L} = \bar{\bar{C}}^{-1}\vec{L}^{o}\bar{\bar{C}} \qquad (1)$$

In these equations, the superscript zero indicates momentum moment matrices associated with the electronic transitions of all of the individual chromophoric groups, i.e., they are the linear and angular momentum moment matrices in the original representation. $\bar{\bar{C}}$ is the unitary matrix which is determined by diagonalization of the Hamiltonian matrix for the molecule under investigation.

One can define a quantity which we choose to call the chiral strength, c, by the matrix equation

$$\bar{\bar{c}} = (\alpha/3)\ \mathrm{Re}\{\vec{P}:\vec{L}'\} \tag{2}$$

where α is the fine structure constant and the factor of 1/3 arises from averaging over all orientations of the molecule. The prime indicates transposition and the symbol : means that scalar products are to be carried out element by element. Thus, for a transition between states I and J, we have

$$c_{I,J} = (\alpha/3)\ \mathrm{Re}\{\vec{P}_{I,J}\cdot\vec{L}_{I,J}\} \tag{3}$$

Substitution of Equations 1 into this equation yields an expression for the chiral strength in terms of the original representation, which includes all possible moments

$$c_{OK} = (\alpha/3)\ \mathrm{Re}\{(\sum_{I^o,J^o} C^{-1}_{OI^o}\vec{P}^o_{I^oJ^o}C_{J^oK})\cdot(\sum_{L^o,M^o} C^{-1}_{KL^o}\vec{L}^o_{L^oM^o}C_{M^oO})\} \tag{4}$$

connecting all possible excited states of the groups.

It is possible to obtain a form of the above equation which is more suitable for practical calculations by restricting calculations to a finite set of singly excited states and excluding ground states from the configuration interaction. These approximations mean that the matrices for $\vec{P}$ and $\vec{L}$ become vectors and only transitions from the ground state need be considered. Thus the following equations can be derived,

$$\vec{P} = \vec{P}^o\bar{\bar{C}} \qquad \vec{L} = \vec{L}^o\bar{\bar{C}}^* \tag{5}$$

$$\vec{P}_{OK} = \sum_{J^o}\vec{P}_{OJ}C_{J^oK} \qquad \vec{L}_{OK} = \sum_{J^o}\vec{L}_{J^oO}C^*_{J^oK} \tag{6}$$

$$c_{OK} = (\alpha/3)\ \mathrm{Re}\{\vec{P}_{OK}\cdot\vec{L}_{KO}\} \tag{7}$$

$$c_{OK} = (\alpha/3)\ \mathrm{Re}\{(\sum_{I^o} C_{I^oK}\vec{P}_{I^o})\cdot(\sum_{I^o} C^*_{I^oK}\vec{L}_{I^o})\} \tag{8}$$

The group moments which are required for the evaluation of the chiral strength from these equations can be determined from ground state spectroscopy. In practice, the chiral strength is calculated from these relationships with momentum expressed in atomic units, $\hbar$ for angular momentum and $me^2/\hbar$ for linear momentum.

It is instructive to derive an expression for the chiral strength in terms of the rotatory strength. Converting to cgs units, the chiral strength becomes:

$$c_{OK} = (1/3\hbar mc)\ \mathrm{Re}\{\vec{P}_{OK}\cdot\vec{L}_{KO}\} \tag{9}$$

Substituting for $\vec{P}$ and $\vec{L}$ by the relationships

$$\vec{L}_{KO} = (2mc/e)\ \vec{m}_{KO} \qquad \text{and} \qquad \vec{P}_{OK} = -(i\omega_{KO}m/e)\vec{\mu}_{OK} \tag{10}$$

one obtains

$$c_{OK} = (2\omega_{KO}m/3\hbar e^2)\ \mathrm{Im}\{\vec{\mu}_{OK}\cdot\vec{m}_{KO}\} \tag{11}$$

or

$$c_{OK} = (2\omega_{KO}m/3\hbar e^2)R_{OK} \tag{12}$$

Upon comparison of the above equation with an expression for the ordinary oscillator strength

$$f_{OK} = (2\omega_{KO}m/3\hbar e^2)D_{OK} \tag{13}$$

significance of the chiral strength becomes obvious. It is seen that the chiral strength is the chiroptical analog of the oscillator strength.

In order to carry out calculations with the momentum formalism one must set up the Hamiltonial matrix for the problem of interest. This array might have the appearance shown below for the dipeptide N-acetyl-L-tyrosine-N-methylamide. This molecule would contain

$$\begin{vmatrix} E_{\alpha_1} & C & \mu m & \mu m & \mu m & \mu m & 0 & \mu m \\ C & E_{\beta_1} & K & K & K & K & \mu m & K \\ \mu m & K & E_{\gamma_2} & 0 & 0 & 0 & \mu m & K \\ \mu m & K & 0 & E_{\delta_2} & 0 & 0 & \mu m & K \\ \mu m & K & 0 & 0 & E_{\epsilon_2} & 0 & \mu m & K \\ \mu m & K & 0 & 0 & 0 & E_{\zeta_2} & \mu m & K \\ \mu m & K & \mu m & \mu m & \mu m & \mu m & E_{\alpha_2} & C \\ \mu m & K & K & K & K & K & C & E_{\beta_2} \end{vmatrix} \tag{14}$$

three groups, two amides and one phenolic chromophore.

The diagonal terms that appear in the above matrix are the unperturbed energies of the electronic transitions of the various groups. The numerical subscripts are group indices. The Greek subscripts refer to the various states; α and β represent the n,π^* and π,π^* states of the amide chromophores, whereas γ, δ, ϵ and ζ refer to the ${}^1L_b$, ${}^1L_a$, ${}^1B_b$ and ${}^1B_a$ transitions which are associated with the phenolic chromophore.

The off-diagonal elements represent different classes of interaction energies which are associated with the various mechanisms for the generation of optical rotatory power. Thus, the elements indicated by the symbol C denote the interaction energies associated with the mixing of the n,π^* and π,π^* states of the same group, i.e., the monopoles of the n,π^* transitions interacting with the static charge distribution of the surrounding atoms. Terms of this type correspond to the one-electron mechanism of Condon, Altar and Eyring (7). The other nonzero off-diagonal elements correspond to the dipole-dipole coupling mechanisms (K) of Kirkwood (8), Kuhn (9) and Moffitt (10), and the mechanism involving the interaction of the quadrupolar charge distribution of the n,π^* transition with the dipolar charge distribution of electrically allowed transitions (μm).

In order to evaluate chiral strengths Equation 14 is diagonalized by standard procedures. This diagonalization yields the elements of the C matrix, the column vectors of which are the eigenvectors required for the evaluation of the components of $\vec{P}$ and $\vec{L}$. Thus, the chiral strength can be calculated by means of the equations presented earlier.

The momentum formalism which has been described above should be superior to techniques based on dipole moments insofar as the calculation of the optical properties of side chain chromophores of proteins is concerned. However, this method mixes the various states in such a way that it is not a straightforward matter to evaluate the contributions of specific chromophoric groups to the overall optical activity of a molecule. This is not the result of any deficiency in the theoretical formalism, but merely reflects the fact that the excited states that contribute to the optical activity may be delocalized over a number of chromophoric groups. In this respect, matrix theories such as this are more realistic than simple pairwise interaction approaches.

Nevertheless, in some cases it is possible to extract information regarding the relative contribution of specific chromophores to the optical activity of molecules. Information pertaining to the extent to which various excited states mix and contribute to the optical activity of a given absorption band may be reflected by the final energies of the various states, the eigenvectors and the optical factors, i.e., the scalar products

$\{P_{OK} \cdot L_{KO}\}$. Meticulous analysis of these elements permits one to approximate the extent to which various states mix and contribute to the optical activity.

METHODS AND RESULTS

The general theoretical approach which was described above has been utilized to calculate the chiroptical properties of numerous molecules over the last few years. Many model compounds which contain various chromophores of biological significance have been investigated (11-14). A preliminary report of calculations involving ribonuclease S has been published (15) and preliminary investigations involving basic bovine trypsin inhibitor, staphylococcal nuclease and lysozyme are under way. Some preliminary results of the lysozyme work will be described here.

Near-UV CD spectra of lysozyme are shown in Figure 1. Data are presented for this molecule in dilute salt solution at neutral and high pH, and at neutral pH in the presence of the inhibitor tri-N-acetyl-glucosamine (tri-NAG). Each spectrum represents the result of repetitive scanning by means of a circular dichrometer which is interfaced to a digital computer.

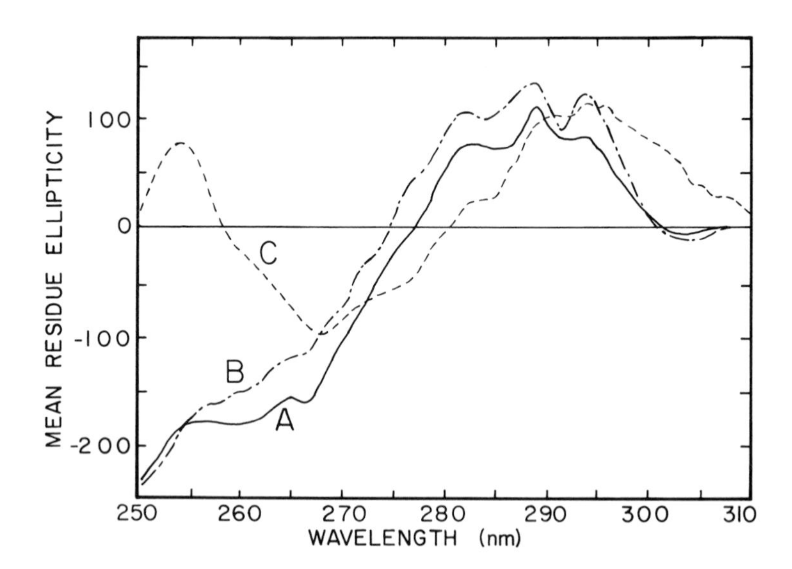

Figure 1: Circular dichroism spectra of lysozyme in dilute aqueous salt solution at neutral pH in the absence (curve A) and presence (curve B) of tri-N-acetylglucosamine, and uninhibited at pH 11.1 (curve C).

There are several obvious differences in the features of these spectra. For example, there is a marked increase in positive ellipticity near 250 nm and the broad positive band near 290 nm is red-shifted when the pH is increased. Furthermore, upon addition of tri-NAG to lysozyme at neutral pH, the positive band near 290 nm is observed to increase in intensity and the magnitude of the small negative feature above 300 nm is more than doubled. These are significant differences, and it would be of interest to determine whether or not they can be accounted for in at least a qualitative fashion by calculations based on the foregoing theoretical formalism.

Theoretical optical calculations were carried out based upon the x-ray coordinates of lysozyme. The M2 coordinate set as refined by Diamond (16) was utilized, and calculations were carried out both in the presence and absence of tri-NAG.

Spectroscopic data for all of the chromophoric groups other than the disulfides have been described in the model compound studies referred to earlier. Optical parameters for the disulfides were derived from the Bergson model as developed by Woody (17). Since these calculations were concerned primarily with the near-UV CD properties, only the two lower energy $n \rightarrow \sigma^*$ transitions (those designated $n_1 \rightarrow \sigma^*$ and $n_4 \rightarrow \sigma^*$ by Woody) were included. The energies were assumed to be degenerate at 250 nm since the crystal structure indicates that all four of the disulfides of lysozyme have dihedral angles fairly close to $\pm 90^o$.

The indole chromophore of tryptophan presents special problems insofar as calculations such as these are concerned. In most indole-containing compounds, including the tryptophan residues of proteins, the ^1L_b and ^1L_a transitions overlap extensively. In many cases the ^1L_b transition appears to give rise to sharp vibrational bands which are superimposed over a broad band which arises from the ^1L_a transition. Furthermore, the fact that the energies of the ^1L_a transitions depend markedly on chromophoric environment introduces severe complications.

The presence of a formal charge near the indole chromophore may shift its ^1L_a transition energy. In an attempt to approximate the magnitude of this effect in lysozyme, electrostatic interaction calculations between the monopoles of the transition dipoles of the six indole chromophores and the atomic static charges which lie within a 10Å radius were carried out. In performing the calculations all carboxyl side chains were assumed to carry unit negative charge and lysine and arginine side chains unit positive charge.

The energies which have been assigned to the ^1L_a transitions in accordance with the calculated interaction energies are listed in Table I. The wavelengths of the bands were assigned from 295 to 300 nm in a manner consistent with the calculated energies. The interaction energies calculated for the ^1L_a transitions of

TABLE I - CALCULATED WAVELENGTHS FOR TRYPTOPHAN ^1L_a TRANSITIONS

Residue	Wavelength	Charged sidechains[a]
W62	295 nm	Asp48,101; Arg61,73
W63	295 nm	Asp52,101,103; Lys97
W123	297 nm	Asp119; Arg5,114,125; Lys33
W28	298 nm	Asp18; Lys96
W111	298 nm	Glu35; Arg112; Lys116
W108	300 nm	Asp52,101,103; Glu35; Arg112

[a]Within 10Å

tryptophans 62 and 63 were of approximately the same magnitude as were those for residues 28, 111 and 123. However, since residue 123 is probably more exposed to the solvent than residues 28 and 111, its ^1L_a transition was placed one nanometer farther to the blue, at 297 nm. All six tryptophan ^1L_b bands were assumed degenerate at 291 nm.

Table II lists the calculated wavelengths and chrial strengths for the near-UV electronic transitions in lysozyme and lysozyme in the presence of tri-NAG. These data were calculated with basis sets which included all aromatic, histidine and disulfide residues, their immediately adjacent amide chromophores, and all other chromophores within 8Å of the aromatics. The results indicate that the near-UV CD spectrum is the result of net positive chiral strength which arises from the tryptophan ^1L_b, the tyrosine ^1L_b and disulfide $n \rightarrow \sigma^*$ bands; negative chiral strength is contributed by the tryptophan ^1L_a transitions. The chiral strength that arises from the tyrosine ^1L_a and tryptophan ^1B_b bands, which are near degenerate at 225 nm, makes a net positive contribution.

The calculations indicate that the tryptophan ^1L_a excited states mix to give rise to the final states listed in Table II. The energies of the tryptophan ^1L_a transitions have been perturbed somewhat as a result of mutual coupling interactions. However, in most cases the ^1L_a transitions which contribute predominantly to the final excited states are distributed over only a few of the residues. These residues and the wavelengths associated with the final excited states to which they make major contributions are presented in the table. The tryptophan ^1L_a transitions, as a result of their distribution of energies, are predicted to give rise to a manifold of bands of mixed sign. The positive band at 293.8 nm is enhanced when the amide transitions associated with tri-NAG are included in the calculation. The net negative contribution to the chiral strength which arises from the remainder of the ^1L_a transitions to the red of this band is also enhanced by the presence of tri-NAG. The only other chiral strengths listed in Table II which

TABLE II - THEORETICAL CHIRAL STRENGTHS x 10^5

Wavelength(nm)	Residues	Transition	Lysozyme	Lysozyme + 3-NAG
291	Tryptophan	^1L_b	0.0180	0.0180
275	Tyrosine	^1L_b	0.0143	0.0126
250	Cystine	$n \rightarrow \sigma^*$	0.0226	0.0209
225	Tryptophan Tyrosine	^1B_b ^1L_a	0.392	0.558

Tryptophan ^1L_a Bands

Wavelength(nm)	Residue(s)	Lysozyme	Lysozyme + 3-NAG
293.8	62 + 63	0.0460	0.0742
296.5	62 + 63	-0.181	-0.206
297.2	123	-0.566	-0.565
298.6	28+108+111	1.65	1.65
299.0	28+108+111	-2.17	-2.18
300.4	28+108+111	1.11	1.09
Total ^1L_a chiral strength		-0.103	-0.125

are changed when the electronic transitions of tri-NAG are included in the basis set are those arising from the tyrosine ^1L_a and the tryptophan ^1B_b bands; the net positive chiral strength which these transitions contribute near 225 nm is predicted to increase.

DISCUSSION

It must be pointed out that the foregoing results are of a preliminary nature, and are therefore subject to future modification. Thus, any conclusions based upon these results must be of a highly tentative nature. The most serious uncertainties at the present time revolve around deficiencies in our knowledge of the disulfide chromophore, uncertainties as to the proper energy assignments for the tryptophan ^1L_a bands and the possible effects of omitting significant portions of the peptide backbone from the calculations.

In spite of these limitations, the calculated chiral strengths that are listed in Table II do appear to account for the near-UV CD spectra of lysozyme in a reasonable fashion. Thus, the net chiral strength which is associated with the tryptophan ^1L_b bands is positive. The positive chiral strength that is predicted for these transitions is in reasonable agreement with the region of positive ellipticity that falls near 288.5 nm in the experimental spectrum, so this feature probably arises from the 0-0 component of the tryptophan ^1L_b transitions.

The situation is somewhat less definitive in the case of the tryptophan ^1L_a bands. Nevertheless, the positive shoulder that falls near 294 nm in the experimental spectrum, and which has been assigned as a component of a tryptophan ^1L_a transition (18), is in reasonable agreement with the calculated results.

The theoretical calculations also predict that the tryptophan ^1L_a transitions should give rise to negative ellipticity to the red of the 294 nm band, as is observed experimentally. On the other hand, a transition with positive chiral strength is predicted to fall farthest to the red, i.e., at 300.4 nm. However, several negative bands, including a relatively strong one at 299 nm, lie just to the blue of this band. As a result, any positive ellipticity that is generated in this region of the spectrum may be so obscured by the adjacent negative bands as to be beyond the capability of present instrumentation to detect. In any event, it appears likely that tryptophan ^1L_a transitions must be responsible for the region of weak negative ellipticity that is observed in the lysozyme CD spectra above 300 nm.

The tryptophan ^1L_a transitions also apparently make significant contributions to the large region of negative ellipticity that appears in the experimental spectra near 265 nm. Although the ^1L_b transitions of both the tryptophan and tyrosine residues contribute positive chiral strength in this region of the spectrum, they are overwhelmed by the much more intense net negative contributions of the tryptophan ^1L_a transitions. This conclusion may be verified by using the data of Table II and assuming gaussian band shapes to compute theoretical spectra. This assignment is also in agreement with the work of Halper et al. (19), who have assigned the ellipticity that is observed in this region to tryptophan ^1L_a transitions.

It should also be pointed out that the tyrosine ^1L_a and the tryptophan ^1B_b transitions are predicted to give rise to relatively strong positive chiral strength near 225 nm. This positive ellipticity, superimposed on the strong negative background which arises at lower wavelengths from the peptide transitions, could easily make significant contributions to the shoulder that is observed near 255 nm in the spectrum that was determined at neutral pH. These conclusions are reinforced by the experimental spectrum that was determined at higher pH. The increase in positive ellipticity near 255 nm is exactly what would be expected if the tyrosine ^1L_a transitions made significant contributions. In addition, the calculations indicate that the disulfide $n \rightarrow \sigma^*$ transitions should make a contribution near 250 nm. However, as a result of the much larger chiral strength that was calculated for the tyrosine ^1L_a and tryptophan ^1B_b transitions, and the experimental evidence that the positive band undergoes a marked red-shift at high pH, one must conclude that present evidence indicates that this spectral feature probably arises primarily from the aromatic chromophores.

The results of the theoretical calculations which were carried out for lysozyme in the presence of tri-NAG also appear to be in reasonable agreement with experiment. The data of Table II indicate that there should be an increase in the positive chiral strength near 225 nm upon binding of tri-NAG. Unfortunately, it is difficult to ascertain whether or not this is in fact the case. The CD data of Figure 1 appear to indicate that near 250 nm the curve for the uninhibited enzyme is approaching that which was determined in the presence of tri-NAG. However, even at lower wavelengths there is no clear evidence for a marked increase in positive ellipticity in the presence of tri-NAG.

The theoretical results for the enzyme-inhibitor complex are in rather good agreement with experiment insofar as the higher wavelength region is concerned. The spectral differences which are observed appear to arise primarily from the interaction of the $\pi \rightarrow \pi^*$ transitions of the N-acetyl groups of tri-NAG with the ^1L_a transitions of some of the tryptophan residues. In the presence of tri-NAG, lysozyme is predicted to show increased positive ^1L_a chiral strength at 293.8 nm relative to the uninhibited enzyme. The increased magnitude of this band is in agreement with the observed increase in ellipticity that is observed at this wavelength in the presence of tri-NAG.

The increase that is observed in the magnitude of the weak negative band that falls near 305 nm upon the binding of inhibitor can also be accounted for by the data of Table II. When one sums the contributions of the tryptophan ^1L_a states that are most likely to make significant contributions in this wavelength region, i.e., those which fall to the red of 294 nm, it is apparent that increases in negative ellipticity are predicted in the presence of tri-NAG. The observed changes occur primarily as a result of decreased positive chiral strength at 300.4 nm and increased negative chiral strength at 296.5 nm.

In summary, the near-UV CD spectrum of lysozyme is predicted to arise primarily from chiral strengths associated with the tryptophan ^1L_a and ^1L_b transitions. The positive maximum in the spectrum at 288.5 nm arises from positive chiral strength associated with the 0-0 component of tryptophan ^1L_b bands. Tryptophan ^1L_a chiral strength contributions of mixed sign are able to account for the negative ellipticity near 305 nm, the positive shoulder at 294 nm and negative ellipticity below 275 nm. The changes which are observed in the CD spectrum of lysozyme upon the binding of tri-NAG can be accounted for in terms of specific electronic interactions between the ^1L_a transitions of some of the tryptophan residues and the $\pi \rightarrow \pi^*$ transitions of the inhibitor.

The fact that the chrioptical calculations, which were based upon atomic coordinates determined from x-ray diffractions studies of crystals, show reasonable agreement with CD spectra determined in solutions may be considered as evidence that the structure of the lysozyme molecule is similar in the two states. However, relatively little attention has as yet been devoted to the effects that small variations in molecular structure might have upon the calculated chiral strengths. Furthermore, it would be especially interesting to compare the results of this investigation with a similar study that is based upon the coordinates of the triclinic form of lysozyme. Such calculations will be attempted as soon as coordinates become available.

ACKNOWLEDGEMENTS

The support of the National Institute of General Medical Sciences of the National Institutes of Health through Grant GM-18092 and the National Science Foundation through Grant CHE-75-13937 is gratefully acknowledged. Equipment purchased with funds from Biomedical Sciences Support Grant S04-RR07099 was utilized in this research. We would like to thank Professor John Rupley for a sample of tri-N-acetylglucosamine. TMH is indebted to Professor John A. Schellman for valuable discussions and significant contributions toward the success of this endeavor.

REFERENCES

1. C. C. F. Balke, G. A. Mair, A. C. T. North, D. C. Phillips and V. R. Sarma, Proc. Roy. Soc., B167, 365 (1967).
2. K. Kurachi, L. C. Sieker and L. H. Jensen, J. Mol. Biol., 101, 11 (1976).
3. J. Mault, A. Yanath, W. Traub, A. Smilansky, A. Podjorny, D. Rabinovich and A. Saya, J. Mol. Biol., 100, 179 (1976).
4. B. L. Vallee, J. F. Riordan, J. T. Johansen and D. M. Livingston, "Cold Spring Harbor Symposia on Quantitative Biology", XXXVI, Cold Spring Harbor Laboratory, 1972, p.517.
5. P. M. Bayley, E. B. Nielsen and J. A. Schellman, J. Phys. Chem. 73, 228 (1969).
6. J. A. Schellman, personal communication.
7. E. U. Condon, W. Altar and H. Eyring, J. Chem. Phys., 5, 753 (1937).
8. J. G.Kirkwood, J. Chem. Phys., 5, 479 (1937).
9. W. Kuhn in "Stereochimie", K. Freudenberg, Ed., Deuticke, Leipzig, 1933, p.317.
10. W. Moffitt, J. Chem. Phys., 25, 567 (1956).
11. T. M. Hooker, Jr., P. M. Bayley, W. Radding and J. A. Schellman, Biopolymers, 13, 549 (1974).
12. J. W. Snow and T. M. Hooker, Jr., J. Amer. Chem. Soc., 97, 3506 (1975).
13. P. E. Grebow and T. M. Hooker, Jr., Biopolymers, 14, 1863 (1975).

14. W. J. Goux, T. R. Kadesch and T. M. Hooker, Jr., Biopolymers, 15, 977 (1976).
15. W. J. Goux and T. M. Hooker, Jr., J. Amer. Chem. Soc., 97, 1605 (1975).
16. R. Diamond, J. Mol. Biol., 82, 371 (1974).
17. R. W. Woody, Tetrahedron, 29, 1273 (1973).
18. K. Ikeda and K. Hamaguchi, J. Biochem. (Tokyo), 71, 265 (1972).
19. J. P. Halper, N. Latovitzki, H. Bernstein and S. Beychock, Proc. Natl. Acad. Sci. U.S., 68, 517 (1971).

DISCUSSION

MICHL :

Am I wrong in believing that the use of the linear momentum instead of (the) position operator in the Rosenfeld formula, which avoids origin-dependence in small-basis set calculations has been well established for many years ?

HOOKER :

It is true that the way in which we introduce the linear momentum moment into our calculations is equivalent to the dipole velocity approach. However, the introduction of linear and angular momentum moments into the matrix formulation of Bayley, Nielsen and Schellman is new. Furthermore, I do not believe that the concept of the chiral strength has been described elsewhere.

PYSH :

Your method will always yield spectra which are conservative, i.e., in which the rotational strength sums to zero. How could your method be modified in order to be successful in cases where the spectrum experimentally is extremely non-conservative, such as poly-L-proline II ?

HOOKER :

The formalism as I have described it is of necessity a conservative one, and is therefore strictly applicable only to conservative spectra. However, additional electronic transitions can be readily included in calculations as data for them becomes available. It is also possible to modify the formalism to include states that are not explicitly included in the basis set by means of the Kirkwood polarizability approximation.

DESCRIPTION OF THE CHIROPTIC PROPERTIES OF SMALL PEPTIDES BY A MOLECULAR ORBITAL METHOD

M. Iseli, R. Geiger* and G. Wagnière

Institute of Physical Chemistry,
University of Zürich,
Rämistrasse 76, CH-8001 Zürich

1. INTRODUCTION

The exciton model has been successful in interpreting the long-wavelength chiroptic properties of both quasi-infinite [1-6] polypeptides and of shorter oligopeptides [7-12]. Initially, only the electric dipole-electric dipole interaction (μ-μ mechanism) of the amide π-π* transitions was considered. After the discovery in the early 1960's of a n-π* band at longer wavelength than the π-π* bands [13-17], several investigations were then undertaken to include magnetic dipole allowed transitions in the exciton scheme (local n-π interaction and nonlocal electric dipole-electric quadrupole interation; μ_i-m_i and μ_i-m_j mechanisms) [7-12].

From another point of view the question may be raised if a molecular orbital approach - of necessity semiempirical - may not also be suited to tackle the problem of describing the lower excited states of oligopeptides. This second approach may prove advantageous in cases where the usual multipole expansion of the interaction is no longer warranted or where spectroscopic knowledge of transitions in individual chromo-

* Present address: Physical Laboratory, Hoffmann-La-Roche, Ltd., Basel, Switzerland

B. Pullman and N. Goldblum (eds.), Excited States in Organic Chemistry and Biochemistry, 137-150.

phores is incomplete; for instance, when small peptide segments interact with long-wavelength absorbing side-chains, such as phenyl, phenol or indole moieties, or with tightly bound pigments [18-20]. However, oligo-peptides of less than a dozen amino acid units contain already several hundred valence electrons, and the application of current all-valence electron MO procedures, such as the CNDO method, to compute these excited states becomes very cumbersome.

The aim of the present investigation is therefore to explore the possibility of applying a simplified MO method to the problem, taking into account only those higher-lying valence electrons which are assumed to take part in the transitions of interest. This paper gives merely an account of the very first steps in this direction. The intended subsequent phases of this investigation will be briefly outlined.

2. METHOD OF CALCULATION

Figures 1 a,b,c, show graphical representations of the two highest occupied and one of the lower unoccupied canonical SCF CNDO orbitals of tetra-L-alanin in a righthanded α-helical conformation. The degree of localization of these orbitals on particular amide groups is striking. They may be qualitatively identified as of π-, n-, π^*-type, respectively. This suggests a simplified approach, starting from the idea of approximate σ-π sequation in individual amide groups within the polymer. Deviations from this local, planar symmetry, due to the asymmetry of the molecule as a whole, is in a first approximation to be expressed by a local parameter Λ_i for each amide monomer i. The lower σ orbitals are to be assigned to an unpolarizable core. The AO basis consists solely of the (pseudo) $2p_\pi$ orbitals of the atoms belonging to the amide groups, of the $2p_n$ orbitals of the oxygen atoms of the amide groups and of the $2p_\pi$ orbitals of atoms belonging to eventual additional unsaturated, long-wavelength absorbing side groups. Each amide group is thus viewed as a 6-electron system. This approach is practically identical to the PPP method [21,22] with the additional inclusion of nonbonding electrons.

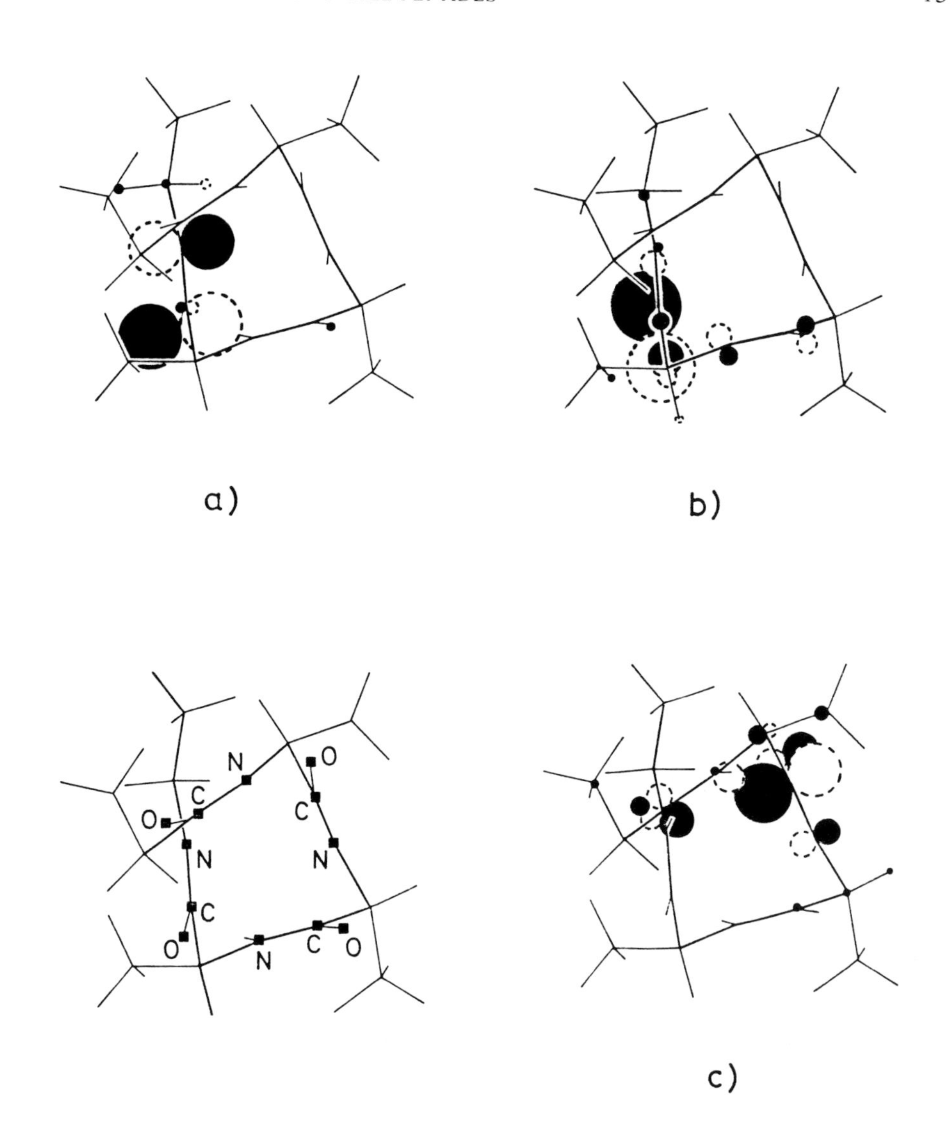

Figure 1. Computer plots [23] of selected higher canonical CNDO SCF orbitals [20,24] of tetra-L-alanine (polar end groups replaced by appropriate nonpolar fragments): a) Orbital 62, predominantly localized on the first amide group, of π-type. b) Orbital 63 (highest filled orbital: the molecule considered has 126 valence electrons), predominantly localized on the first amide group, of oxygen-n-type. c) Orbital 64, predominantly localized on the third amide group, of π*-type.

Within a given monomer i the parameter Λ_i couples the $2p_{nO}$ ($\equiv$ n) and $2p_{\pi O}$ ($\equiv$ O) orbitals. The corresponding Fock matrix element reads, in usual notation:

$$F_{nO}^{(i)} = \Lambda_i - \frac{1}{2} P_{nO} \gamma_{nO}$$

Core matrix elements within the same amide group between n and the orbitals $2p_{\pi C}$ ($\equiv$ C) and $2p_{\pi N}$ ($\equiv$ N) are, for simplicity, neglected. Core matrix elements between different monomers are neglected in a first approximation, but may be selectively introduced at a later stage (see Section 5). All electron repulsion integrals γ between all relevant atomic orbitals in the polymer enter the calculation in the frame of the ZDO approximation.

The semiempirical parameters are calibrated on the electronic properties of the isolated amide group and carbonyl group [25-27]. At the same time, care is taken to stay as close as possible to current PPP parametrization [19-22]. Critical quantities for the amide group are the effective valence state ionization potential for the nonbonding oxygen electrons I_n' and the electron repulsion integrals $\gamma_{nn} \equiv \langle nn|nn \rangle$, $\gamma_{nO} \equiv \langle nO|nO \rangle$ and $\gamma_{nC} \equiv \langle nC|nC \rangle$ (see Table 1). Starting from a given molecular geometry the transition energies and wavefunctions are then computed by the standard SCF-single excitation CI procedure. All amide $n\pi^*$ and $\pi_2\pi^*$ (i.e. $\pi^0\pi^-$) configurations are taken into account.

From the approximate wavefunctions the electric dipole transition moments and the magnetic dipole transition moments are computed without further approximations, taking into account all one- and multi-center terms [28]. The axis of the $2p_\pi$ function on a given atom in a given monomer (amide group, phenyl group, etc.) is assumed to be perpendicular to the plane of that particular chromophore [19,20]. The axis of a given oxygen nonbonding orbital $2p_{nO}$ in an amide group is assumed to lie in the plane of the chromophore and to be perpendicular to the C–O axis. These different p orbitals enter the calculation of the transition moments as Slater functions with standard exponents (C 1.625, N 1.950, O 2.275) and must then be expressed in terms of their components in the molecular frame of reference. To ensure the origin-independence of the rotatory

Table 1a. Core matrix elements for the amide chromophore in eV. $N \equiv 2p_{\pi N}$, etc. $n \equiv 2p_{nO}$.

	N	C	O	n
N	-20.90			
C	- 3.15	-10.45		
O	0.	- 3.30	-15.65	
n	0.	0.	Λ	-26.80

Table 1b. Electron repulsion integrals in the isolated amide chromophore in eV.

	N	C	O	n
N	12.27			
C	7.65	10.53		
O	5.91	8.42	14.50	
n	5.91	7.91	10.14	14.50

strengths, the electric dipole transition moments are computed in the dipole velocity form.

For a given transition $a \rightarrow b$ the rotatory strength is then obtained from the formula [20]

$$R_{ab} = \frac{e^2 \hbar^3}{m^2 c (E_b - E_a)} \sum_{ik} \sum_{i'k'} B_{ik} B_{i'k'} \langle \varphi_i | \vec{\nabla} | \varphi_k \rangle \langle \varphi_{i'} | \vec{r} \times \vec{\nabla} | \varphi_{k'} \rangle$$

in which φ_i, φ_k are appropriate SCF MO's and B_{ik} designate single-excitation CI coefficients.

3. PRELIMINARY RESULTS

As a first working hypothesis we assume the parameter Λ of a given amide group to depend strongly on the absolute configuration at the C_α atom adjacent to the amide N atom. In the following calculations concerning poly-L-amino acids, Λ is always given the same value, calibrated to be 0.5 eV. This is admittedly an oversimplification, but at least it serves to ascertain if correct orders of magnitude are obtainable.

We designate the transitions appearing in the region of 230-250 nm collectively as $(n-\pi^*)$, those below 200 nm, according to the sign of the rotatory strength, collectively as $(\pi-\pi^*)_+$ and $(\pi-\pi^*)_-$, respectively. In general, we notice that the computed $n-\pi^*$ transitions lie too far to the red, the $\pi-\pi^*$ transitions too far to the blue. A relatively minor modification of the parametrization should be able to correct this in future calculations. If within a computed composite band we add the rotatory strengths R and divide by the number of monomers, we find the values given in Table 2. The relation to $\Delta\epsilon_{max}$ is given by the approximate formula [29]

$$R \approx 23 \cdot 10^{-40} \sqrt{\pi} \frac{\Delta}{\lambda_0} \Delta\epsilon_{max}$$

in which Δ/λ_0 is set equal to 0.06 for the composite $(n-\pi^*)$ band and 0.09 for the two $(\pi-\pi^*)$ bands separately.

It is, however, difficult to compare the computed spectrum of the α-helical oligomer with the experimental one for the polymer. In the polymer the negative $(\pi-\pi^*)_-$ "parallel" band is at longer wavelength than the positive $(\pi-\pi^*)_+$ "perpendicular" band. In the computed spectra we notice that on going from smaller to larger α-helical systems the shorter-wavelength $(\pi-\pi^*)_-$ band is gradually shifted to the red with respect to the $(\pi-\pi^*)_+$ band. If we consider the computed negative $\pi-\pi^*$ transition at longest wavelength in each calculation, we notice that as the helix grows the absolute value of the rotatory strength increases, as well as the relative component of the electric dipole transition moment parallel to the helix axis (Table 3).

A striking feature of the CD spectrum of poly(L-proline)I is the appearence of a large positive Cotton effect at relatively long wavelength [32,35]. This feature is qualitatively borne out by the calculation on the hexamer. The $\pi-\pi^*$ transition in question is polarized roughly parallel to the helix axis. There appears to be an analogy in the sign and position of the long-wavelength CD bands of poly(L-proline)I and of cyclo(tri-L-proline) [30]. This analogy is also reflected by the calculation (see Figure 4).

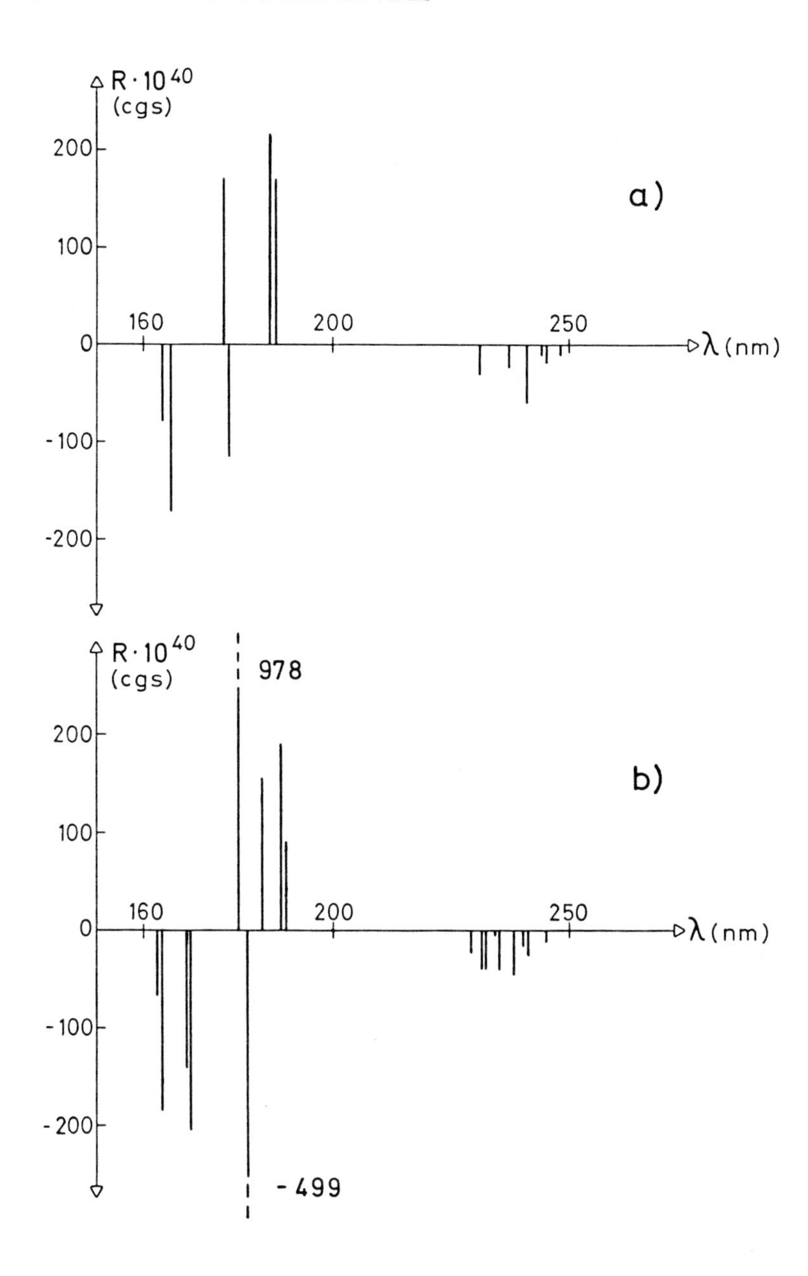

Figure 2. Computed rotatory strengths for the righthanded α-helix. a) Hexamer; b) nonamer.

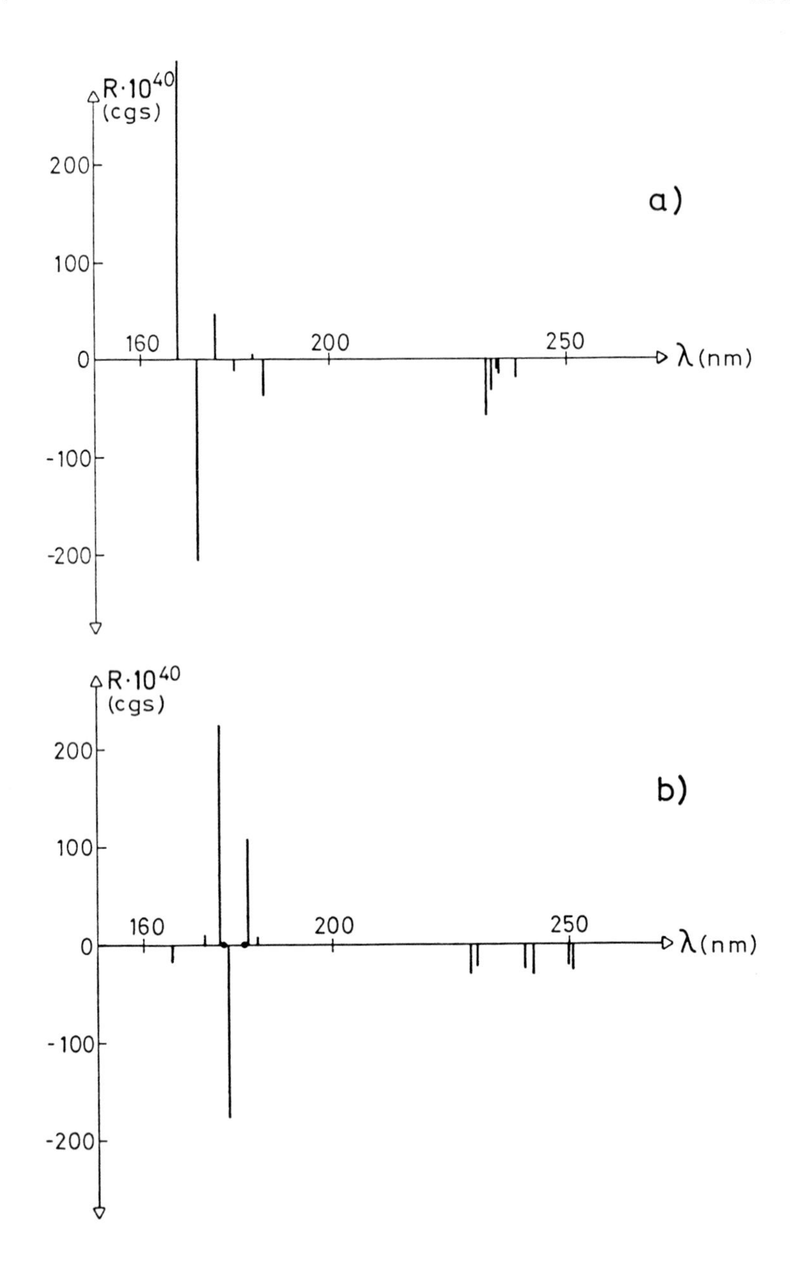

Figure 3. Computed rotatory strengths for a) the hexamer of the PC pleated sheet with $\Phi = -119^{\circ}$, $\psi = 113^{\circ}$, $\omega = 180^{\circ}$; b) the cyclo-hexamer in the conformation given in [31].

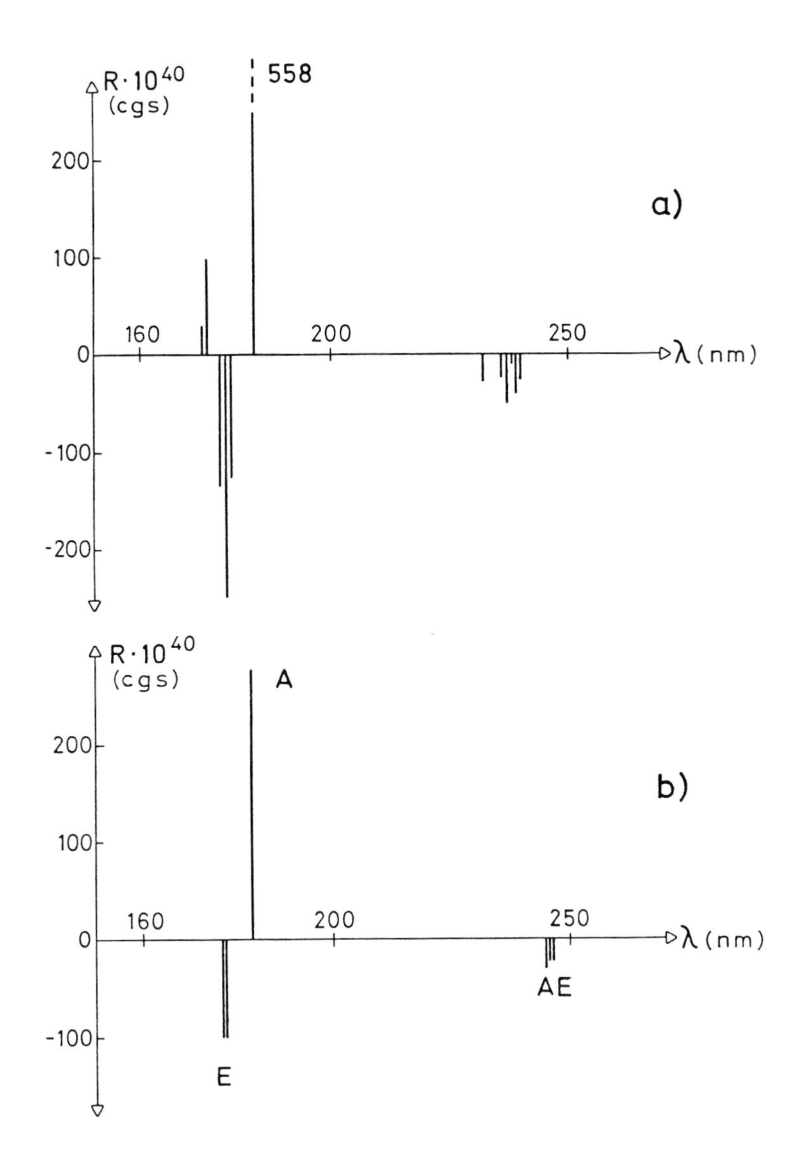

Figure 4. Computed rotatory strengths for a) the hexamer of poly(L-proline)I with $\Phi = -83^o$, $\psi = 158^o$, $\omega = 0^o$; b) cyclo(tri-L-proline) in the conformation of symmetry C_3 described in [32]. See also [33].

Table 2. Rotatory strength per monomer in cgs units and $\Delta\epsilon_{max}$ values, as computed from the hexamer.

α-Helix righthanded

$R(n-\pi^*)$	$= -25.7 \cdot 10^{-40}$	$\Delta\epsilon_{max}\ (n-\pi^*)$	$= -10.5$
$R(\pi-\pi^*)_+$	$= 93.$	$\Delta\epsilon_{max}\ (\pi-\pi^*)_+$	$= 25.$
$R(\pi-\pi^*)_-$	$= -60.$	$\Delta\epsilon_{max}\ (\pi-\pi^*)_-$	$= -16.$

PC Pleated sheet

$R(n-\pi^*)$	$= -22.7 \cdot 10^{-40}$	$\Delta\epsilon_{max}\ (n-\pi^*)$	$= -9.3$
$R(\pi-\pi^*)_+$	$= 60.$	$\Delta\epsilon_{max}\ (\pi-\pi^*)_+$	$= 16.$
$R(\pi-\pi^*)_-$	$= -42.$	$\Delta\epsilon_{max}\ (\pi-\pi^*)_-$	$= -11.$

Poly(L-proline)I

$R(n-\pi^*)$	$= -29.3 \cdot 10^{-40}$	$\Delta\epsilon_{max}\ (n-\pi^*)$	$= -12.$
$R(\pi-\pi^*)_+$	$= 115.$	$\Delta\epsilon_{max}\ (\pi-\pi^*)_+$	$= 31.$
$R(\pi-\pi^*)_-$	$= -84.$	$\Delta\epsilon_{max}\ (\pi-\pi^*)_-$	$= -23.$

Table 3. The π-π* transition with negative rotatory strength appearing at longest wavelength in the right-handed α-helix. Computed components of the electric dipole transition moment in a.u., rotatory strength R and wavelength λ. The z-axis coincides with the helix axis. Notice the relative increase of $\langle\nabla\rangle_z$ as the helix grows, as well as the concomitant redshift.

	$\langle\nabla\rangle_x$	$\langle\nabla\rangle_y$	$\langle\nabla\rangle_z$	$R \cdot 10^{40}$(cgs)	λ(nm)
Trimer	-0.0022	0.1410	0.3147	-220.	169.9
Hexamer	0.2378	-0.0622	0.3562	-115.	177.8
Nonamer	0.1415	-0.0003	0.5396	-499.	181.5

4. MAGNITUDE OF THE nπ*-ππ* INTERACTION

Section 2 describes the way in which the local interactions Λ_i couple n and π at the SCF level. In the frame of the single-excitation CI the following non-vanishing nondiagonal matrix elements occur:

a) $\langle n_i \to \pi_i^* | \mathcal{H} | \pi_i \to \pi_i^* \rangle = -\langle \pi_i \pi_i^* | n_i \pi_i^* \rangle + 2 \langle \pi_i \pi_i^* | \pi_i^* n_i \rangle$

b) $\langle n_i \to \pi_i^* | \mathcal{H} | \pi_j \to \pi_j^* \rangle = 2 \langle \pi_j \pi_i^* | \pi_j^* n_i \rangle$

c) $\langle n_i \to \pi_i^* | \mathcal{H} | n_j \to \pi_j^* \rangle = 2 \langle n_j \pi_i^* | \pi_j^* n_i \rangle$

d) $\langle \pi_i \to \pi_i^* | \mathcal{H} | \pi_j \to \pi_j^* \rangle = 2 \langle \pi_j \pi_i^* | \pi_j^* \pi_i \rangle$

In comparison with the exciton model, these matrix elements may be interpreted in the way indicated below and lead to the following types of contributions to the rotatory strength:

Type of interaction	Rotatory strength contribution
a) local	$\mu_i - m_i$
b) dipole-quadrupole	$\mu_i - m_j$
c) quadrupole-quadrupole	
d) dipole-dipole	$\mu_i - \mu_j$

In the numerical calculations one finds for these matrix elements values in eV ranging from a) 0.7-1.0, b) 0.001-0.06, c) 0.0001-0.02. The $(n_i \to \pi_i^*) - (n_j \to \pi_j^*)$ interaction, while significantly smaller than the $(n_i \to \pi_i^*) - (\pi_j \to \pi_j^*)$ interaction, is nevertheless non-negligible.

5. CONCLUSIONS. FURTHER INVESTIGATIONS

The simplified MO procedure described here gives a consistent, though admittedly still crude description of the chiroptic properties of oligopeptides. To become a useful tool, the method must still be refined. Foremost, the $n\pi^*$-$\pi\pi^*$ interaction should be treated in a more differentiated way. If this interaction is only taken into account by the parameters Λ_i, it is to be apprehended that the local μ_i-m_i mechanism will outweigh the μ_i-m_j mechanism in an unrealistic fashion. Furthermore, it is not apparent how the interaction parameters Λ_i may be consistently calibrated to reflect changes in local and overall conformation. It is therefore suggested to introduce, in addition, a conformation-dependent resonance integral

$$\omega_{i(i+1)} = \eta \cdot \langle N_i | n_{i+1} \rangle$$

where η is a proportionality constant and $\langle N_i | n_{i+1} \rangle$ the overlap integral between the N orbital on the amide group i and the n orbital on the amide group i+1:

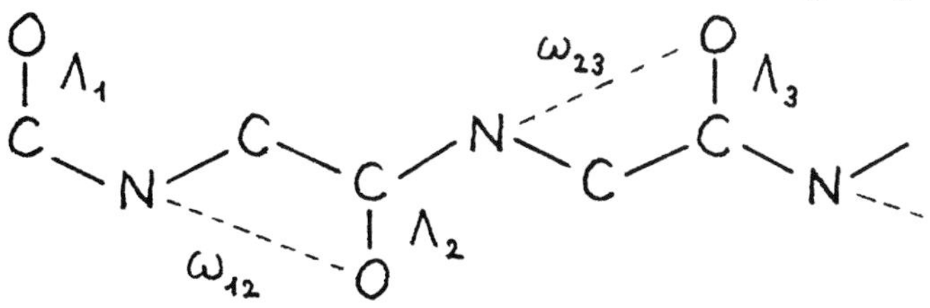

The calibration of the parameters Λ_i and the constant η will have to be performed on a series of oligopeptides of known, relatively rigid conformation. For instance, the cyclo-hexapeptides containing different combinations of glycine, L- and D-alanine [31,36,37] might be well-suited for this purpose.

REFERENCES

1 W.Moffitt, J.Chem.Phys. 25, 467 (1956).
2 D.D.Fitts and J.G.Kirkwood, Proc.Natl.Acad.Sci. USA 42, 33 (1956).
3 W.Moffitt, D.D.Fitts and J.G.Kirkwood, Proc.Natl. Acad.Sci. USA 43, 723 (1957).
4 F.M.Loxsom, J.Chem.Phys. 51, 4899 (1969).
5. M.R.Philpott, J.Chem.Phys. 56, 683 (1972).
6 A.E.Hansen and J.Avery, Chem.Phys.Letters 13, 396 (1972).
7 I.Tinoco,Jr., Advan.Chem.Phys. 4, 113 (1962).
8 R.W.Woody and I.Tinoco, J.Chem.Phys. 46, 4927 (1967).
9 R.W.Woody, J.Chem.Phys. 49, 4797 (1968).
10 P.M.Bayley, E.B.Nielsen and J.A.Schellman, J.Phys. Chem. 73, 228 (1969).
11 V.Madison and J.Schellman, Biopolymers 11, 1041 (1972).
12 E.S.Pysh, J.Chem.Phys. 52, 4723 (1970).
13 N.S.Simmons and E.R.Blout, Biophys.J. 1, 55 (1960).
14 N.S.Simmons, C.Cohen, A.G.Szent-Gyorgyi, D.B.Wetlaufer and E.R.Blout, J.Am.Chem.Soc. 83, 4766 (1961).
15 S.Beychok and E.R.Blout, J.Mol.Biol. 3, 769 (1961).
16 G.Holzwarth, W.B.Gratzer and P.Doty, J.Am.Chem.Soc. 84, 3194 (1962).
17 G.Holzwarth and P.Doty, J.Am.Chem.Soc. 87, 218 (1965).
18 G.Blauer, in Structure and Bonding, Vol. 18, Springer, New York 1974.
19 G. Blauer and G.Wagnière, J.Am.Chem.Soc. 97, 1949 (1975).
20 G.Wagnière and G.Blauer, J.Am.Chem.Soc. 98, 7806 (1976).
21 R.Pariser and R.G.Parr, J.Chem.Phys. 21, 466, 767 (1953); J.A.Pople, Proc.Phys.Soc. A 68, 81 (1955).
22 H.Labhart and G.Wagnière, Helv.Chim.Acta 46, 1314 (1963).
23 Orbital plot program by E.Haselbach and A.Schmelzer, Helv.Chim.Acta 54, 1299 (1971).

24 J.A.Pople, D.P.Santry and G.A.Segal, J.Chem.Phys. 43, S 129 (1965); J.A.Pople and G.A.Segal, J.Chem.Phys. 44, 3289 (1966).
25 D.W.Turner, C.Baker, A.D.Baker and C.R.Brundle, "Molecular Photoelectron Spectroscopy", Wiley Interscience, London 1970, p. 132-138.
26 J.N.Murrell, "The Theory of Electronic Spectra of Organic Molecules", Methuen & Co., London 1963, Chap.8.
27 M.B.Robin, "Higher Excited States of Polyatomic Molecules", Vol. II, Academic Press, New York 1975, p. 75-106, p. 121-160.
28 W.Hug and G.Wagnière, Theoret.Chim.Acta 18, 57 (1970).
29 A.Moscowitz, in C.Djerassi, Optical Rotatory Dispersion, McGraw Hill, New York 1960, p. 165.
30 M.Rothe, cited in R.E.Geiger, Doctoral dissertation, University of Zurich, 1974, p. 95.
31 I.L.Karle, J.W.Gibson and J.Karle, J.Am.Chem.Soc. 92, 3755 (1970).
32 R.E.Geiger, Doctoral dissertation, University of Zurich, 1974, p. 117.
33 M.E.Druyan, C.L.Coulter, R.Walter, G.Kartha and G.K.Ambady, J.Am.Chem.Soc. 98, 5496 (1976).
34 C.M.Deber, A.Scatturin, V.M.Vaidya and E.R.Blout, in Peptides: Chemistry and Biochemistry, B.Weinstein and S.Lande, Eds., Dekker, New York 1970, p. 166.
35 F.A.Bovey and F.P.Hood, Biopolymers 5, 325 (1967).
36 H.Gerlach, J.A.Owtschinnikow and V.Prelog, Helv. Chim.Acta 47, 2294 (1964).
37 H.Gerlach, G.Haas and V.Prelog, Helv.Chim.Acta 49, 603 (1966).

DISCUSSION

PYSH :

Your method will always yield spectra which are conservative, i.e. in which the rotational strength sums to zero. How could your method be modified in order to be successful in cases where the spectrum experimentally is extremely non-conservative, such as poly-L-proline II ?

WAGNIERE :

Our predicted spectra are indeed conservative and thus primarily applicable to systems which show conservative spectra in the wavelength range concerned. To predict spectra which are non-conservative in this range, one would conceivably have to include higher σ-σ^* configurations in the calculation. At present I do not see how this can be done in a clear-cut way without reverting to the difficulties encountered in the all-valence approach.

SNATZKE :

Is there a simple regularity on which peptide unit HOMO and LUMO are "localized" (n = 24 and n = 24 + 1)^2.

WAGNIERE :

We have only performed a limited number of such CNDO calculations and have not noticed such a regularity. In contrast to open-ended peptides, to which this question specifically refers, the CNDO SCF MO's in cyclic peptides of symmetry C_N are of course not localized on individual amide groups; however, the higher occupied MO's can still be classified as "pseudo " or "pseudo n" and the lower unoccupied ones as "pseudo π^*".

INFLUENCE OF 3-SUBSTITUTION ON EXCITED STATE PROPERTIES OF INDOLE IN AQUEOUS SOLUTIONS

Gilbert LAUSTRIAT, Dominique GERARD et Claude HASSELMANN

Laboratoire de Physique de la Faculté de Pharmacie,
Equipe de recherche associée au CNRS (ERA 551),
Université Louis Pasteur, 67083 Strasbourg Cedex.

INTRODUCTION

Considerable attention has been paid to the excited states of the indole ring in the past two decades, since as, the aromatic moiety of tryptophan, it is the major near-ultraviolet chromophore of proteins where it can play the role of a fluorescent probe. Nevertheless photophysical and photochemical properties of indole are rather particular and remain largely unelucidated (Longworth, 1971).

In the course of a study on molecular models of tryptophyl residues, we examined a number of indole derivatives and observed that the ring excited state properties were sensitive to substitution at position 3, which in tryptophan is the attachment site of the alanine chain. This natural substituent exerts on the

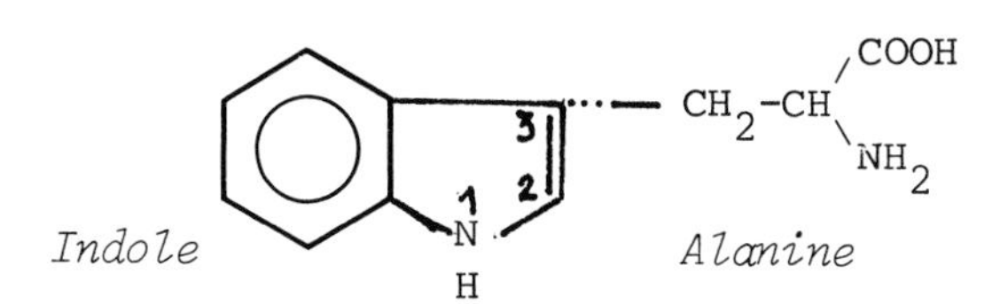

indole ring a mesomeric and an electron-repelling (inductive) effect able to perturb π-electron density distribution. We therefore decided to investigate, by means of various 3-substituents, these chain effects on the spectral behavior of indole in water and on two particular properties of its excited state : the N-H bond lability and the electron ejection.

B. Pullman and N. Goldblum (eds.), Excited States in Organic Chemistry and Biochemistry, 151-162.

The aim of this paper is to discuss the main results of this study, the different parts of which have been — or will be — presented in more detail elsewhere.

EXPERIMENTAL

1. Materials and solutions.

Compounds of the purest quality available (Sigma) were recrystallized from ethanol or water and dissolved in bidistillated water without buffer. The pH was adjusted to the desired value with sodium hydroxide or hydrochloric acid. All measurements were performed at room temperature and on air-saturated solutions unless otherwise indicated.

2. Methods.

Absorption and fluorescence spectra were recorded with a Cary 15 spectrophotometer and an absolute (Fica 55) spectrofluorimeter respectively.

Fluorescence quantum yields (ϕ) were determined from the areas under emission spectra plotted in wavenumbers at constant wavenumber resolution, taking an L-tryptophan neutral solution as the reference (ϕ = 0.14 at 25°C ; Chen, 1967).

Fluorescence decay times were determined by the monophoton sampling technique, as previously decribed (Pfeffer *et al*, 1963).

Irradiations were performed at 254 nm, on nitrogen-saturated equimolar solutions of indole and imidazole (2×10^4 M), by means of an immersed low-pressure mercury lamp (Hanau TNN 15/32). The small 185 nm component of the lamp emission was filtered out with an acetic acid aqueous solution (20 % v/v). The method of Hatchard and Parker (1956) was used to measure the light intensity within the reaction cell (3.5×10^{18} quanta/sec). After irradiations (5 min), imidazole was titrated by the Pauly reaction (Hasselmann, 1976).

RESULTS AND DISCUSSION

1. Choice of indole derivatives.

Compounds have been chosen so as to present at position 3 of the indole ring side chains of various electron-repelling strengths. The most intense repelling effect was attained with the methyl substituent (skatole). Starting from this derivative, decreasing repelling capacities of the chain were obtained with substituents composed of an ethylene bridge (-Et-) and various terminal groups of increasing electron-attracting properties, which progressively compensate the repelling effect of the ethylene bridge (Tournon *et al*, 1972). As increasingly attracting terminal groups, we used : $-COO^-$ (which has very little if any effect), $-NH_2$, $-COOH$ and $-NH_3^+$ (which exhibits the strongest effect due to its extra positive charge). By means of skatole, indole propionic acid, tryptamine and tryptophan, in aqueous solutions at various pHs, the following side chains of decreasing electron-repelling effects could therefore be studied :

$-CH_3$ > $-Et\text{-}COO^-$ > $-Et\text{-}NH_2$ > $-Et(COO^-, NH_2)$ > $-Et\text{-}COOH$ >

$-Et\text{-}NH_3^+$ > $-Et(COO^-, NH_3^+)$ > $-Et(COOH, NH_3^+)$.

With respect to the inductive effect of the chain on the indole chromophore, it is particularly expected that the spectral behavior of ionized propionic acid ($Et\text{-}COO^-$) be close to that of skatole, whereas the behavior of the acidic form of tryptophan $Et(COOH, NH_3^+)$, where the presence of two strongly attracting terminal groups should merely counterbalance the repelling effect of the ethyl bridge, be close to that of unsubstituted indole.

2. Absorption spectra.

The first absorption bands of indole and skatole, which in agreement with the previous considerations exhibit the most important differences among the studied compounds, are shown on figure 1a.

The indole spectrum is known to be mainly composed of a broad band corresponding to the $^1L_a \leftarrow A$ transition and peaking at 268 nm. To this band, approximately represented on the figure by the dashed line, are superimposed two smaller bands pertaining to the $^1L_b \leftarrow A$ overlaping transition and responsible for the secondary maxima at 276 and 286 nm (Bernardin, 1970 ; Strickland and Billups, 1973 ; Valeur and Weber, 1977).

The spectrum of skatole presents the same general structure as that of indole, with definite differences. These differences, however, are seen to result mainly from a decrease in intensity and a red shift of the ^1L_a band, without important variations of the ^1L_b band. Such modifications of the ^1L_a band explain that on the whole spectrum the red shift is greater for the short-wavelength peak ($\Delta\lambda$ = 7 nm, $\Delta\lambda$ = 950 cm^{-1}) than for the long-wavelength peak ($\Delta\lambda$ = 2 nm, $\Delta\lambda$ = 245 cm^{-1}), and also the fact that the maximum of the spectrum now appears on the intermediate band (280 nm).

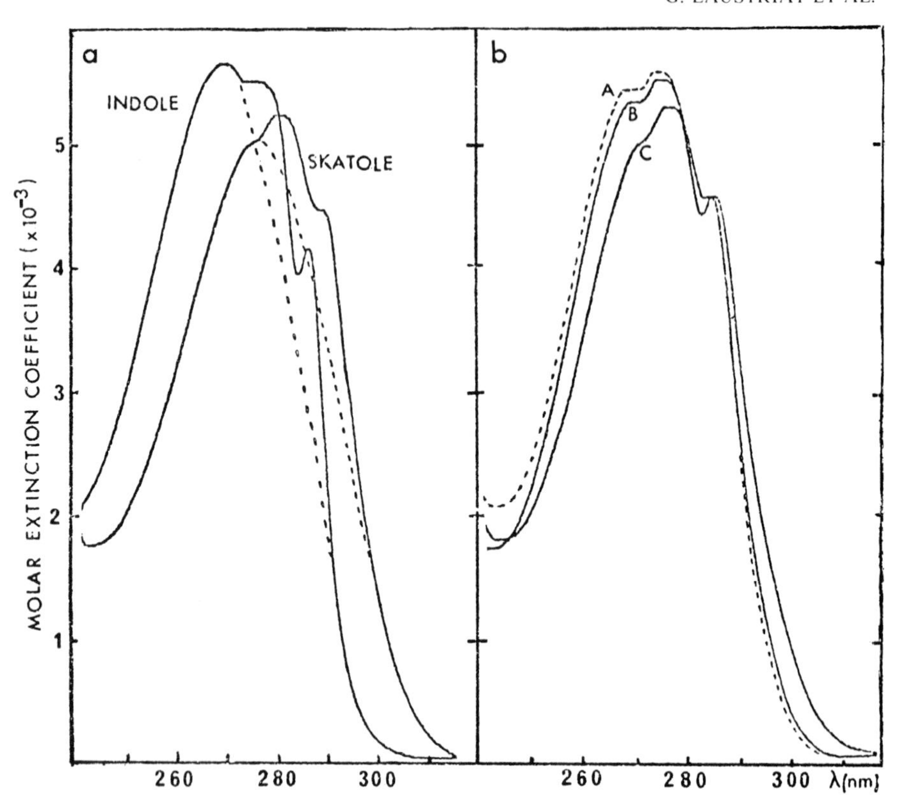

Figure 1. Absorption spectra in aqueous solutions. a) indole and skatole (dashed lines : approximate shape of the ^{1}La bands). b) tryptophan : pH 0.5 (curve A), pH 7 (curve B), pH 12 (curve C).

That the change in the spectrum shape is partly due to an inductive effect of the substituent on the indole chromophore is illustrated in figure 1b, where similar variations are seen to progressively appear in the absorption spectrum of tryptophan when passing from Trp(COOH, NH_3^+) to Trp(COO^-, NH_3^+) and Trp(COO^-, NH_2), that is when the electron-repelling effect of the chain is increased. It is also interesting to note that, as expected, the spectrum of Trp(COO^-, NH_2) is almost identical to that of skatole whereas between indole and Trp(COOH, NH_3^+) there remains differences which are probably ascribable to the mesomeric effect of the substituent.

As to the red shift of the ^1L_a band on increasing electron-repelling effect of the chain, it could reflect an increase in the solvent-solute interaction due to an increase in the dipole moment change $\Delta\vec{\mu}$ upon excitation to the ^1L_a state (Tinoco *et al*, 1962). As a matter of fact, this change has been shown to be important for the $^1L_a \leftarrow A$ transition (while very small for the $^1L_b \leftarrow A$ transition) (Sun and Song, 1977) and could thus be sensitive to inductive effects of the substituents. Ulterior discussion of emission spectra will corroborate this interpretation.

It thus appears that the electron-repelling effect of the chain at position 3 of the indole ring mainly affects the $^1L_a \leftarrow A$ transition, by reducing the oscillator strength and increasing the dipole moment change. It is hoped that calculations in progress in this laboratory will sustain this experimental conclusion.

3. Fluorescence parameters.

In aqueous solutions, where the indole fluorescence is attributed to the ^1L_a state (Konev, 1967), the emission spectrum consists in a well-known broad and structureless band, exhibiting an unsually large Stokes shift. This shift has been ascribed to the formation of a specific excited state complex (exciplex) between solvent and solute molecules (Walker *et al*, 1966, 1967) or to a solvent shell relaxation during the lifetime of the excited state (Mataga *et al*, 1964 ; Eisinger and Navon, 1969), with a possible contribution of hydrogen bonding at the ring nitrogen (Van der Donckt, 1969). Recent calculations (Sun and Song, 1977) showed that the Stokes shift can be accounted for by the solvent relaxation process, owing to the large change in dipole moment upon excitation to the ^1L_a state.

Two fluorescence parameters of the studied compounds are given in Table I : the emission maximum wavelength (λ^e_{max}) and the ratio of the quantum yield (ϕ) to the lifetime (τ).

λ^e_{max} is seen to progressively increase from indole to skatole, the variation paralleling exactly that of the electron-repelling effect of the 3-substituent. With the exception of Trp (COOH, NH_3^+), the red shift of the emission spectrum upon indole substitution exceeds the corresponding red shift of the ^1L_a absorption band, indicating that the phenomenon is due to an increase of the Stokes shift. This Stokes shift increase is given by :

$$\Delta(\Delta\bar{\nu}) = \Delta\bar{\nu}_{In-S} - \Delta\bar{\nu}_{In}$$

where $\Delta\bar{\nu}$ is the wavenumber difference between the ^1L_a absorption band and the emission spectrum maxima (In an In-S refering to indole and substituted indole respectively). Values of $\Delta(\Delta\bar{\nu})$, indicated in Table I, are seen to increase, as those of λ^e_{max}, from Trp(COOH, NH_3^+) to ionized indole propionic acid and skatole. With the most electron-repelling substituents, the Stokes shift

Table I. Fluorescence parameters of indoles.

3-Substituent	λ^{e}_{max} (nm)	$\Delta(\Delta\bar{\nu})$ (cm^{-1})	$\phi/\tau \times 10^{-7}$ (s^{-1})
-H (indole)	343	0	6.5 ± 0.5
$-CH_2-CH(COOH)(NH_3^+)$	344	- 300	5.5 ± 1
$-CH_2-CH(COO^-)(NH_3^+)$	351	+ 120	4.6 ± 0.3
$-CH_2-CH_2-NH_3^+$	351	+ 120	5.4 ± 0.4
$-CH_2-CH_2-COOH$	363	+ 810	-
$-CH_2-CH(COO^-)(NH_2)$	365	+ 810	4.5 ± 0.2
$-CH_2-CH_2-NH_2$	366	+ 810	-
$-CH_2-CH_2-COO^-$	367	+ 960	3.9 ± 0.3
$-CH_3$ (skatole)	369	+1110	4.0 ± 0.3

enhancement due to 3-substitution is very important (λ^{e}_{max} = 26 nm, $\Delta(\Delta\bar{\nu})$ = 1100 cm^{-1}) ; in contrast with Trp (COOH, NH_3^+) the Stokes shift enhancement is about nil, meaning that the electron-attracting effects of the two terminal groups cancel the repelling and mesomeric effects of the ethylene bridge.

According to the solvent relaxation mechanism, the observed correlation would indicate that the electron-repelling effect of the chain promotes an increase in the dipole moment change upon excitation to the ^1L_a state. The fact that the same conclusion was derived from the analysis of absorption spectra strengthens this interpretation, which in turn sustains the solvent relaxation process as the main cause of the Stokes shift of indoles in water.

Another consistency between absorption and emission data lies in the variation of the ϕ/τ ratio for the studied compounds. As a matter of fact, this ratio is a measure of the rate constant k_f of the radiative $^1L_a \rightarrow A$ transition and it should therefore be correlated to the oscillator strength of the corresponding $^1L_a \leftarrow A$ absorption transition. Despite the uncertainties in the experimental values of ϕ/τ, this parameter is seen (Table I) to decrease from indole to skatole, that is to vary in the same way as the oscillator strength of the $^1L_a \leftarrow A$ transition.

The two independent correlations (spectral shifts and oscillator strengths) observed between variations of the ^1L_a absorption band and of fluorescence parameters confirm that the emission of indoles in water originates from the ^1L_a state and shows that the $\Delta\vec{\mu}$ change in going from the ground to the ^1L_a state is particularly sensitive to inductive effects at position 3 of the ring.

4. Deprotonation of the indole nitrogen.

Recent calculations (Song and Kurtin, 1969 ; Sun and Song, 1977) have shown that the π-electronic charge on the indole nitrogen N_1 decreases upon excitation to the ^1L_a state, implying that the N_1-H bond strength is weaker in the excited than in the ground states. This is experimentally verified by the facts that the nitrogen deprotonation pK of indole is very high (17.0) in the ground state (Dross, 1976), while the fluorescence quenching of indole by OH^- ions — which results from deprotonation in the excited state, leading to the non-fluorescent indolate anion — occurs in the pH range 11-13 (Schang, 1975).

Calculations have also shown that substitution of indole could modify — in various manners according to the position and the nature of the substituent — the N_1 electronic charge (e_N^*) in the ^1L_a state (Sun and Song, 1977). It could then be expected that an electron-donating substituent on carbon 3 would increase e_N^* and therefore strengthen the N-H bond in the excited state. This eventuality has been tested by analyzing the alkaline fluorescence quenching of indole, skatole and tryptophan Trp(COO^-, NH_2) . To this end, the Stern-Volmer constant K_{sv} of the quenching (deprotonation) reaction :

$$InNH^* + OH^- \rightarrow H_2O + InN^{-*} \quad (\rightarrow InN^- + \text{heat})$$

was determined. This constant may be written as :

$$K_{sv} = k_r \tau_o p$$

Table II. Kinetic parameters of the alkaline quenching of some indoles.

Compound	K_{sv} (M^{-1})	k_r (M^{-1} s^{-1})	τ_o (ns)	p
Indole	5.8		4.3	0.9 ± 0.15
Skatole	6.0	1.5×10^9	9.0	0.5 ± 0.1
Trp(COO^-,NH_2)	6.1		9.0	0.5 ± 0.1

where τ_o is the fluorescent state lifetime at very low concentration in OH^- ions ($[OH^-] < 10^{-5}$ M), k_r is the rate constant of the diffusion-controlled reaction, which can be evaluated by means of Smoluchovski's equation, and p is the probability of the indole ring deprotonation during a molecular encounter.

Results presented in Table II show that the deprotonation probability p (determined from the above relation) is close to unity for indole, reflecting the relative lability of the nitrogen proton in the excited state, but is significantly smaller for skatole and tryptophan, indicating that the electron-donating 3-substituent indeed strengthens the N-H bond.

It is to be noted that this phenomenon could account for the unexplained increase in fluorescence quantum yield from indole (ϕ = 0.28) to skatole (ϕ = 0.36) and alkaline tryptophan (ϕ = 0.40). As a matter of fact, deprotonation — or partial deprotonation — of the ring nitrogen has been shown to be an important cause of the non-radiative deexcitation of indole in water (Stryer, 1966 ; Ricci, 1970). The enhancement of the N-H bond strength by the substituent, by lowering the efficiency of this process, would increase the fluorescence quantum yield.

5. Electron photoejection.

Electron photoejection from indoles in aqueous solutions, first observed in flash photolysis experiments (Grossweiner and Joschek, 1965), has been shown to originate from the excited singlet state (Feitelson, 1971 ; Bent and Hayon, 1975). Under monophotonic (steady state) excitation, the quantum yield of this process is relatively low (less than 0.1 ; Feitelson, 1971 ; Steen, 1974) and is constant over the first absorption band (Steen, 1974 ; C. Hasselmann and G. Laustriat, unpublished results). The admitted

Table III. Electron photoejection quantum yield ϕ_e (x 10^2) and rate constant k_e (x 10^6) of some indoles (pH 6).

	Indole	Tryptophan	Tryptamine	Skatole
ϕ_e	3.2	0.8	1.4	2.1
k_e	7.5	2.5	2.3	2.3

reaction scheme is (Bent and Hayon, 1975 ; Evans *et al*, 1976) :

$$^1InNH^{*} \rightarrow e^-_{aq} + InNH^{+\bullet} \xrightarrow{(pK = 4.3)} InN^{\bullet} + H^+$$

Influence of 3-substitution on indole electron photoejection efficiency was investigated using protonated imidazole (ImH^+) as a hydrated electron (e^-_{aq}) scavenger in deareted solutions at pH 6 (Hasselmann, 1976). e^-_{aq} is known to be very reactive toward ImH^+ and to provoke the imidazole ring disruption (Bazin *et al*, 1975), so that in fixed experimental conditions and for short irradiation times the imidazole degradation rate is proportional to the indole photoejection quantum yield ϕ_e.

Values of ϕ_e and of the rate constant $k_e = \phi_e \tau_o^{-1}$ obtained by this method for indole and some 3-substituted derivatives are indicated Table III. They show that the presence of an electron-donating substituent at position 3 of the indole ring promotes a decrease of the electron photoejection efficiency.

No straightforward interpretation of this result may be offered, since an opposite trend could have been expected *a priori*, due to a lowering of the ring ionization potential by the substituent. It could however be consistent with the partial proton transfer process mentioned above. As a matter of fact, the anionic form of excited indole should participate in electron ejection due to the increased electronic charge of the ring. The enhancement of the N-H bond strength by the substituent, by reducing the proton transfer efficiency, would then decrease the electron ejection quantum yield. Such a mechanism however, if existant, should not be unique since hydrated electrons formation was also evidenced with N-methyl indole, but to a much lesser extent (ϕ_e = 1.3 x 10^{-2}) than with indole or skatole.

CONCLUSION

This experimental study shows that the excited state properties of the indole ring are modified by perturbing its π-electronic system with electron-withdrawing and donating 3-substituents. The main effects and their implications are :

- a decrease in the oscillator strength and in the energy of both ($^1L_a \leftarrow A$) and emission transitions, confirming the assignment of indole fluorescence in water to the ^1L_a state ;

- an increase in the Stokes shift, implying that the dipole moment change upon excitation to the ^1L_a state is very sensitive to 3-substitution and that the unusually large Stokes shift of indoles in water is mainly due to a solvent relaxation process ;

- an increase in the N_1-H bond strength, accounting for the enhancement of the fluorescence quantum yield by 3-substituents devoid of terminal quenching group(s), and possibly for the variations of the electron photoejection yield upon substitution.

Acknowledgement. This work was supported by the Institut National de la Santé et de la Recherche Médicale (Grant 74.1.225.3).

REFERENCES

Bazin, M., Hasselmann, C., Laustriat, G., Santus, R. and Walrant, P. (1975) Chem. Phys. Lett. 36, 505.

Bent, R. and Hayon, E. (1975) J. Am. Chem. Soc. 97, 2612.

Bernardin, J.E. (1970) Ph. D. Thesis, University of Oregon.

Chen, R.F. (1967) Anal. Lett. 1, 35.

Dross, C. (1976) Thèse (Pharmacie), Université Louis Pasteur, Strasbourg.

Eisinger, J. and Navon, G. (1969) J. Chem. Phys. 50, 2069.

Evans, R.F., Ghiron, C.A., Volkert, W.A. and Kuntz, R.R. (1976) Chem. Phys. Lett. 42, 43.

Feitelson, J. (1971) Photochem. Photobiol. 13, 87.

Grossweiner, L.J. and Joschek, H.J. (1965) Advan. Chem. Ser. n° 50, p. 279, American Chemical Society, Washington D.C.

Hasselmann, C. (1976) Thèse Doct. Sci., Université Louis Pasteur, Strasbourg.

Hatchard, C.G. and Parker, C.A. (1956) Proc. Roy. Soc. A 235, 518.

Konev, S.V. (1967) Fluorescence and Phosphorescence of Proteins and Nucleic Acids, Plenum Press, New-York.

Longworth, J.W. (1971) in Excited States of Proteins and Nucleic Acids (Ed. Steiner, R.F. and Weinryb, I.), Plenum Press, New-York, p. 319.

Mataga, N., Torihashi, Y. and Ezumi, K. (1964) Theoret. Chim. Acta 2, 158.

Pfeffer, G., Lami, H., Laustriat, G. and Coche, A. (1963) Colloque International d'Electronique Nucléaire de Paris, p. 93.

Ricci, R.W. (1970) Photochem. Photobiol. 12, 67.

Schang, J. (1975) Thèse (Pharmacie), Université Louis Pasteur, Strasbourg.

Song, P.S. and Kurtin, W.E. (1969) Photochem. Photobiol. 9, 175.

Steen, H.B. (1974) J. Chem. Phys. 61, 3997.

Strickland, E.H. and Billups, C. (1973) Biopolymers 12, 1989.
Stryer, L. 1966 J. Am. Chem. Soc. 88, 5708.

Sun, M. and Song, P.S. (1977) Photochem. Photobiol. 25, 3.

Tinoco, I., Halpern, A. and Simpson, W.T. (1962) Polyamino acids Polypeptides and Proteins, Univ. Wisconsin, Press Madison, p. 147.

Tournon, J., Kuntz, E. and El-Bayoumi M.A. (1972) Photochem. Photobiol. 16, 425.

Valeur, B. and Weber, G. (1977) Photochem. Photobiol., in press.

Van der Donckt, E. (1969) Bull. Soc. Chim. Belg. 78, 69.

Walker, M.S., Bednar, T.W. and Lumry, R.W. (1966) J. Chem. Phys. 45, 3455.

Walker, M.S., Bednar, T.W. and Lumry, R.W. (1967) J. Chem. Phys. 47, 1020.

THE SUDDEN POLARIZATION EFFECT

L. SALEM

Laboratoire de Chimie Théorique (ERA n° 549)
Université de Paris-Sud, Centre d'Orsay
91405 Orsay (France)

I - INTRODUCTION

In 1970, Dauben obtained[1] highly stereospecific products in the photocyclization of trans-3-ethylidene cyclooctene :

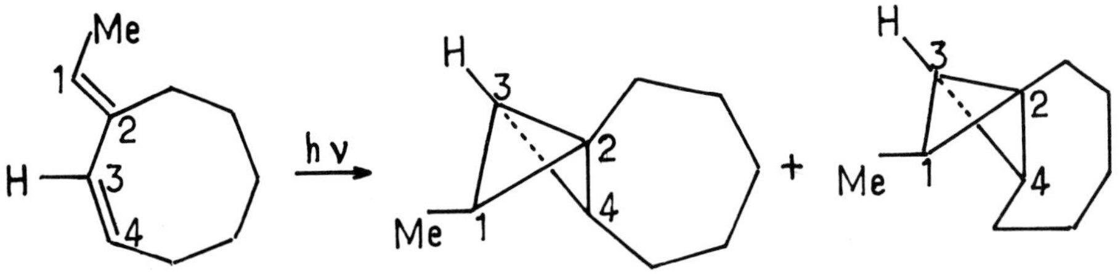

He postulated that the exclusively conrotatory closing of the extra-annular 3-membered ring could be ascribed to the transient formation of an allyl anion system, viz. :

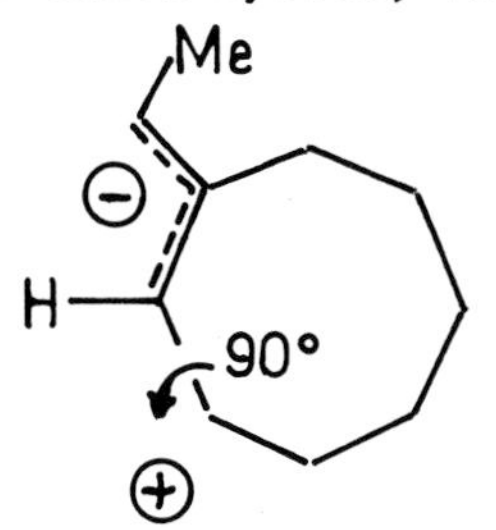

where the charge separation results from rotation around the ring double bond. Three years later Dauben extended his proposal[2] to the photochemistry of s-cis, s-trans hexatrienes. The general scheme

B. Pullman and N. Goldblum (eds.), Excited States in Organic Chemistry and Biochemistry, 163-174.

was suggested for the formation of bicyclo-[3.1.0]-hexenes. Dauben 's zwitterionic mechanism was a rather daring proposal since only one reaction had been performed[3] with sufficient labeling (R=Ph, R'=H, with a phenyl substituent in the 1 position and another Ph group on carbon 2) to observe stereospecific conrotatory closure of the three-membered ring.

In an apparently totally unrelated theoretical study, and at a similar time, Wulfman and Kumei emphasized[4] the highly peculiar physical characteristics of the excited states of alkenes. They showed that the two lowest excited states of twisted ethylene -which were known, since the work of Mulliken, to be purely ionic[5]- should be also highly polarizable. The wave functions for these two states are :

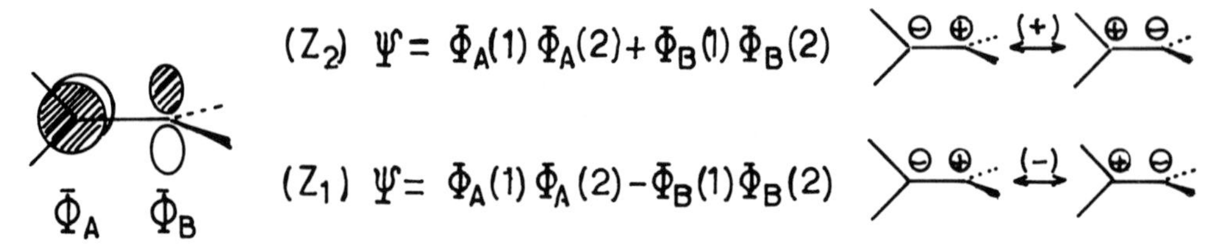

with the "ion-pair" state Z_2 lying higher than Z_1. The very close energy of these states, and the existence, in either one, of identical subcomponents, should allow any internal or external perturbation -according to Wulfman and Kumei- to mix them. Thus either state can become polarized by increasing its share (via borrowing from the other state) of one ionic structure :

Z_1 ⊕——⊖ ⊖——⊕ Z_2

us to a keen interest in their excited zwitterionic states. Intrigue by Dauben's suggestion, we decided to investigate the charge distri- bution in the ground diradical and excited zwitterionic states of the diallyl system corresponding to s-cis, s-trans hexatriene twisted about the central 34 bond :

The open-shell restricted Hartree-Fock technique, complemented by 3-by-3 configuration interaction, appeared to be a reasonable starting point for such a calculation. To our great surprise we found both zwitterionic states to be strongly polarized[7] :

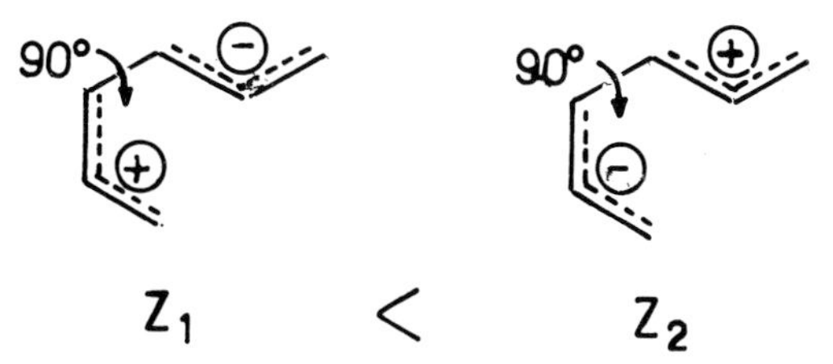

The calculated net charges, in both cases, are roughly equal to 0.8e. Such a charge separation, in a non-polar neutral hydrocarbon, is enormous by conventional standards. It was clear that this unconventional phenomenon, which appeared to confirm both Wulfman and Kumei's, and Dauben's predictions, required further investigation.

II - POLARIZED ZWITTERIONIC STATES IN VARIOUS SYSTEMS

We first pursued our study of diallyl by investigating whether the polarization subsisted at twist angles θ other than 90°. Figure 1 shows that the charge separation peaks very sharply for θ 90°, and drops to practically zero outside a narrow 2° region surrounding this sharp maximum. This result indicates that incipient overlap between the two allylic fragments rapidly destroys the polarisation. Overlap between the radical sites must be very small. We next verified that in s-cis, s-cis diallyl and s-trans, s-trans diallyl - which both have an element of symmetry- the zwitterionic states are not polarized :

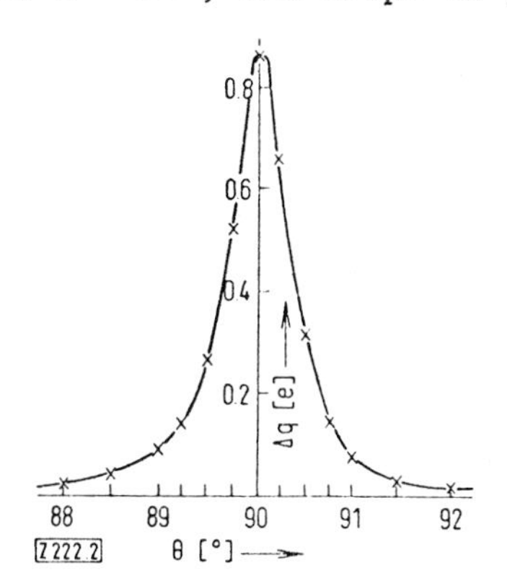

Fig. 1. Charge separation Δq in Z_1 excited state of s-cis, s-trans-diallyl (1,3,5-hexatriene) as a function of twist angle θ[7].The curve is not symmetric about 90°.

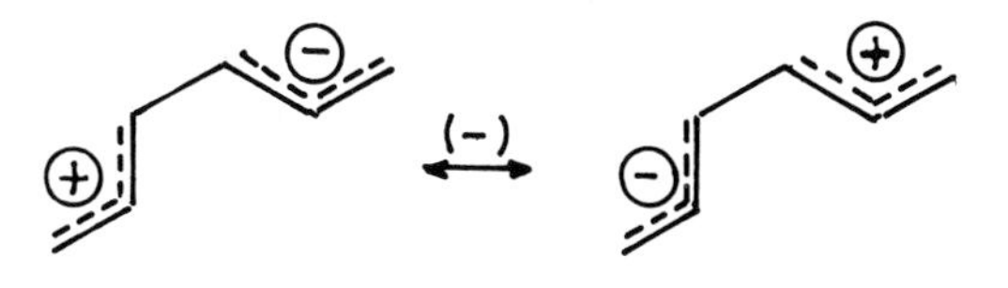

Hence dissymmetry between the two sites of s-cis, s-trans diallyl, however slight, seems also to be required for polarization.

The prerequisite of dissymmetric sites is confirmed by a calculation on 90°-twisted ethylene in which one terminal group is progressively slightly pyramidalized :

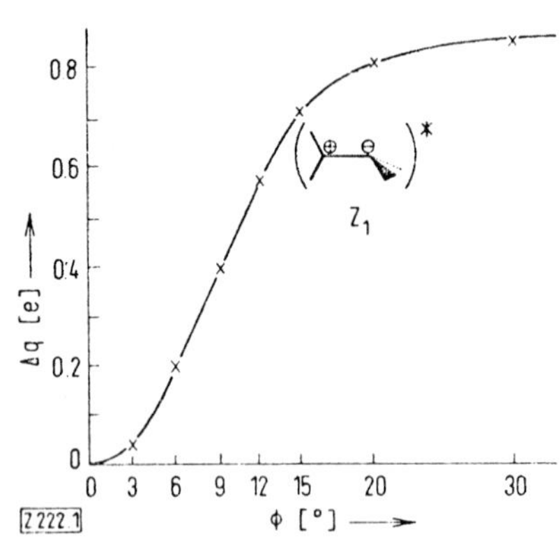

Fig. 2. Charge separation Δq in Z_1 excited singlet state of 90°-twisted ethylene as a function of the pyramidalization angle φ[7]. The number Δq has been corrected for the intrinsic charge separation in the ground state (which never exceeds 0.04e).

The charge separation increases rapidly with angle of pyramidalization, reaching a plateau for ∿ 12° (Figure 2).

Other diradicaloid systems in which we found polarized zwitterionic states are methylene-allyl I (where the direction of polarization depends on the basis set), edge-to-face trimethylene II, and 1, 2 cyclopropylcarbinyl III with one methylene group bisected and the other group perpendicular relative to the ring. This last result may have implications on the mechanism of the photochemical di-π methane rearrangement[8]. In edge-to-face trimethylene the polarization subsists for a rather wide range

(STO-3G) (4-31G)

I II III

of angles of twist of the edge group away from the horizontal symmetry plane. This result is consistent with the requirement of very small overlap. Indeed in trimethylene the 1-3 orbital overlap remain minute untill both methylene groups come near to the face-to-face conformation.

III - THE THEORY UNDERLYING THE PHENOMENON

(a) Our "experimental" calculations seem to indicate two requirements for polarization of Z_1 and Z_2 :

(1) Small to vanishing overlap between radical sites (orbitals)

(2) Dissymmetry between radical sites

The quantum-mechanical basis for these conditions is best illustrated for a case with zero-overlap between site orbitals, for which we consider the two polar forms as starting point. For 90°-twisted ethylene the two-electron energies of the two ionic forms are respectively J_{aa} and J_{bb} :

For zero pyramidalization, $J_{aa} = J_{bb}$. But as soon as the C_aH_2 group is pyramidalized $J_{aa} < J_{bb}$ (center a has a more expanded orbital, due to hybridization acquired <u>via</u> pyramidalization). Now resonance between these two structures, in the classical sense -or mixing, in the quantum-mechanical sense- is possible via the exchange integral K_{ab}. This integral accounts for the fact that the charge, and the counter-hole, can move back and forth from center a to center b at a frequency[9]

$$\frac{K_{ab}}{\pi}$$

For 90°-twisted ethylene, $K_{ab} = 0.0021$ a.u. and the time for exchange is of the order of 10^{-14} seconds.

Whether or not the two states of the system are polarized structures or are non-polar resonance mixtures depends on the eigenfunctions of the coupling matrix

$$\begin{vmatrix} J_{aa} - E & K_{ab} \\ K_{ab} & J_{bb} - E \end{vmatrix} = 0 \qquad (1)$$

Essentially, the eigenstates are ionic, as drawn above, if

$$K_{ab} \ll J_{aa} - J_{bb} \qquad (2)$$

Since the exchange integral is non-zero the r.h.s. of (2) must be larger than zero : whence the necessary <u>dissymmetry</u> between centers a and b. If the dissymmetry is introduced progressively (as in the pyramidalization of 90°-twisted ethylene) the polarization will occur suddenly when (2) is verified. Whence the term <u>"sudden" polarization</u> for the phenomenon. At the same time, $J_{aa}-J_{bb}$ will never be very large and the exchange integral must remain small. If we now consider a system in which the two orbitals ϕ_a and ϕ_b can overlap, this condition is equivalent to requiring <u>small or zero overlap</u>.

(b) Normally, then, the variation of energy of a Z state with increasing $J_{aa}-J_{bb}$ should be monotonic and relatively slow. This seems indeed to be the case in 90°-twisted ethylene as a function

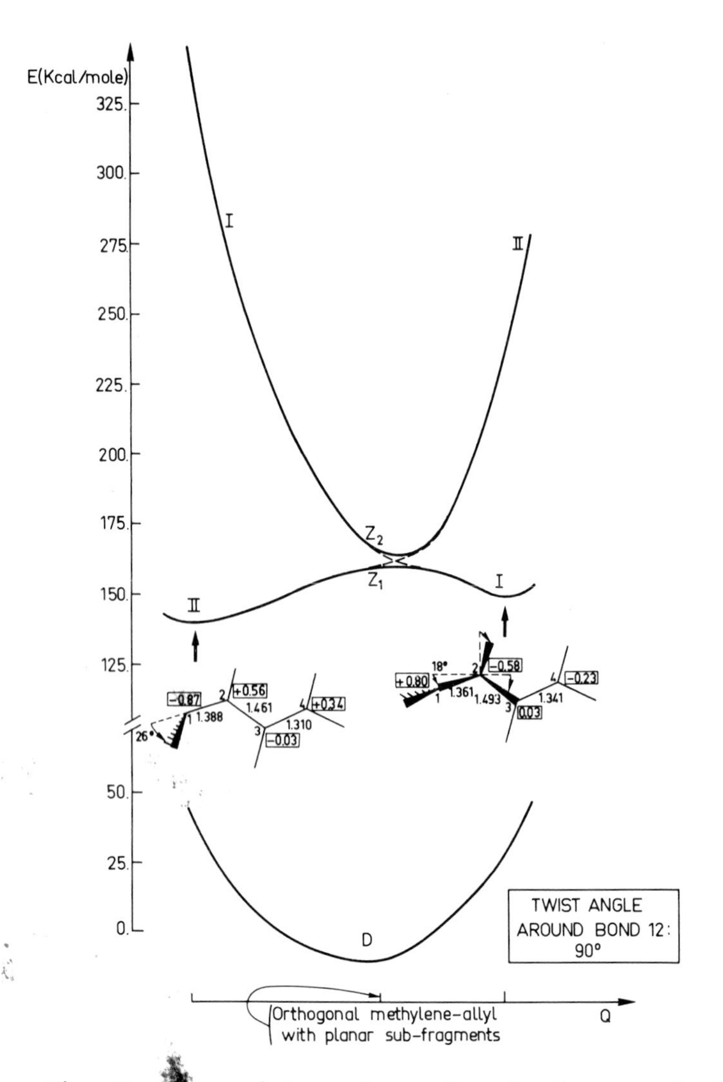

Fig. 3. Potential surfaces for the lowest singlet states of 90°-twisted methylene-allyl (minimal basis set)[13]. The coordinate Q is obtained by linear interpolation from the geometry of II to an intermediate geometry with planar subfragments, and from this intermediate geometry to the geometry of I. Bond lengths in this "half-way" skeleton are C_1C_2=1.375 Å, C_2C_3=1.475 Å, C_3C_4=1.325 Å (for this geometry II lies below I). Dotted lines show the avoided crossing between the configurations corresponding respectively to I and II. Note that the diradical D is a maximum, or a near-maximum, along the coordinate for twisting around bond 12 (i.e., vertical excitation does not occur from D, but from the untwisted butadiene).

of pyramidalization[6] (see the Table). Berthier has raised[10] the interesting point that the nearly-discontinuous nature of the charge polarization might reflect a discontinuous energy solution in the Hartree-Fock method, similar to an instability. A calculation, for 90°-twisted ethylene at various angles of pyramidalization, of the zwitterionic states can be made directly by imposing

Pyramidalization angle	Energy (Kcal/mole)
0°	74.6
3°	74.6
10°	74.6
20°	73.1
30°	71.4
40°	69.4
54°7	66.7

Table. Energies of Z_1 state relative to Ground State Coplanar Ethylene. Calculation including 271 configurations.

a closed-shell restriction on the wave function and using a direct minimization technique. In addition to the energies of Z_1 and Z_2, a third state of lower energy is calculated by Berthier at $\approx$ 35 Kcal/mole below Z_1 and Z_2. It appears, however, that the third solution is simply an approximation, given poorly by a closed-shell wave function, to the diradical state ^1D which lies at $\approx$ 80 Kcal/mole[11] below Z_1 and Z_2[12].

(c) Another question which may be addressed is the variation in energy of Z_1 and Z_2 as a function of geometry, in particular if the geometries of the two oppositely polarized forms are optimized. The excited molecule, in either of the polarized forms, will seek to optimize its energy by adopting the geometry most appropriate to that form. We find that, in its optimized geometry, each zwitterionic form lies <u>below</u> the form with the opposite polarization. Hence the optimally distorted molecules, with <u>either</u> polarization, lie on the <u>same</u> (lowest excited) singlet surface. The resulting double minimum potential is illustrated in Figure 3 for terminally twisted butadiene. The two minima do not have the same height ; moreover their relative height will depend on the pattern of substitution. In this manner <u>reversal of stereochemistry</u> is predicted for ring closure in polarized zwitterionic states[13,14] :

Me ⊖ ⊕ < Me ⊕ ⊖ (Conrotatory closure)

(Disrotatory closure)

In practice one (or even both) of the polarized forms may have a reaction channel faster than electrocyclic ring closure. This may be the case in Havinga's recent experiments[15], where at least partial confirmation of the dual photochemistry associated with the oppositely polarized zwitterionic forms of s-trans,-s-cis diallyl seems to be forthcoming. The photochemistry of previtamin D (P)

leads to two toxisterol products, a bicyclohexene and a spiro compound :

I

II

The stereochemistry of I is unambiguously $2\pi a + 4\pi a$, which is consistent with a zwitterionic precursor :

where the 3-membered ring closes first. Compound II is obtained <u>via</u> H abstraction from 4 to 9, followed by 4-8 bond formation. A possible mechanism for such hydrogen abstraction is passage through the <u>oppositely</u> polarized zwitterionic form :

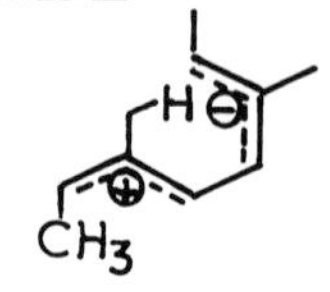

where everything conspires to pull the proton off from 4 to 9. This hydrogen abstraction would occur preferentially to disrotatory three-membered ring closure. It is interesting that the terminal-methyl s-trans, s-cis diallyl skeleton is precisely that for which V. Koutecky calculates[14] the smallest (1 Kcal/mole) energy difference between the two opposite zwitterionic forms.

(d) A final problem concerns the actual extent of charge separation. In some recent important work, Malrieu at all. have shown[16] that the charge separation (and concomitant large depole moment) is generally smaller than that calculated by a limited, three-configuration interaction technique. If sufficient configuration interaction is introduced (in particular with covalent states) to allow total freedom of distribution of the charge distribution in Z_1 and Z_2, the charge is found to concentrate in the regions closest to the border between the two odd orbital sites. Malrieu's charge distribution for twisted butadiene is

- 0.17 +0.17
+0.43
-0.43

Malrieu's effect decreases the overall depole moment created by sudden polarization, whatever the size of the system, to approximately 4 Debyes.

IV - SUDDEN POLARIZATION IN THE N-RETINYLIDENE CHROMOPHORE

A rough description of a rod cell of the eye is shown in Figure 4. The arrows indicate the direction of flow of sodium ions, from outside the cell into the outer segment, through the inner segmant and ouside again. When light impacts on the cell, the plasma membrane becomes less permeable to the ions. Apparently,

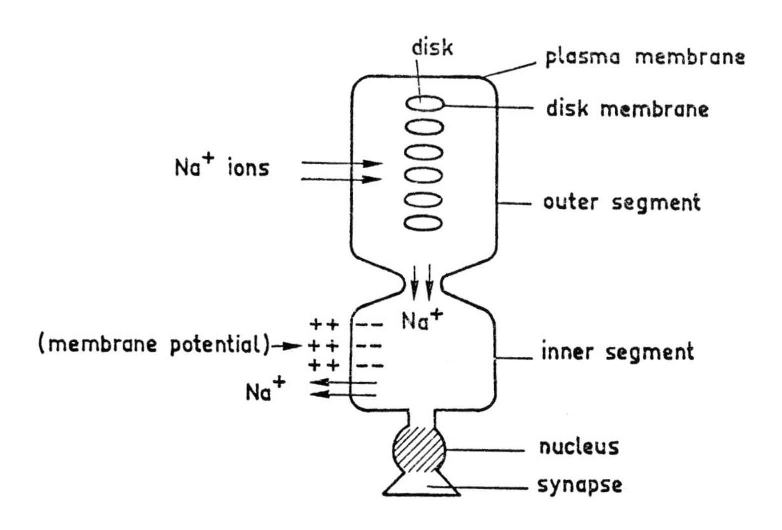

Fig. 4. Schematic description of rod cell.

this arises from a change in the membrane potential, itself triggered by a change in permeability to ions of the disk membranes. These disk membranes contain rhodopsin, which is the crucial molecule in the phenomenon of vision.

The rhodopsin molecule is a protonated Schiff base of retinal, whose skeleton is probably in the 11-cis, 12-s-cis form. The primary excitation of the retinal skeleton leads to the all-trans form, via rotation around the 11-12 double bond. If we restrict this rotation to its half-way point at 90°, we notice that the conjugated carbon skeleton is transformed into two pentadienylic moieties (for simplicity we neglect the cyclohexene fragment of

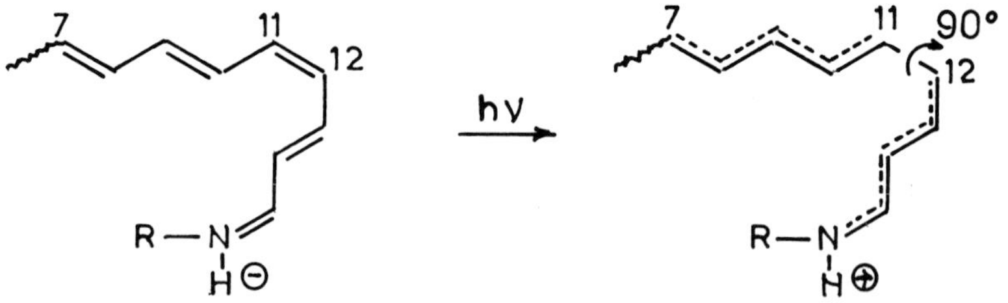

retinal, whose double bond seems to be well out-of-plane relative to the remaining carbon skeleton). Since these fragments are not equivalent (because of the imino group), polarization of the lowest excited state should occur. A negative charge should move towards the positive end (giving a neutral 12-16 fragment) while a positive charge is created on the 7-11 fragment. These results, postulated and demonstrated earlier[17] by calculation, are shown in Figure 5.

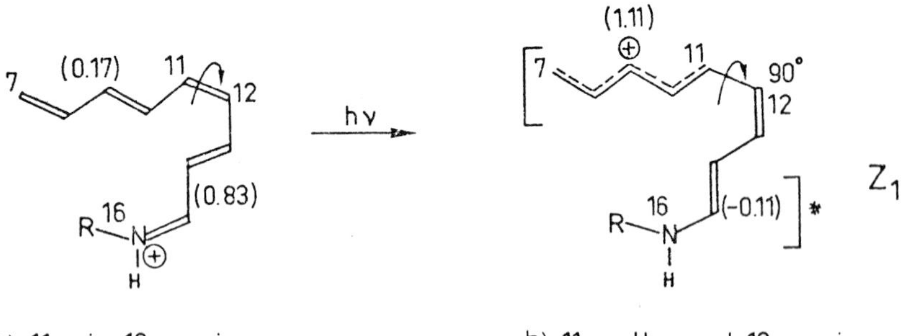

a) 11-cis, 12-s-cis

b) 11-orthogonal, 12-s-cis

Fig. 5. Sudden polarization in N-retinylidene chromophore[17]. The indicated net charges correspond to a calculation for $R=CH_3$.

The net result of excitation is the transformation of a photon into an electrical signal, as the positive charge, originally in position 16, migrates to positions 9, 7 and 11 ! This sudden polarization may induce a conformational change in the protein, or even cause directly, via electrostatic interactions, provisional changes in permeability.

In relation to Malrieu's calculations[16], it is important to note that in the retinyledene system the requirements for charge concentration near the twisted bond and reduced polarization dipole do not exist. Since, in the excited state, only one fragment is charged, no counter-charge exists to pull this positive charge near the center. The change in dipole moment induced by the excitation of rhodopsin is therefore enormous ($\Delta\mu \sim 35$ D).

V - CONCLUSION

Starting from the suggestive interpretation by Dauben of some experimental evidence, we have been led to predict a general phenomenon, intervening in numerous systems. Some fragmentary experimental evidence gives indications, but not yet proof, of the existence of the phenomenon :

(a) Havinga's aforementioned observed dual photochemistry for previtamin D.

(b) The recent observation by Mathies and Stryer[18] of a very large dipole moment in the vertically excited state of the retinylidene chromophore (our calculations, however, refer to the twisted excited state).

(c) The recent observation, by Grabowski and his collaborators[19], of a fully polarized fluorescent state of para N-dimethyl cyanobenzene

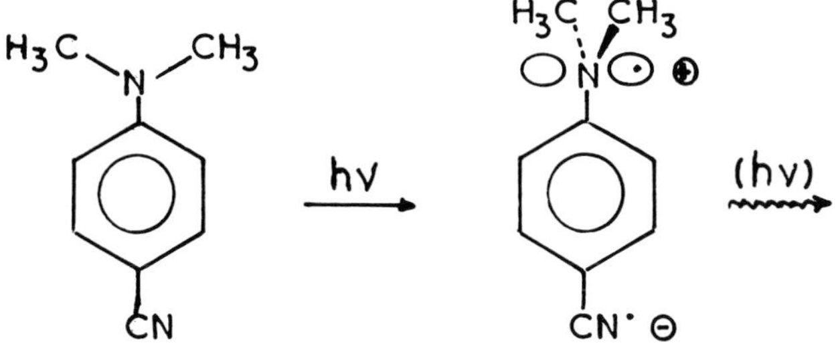

Here the polarization is ascribed to rotation of the $N(CH_3)_2$ group about the CN bond. Since this bond has partial double bond character in the molecular ground state, this unique process bears some similarity to the sudden polarization predicted for the twist of olefins.

REFERENCES

1. Dauben, W.G. and Ritscher, J.S., (1970), J.Am.Chem.Soc. 92, 2925
2. Dauben, W.G., Kellogg, M.S., Seeman, J.I., Wietmeyer, N.D., Wendschuh, P.H., (1973), Pure Appl.Chem.33, 197
3. Padwa, A., Brodsky, L., and Clough, S., (1972), J.Amer.Chem. Soc. 91, 6767
4. Wulfman, C.E. and Kumei, S.E., (1971), Science 172, 1061
5. Mulliken, R.S., (1932), Phys.Rev. 41, 751
6. Salem, L., and Rowland, C., (1972), Angew.Chemie Int.Ed. 11, 92
7. Bonacic-Koutecky, V., Bruckmann, P., Hiberty, P., Koutecky, J., Leforestier, C., and Salem, L., (1975), Angew.Chemie Int.Ed. 14, 575
8. Salem, L., (1975), Isr.J.Chem.14, 89
9. Bethe, H.A., and Salpeter, E.E., (1957), Quantum Mechanics of One and Two-electron Systems, Springer-Verlag, p.132
10. Berthier, G., private communication to the author (1976)
11. Buenker, R.J. and Peyerimhoff, S.D., (1975), Chem.Phys.9, 75
12. Indeed the three wave functions are all combinations of $\phi_a(1)\phi_a(2)$, $\phi_a(1)\phi_b(2)$, $\phi_b(1)\phi_a(2)$, $\phi_b(1)\phi_b(2)$. Hence the overall character in the three states is 200% ionic and 100% covalent - the latter predominating in the lower solution
13. Bruckmann, P. and Salem, L., (1976), J.Am.Chem.Soc.,98,5037
14. Bonacic-Koutecky,V., (1976), submitted for publication to J.Amer.Chem.Soc.
15. Havinga, E., (1976), Chimia 30, 27
16. Bruni, M.C., Daudey, J.P., Langlet, J. Malrieu, J.P. and Momicchioli, F., (1977), J.Am.Chem.Soc., in press
17. Salem, L.and Bruckmann, P., (1975), Nature, 258, 526
18. Mathies, R. and Styer, L., (1975), Proc.Nat.Acad.Sci. 73, 2169
19. Grabowski, Z., Lecture at Ecole de Physique et Chimie,Paris, November 1976.

DISCUSSION

WAGNIERE :

If the polar structures are as important as you imply, shouldn't there be detectable solvent effects?

SALEM :

If the actual reaction mechanism is ionic, *via* an excited Z state, already in non-polar solvents, it is not clear what the effec of a polar solvent will be (except to lower the energy of Z and henc eventually increase quantum yields). Polar solvents would influence *competition* between a Z-mechanism and a non-polar mechanism.

PHOTORECEPTORS AND PHOTOPROCESSES IN THE LIVING CELL

Jerome J. Wolken

Carnegie-Mellon University
Pittsburgh, Pa. U.S.A.

Living organisms from bacteria to man exhibit various kinds of sensitivity to the visible part of the electromagnetic spectrum of energy. This is seen in behavior, for an organism will bend, move, or swim toward or away from a light source. Such behavior is described as phototropism and phototaxis. Plant cells utilize solar radiation directly in photosynthesis. In animals, photosensory cells evolved, giving rise to eyes and to vision. Other photobiological phenomena are now known, such as photoperiodism and photomorphogenesis, which control many developmental growth processes as well as hormonal stimulation of the sexual cycles, the timing of the flowering of plants, and the color and shade changes in the skin of animals. Also, there is photodynamic action, the photosensitization by a molecule which becomes activated by light and causes destructive photooxidation in the cell, and photoreactivation, the recovery of ultraviolet damage by visible radiation.

In all of these photobiological phenomena, receptor molecules are necessary to absorb the energy. The photoreceptor molecule, or chromophore, is bound or complexed with a specific protein which resides in the cell membrane or in the membranes of the photoreceptor structure. The photoreceptors are the cell's transducers, converting one form to another in the process. For example, light energy → mechanical energy ↔ chemical energy ↔ electrical energy.

The photochemical behavior of photoreceptor molecules (e.g. chlorophyll, flavins) are describable in terms of their excited states. In some molecules in the excited state, the high energy electrons do not escape from the molecule, but return to their original low energy ground state, and in the process, some of the

B. Pullman and N. Goldblum (eds.), Excited States in Organic Chemistry and Biochemistry, 175-186.

energy appears as fluorescence. When the photoreceptor molecule is in solution it dissipates its excitation energy to the surrounding medium, resulting in no utilization of the energy by the cell. But, when the excited photoreceptor molecule is bound to a protein in or on a membrane, it dissipates its excitation energy to the protein as it relaxes to the ground state. As a result, the protein is locally excited, and the excitation coupled to the electronic relaxation of the photoreceptor molecule leads to a conformational change of the protein in the membrane. This affects the trans-membrane potential and triggers the transducing system. The understanding of such behavior is of great importance in photoreceptor processes and, therefore, photobiological phenomena.

PHOTOTROPISM

Phototropism is defined as the directed bending of fixed organisms in response to light. Most fungi are phototropic, and the sporangiophore of *Phycomyces blakesleeanus* is a good model. The sporangiophore is a large single cell which can grow to more than 10 cm in length and is about 0.1 mm in diameter. Its life cycle Stages I - IVb is complete in < 100 hours. The sporangiophore exhibits phototropism when it is illuminated unilaterally. It is positively phototropic to wavelengths from 300 to 510 nm, with a peak sensitivity at about 280 nm. The action spectra for *Phycomyces* phototropism show absorption peaks around 280, 365-385, 420-425, and 445-485 nm [3,5].

The growth-zone is the sensory region of the sporangiophore. It extends from 0.1 mm to about 3 mm below the sporangium and occupies a surface area of about 6×10^{-3} cm^2. The sporangiophore is most photosensitive in Stage I and IVb. In scanning down the growth zone of the sporangiophore with a microspectrophotometer, the absorption spectrum gradually shifts from that of a carotenoid with absorption peaks at 430, 460, and 480 nm to that of a flavoprotein with absorption peaks near 280 and 370 nm and around 460 nm. These absorption peaks taken together at 280, 370, 435, 460, and 480 nm are similar to those for the phototropic action spectrum. Although there is little experimental evidence for a carotenoid, the spectral data suggests that flavins or flavoproteins are involved in the photoprocess [1,2,20,22]. Flavins isolated from the sporangiophores were identified as riboflavin, lumiflavin, flavin adenine dinucleotide (FAD) and flavin mononucleotide (FMN). The number of flavin molecules was estimated to be 1.3×10^{13} per sporangiophore. In addition to the flavins, cytochrome *c* was isolated from the sporangiophores, whose reduced absorption peak was at 552 nm. Indolacetic acid (IAA) was also isolated and was found to be 0.05 μg/g of fresh weight sporangiophores. IAA is a plant growth hormone and has been shown to participate in the phototropism of the *Avena* coleoptile [15].

In search of a photochemical system to account for *Phycomyces* phototropism, it was of interest to determine how IAA could function upon light excitation with a flavin and cytochrome *c*. A possible biochemical scheme is the dehydrogenation of IAA catalyzed by a flavin system which is capable of transferring reducing equivalents to an intermediate carrier cytochrome *c* and finally by a series of steps to the reduction of oxygen and to water (Fig. 1). The reaction was tested *in vitro* with a solution containing 2×10^{-4} M IAA and 5×10^{-5} M riboflavin buffered with 2×10^{-2} M Tris at pH 7.4. The solution was degassed and then irradiated in a quartz reaction cell with 436 nm (isolated from a pressure mercury arc lamp through a liquid filter). From the amount of IAA photolyzed and the amount of photons absorbed by the system, which was measured by potassium ferrioxalate chemical actinometry [7], the quantum yield of the riboflavin sensitized photodecomposition of IAA was 0.65. It was proposed that the reaction mechanism involves two molecules of riboflavin [10]. If so, the quantum yield should be less than 0.5. Photolysis of the solution was repeated as above but 1 mg/ml of cytochrome *c* from *Phycomyces* was added, and the quantum yield was 0.63. The optical density at 552 nm increased with irradiation time, indicating reduction of cytochrome *c*. The reaction cell was then connected to the vacuum line again and 1 atmosphere of oxygen was introduced. The absorption spectra taken after mixing the solution showed disappearance of the 552 nm peak and a shift to the oxidized cytochrome *c*. The photodestruction of IAA with blue-violet light in the presence of riboflavin and cytochrome *c* is possible. This suggests that a similar photochemical system may function in *Phycomyces* phototropism.

The Photoreceptor

What is the photoreceptor and where is it located? One of the first observations made with polarization microscopy revealed that birefringent rod crystals, from 1 to 2 μm wide and from 5 to 10 μm in length, were aligned near the vacuole in the light growth-zone, the region where the photoreceptor is believed to lie [20].

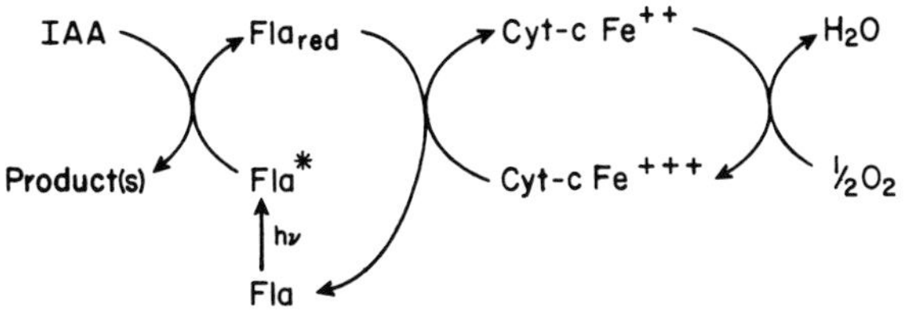

Figure 1. Flavin-sensitized photodecomposition of indoleacetic acid (IAA). Fla, flavin; Fla* excited; Fla_{red}, reduced; cyt-*c* Fe^{+++}, cytochrome *c* oxidized; cyt-*c* Fe^{++}, reduced.

Octahedral crystals were also observed which were fluorescent but not birefringent. The octahedral crystals were isolated from the sporangiophore in a relatively pure state. From 1 gm of fresh weight sporangiophores, 1 to 2 mg of crystals were obtained. The crystals can be solubilized only at Phs below 2.5 and greater than 9.5. They are insoluble in 2% digitonin solution, but are soluble in 1% sodium dodecyl sulphate (SDS), pH 6.0 to 7.1. Chemical analyses of the crystals identified them as protein (95%) of which 1% by weight is tryptophan. In some analyses, lipid, flavin and retinal was found associated with the crystals [24].

The finding that there are crystals in the growth-zone of Phycomyces sporangiophores and that they could be the photoreceptors required further investigation. A sporangiophore from growth stages I - IVb was centrifuged at high speed and three crystal layers became stratified according to their densities in the sporangiophore. The absorption spectra of these crystal layers were obtained with the microspectrophotometer. The crystal layer (a) had an absorption maximum near 460 nm (flavin), the crystal layer (b) around 520 nm (carotenoid), and the crystal layer (c) at 412 nm, and around 550 nm (cytochrome); these spectra suggest that they are involved in the photoprocess [20,22]. The absorption spectrum for a single octahedral crystal (Fig. 2) showed absorption peaks around 280 nm, 350 nm, and 460 nm, that of a flavoprotein [13]. The spectrum of the isolated crystals packed in a special absorption cell indicated an oxidized flavoprotein; when irradiated, the spectrum shifted to that of a flavin semiquinone (Fig. 3).

PHOTOTAXIS

Phototactic behavior can be classified into photokinesis and phototaxis. Photokinesis is the change in velocity or rate of

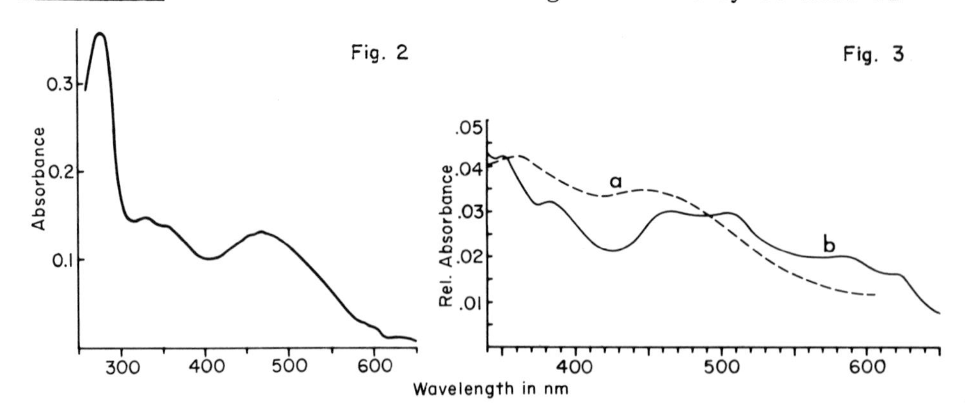

Figure 2. Absorption spectrum of octahedral crystal.
Figure 3. Packed crystals. Spectrum obtained (a) in the dark, (b) after irradiation.

motion upon illumination, without regard to orientation, whereas phototaxis is the orientation of the entire organism with reference to light. We can use light then to "communicate" with the organism to the extent that its movement is directed. The action spectrum for such behavior should correspond to the absorption spectrum of the photoreceptor molecule.

Euglena, a protozoan algal flagellate, has a photoreceptor system for location, which consists of the eyespot, paraflagellar body, and flagellum. The eyespot consists of numerous orange to red granules, from 0.1 to 0.3 μm in diameter. They are thought to be the organism's shading device and filter for the photoreceptor. The paraflagellar body is a crystalline structure attached to the flagellum; it is regarded as the photoreceptor for phototaxis. The Euglena cell can be likened in structure to that of a photo-neuro-sensory cell [24].

The photokinesis action spectrum plotted for the rate of swimming (mean velocity in mm/sec versus wavelength at 6 μwatts/cm^2 intensity) is illustrated in Fig. 4a. It will be observed that there is a major peak at 465 nm and another peak near 630 nm [25]. This action spectrum resembles the absorption spectrum of protochlorophyll, chlorophyll b, and/or a carotenoid. The action spectrum for phototaxis shows a major peak around 495 nm and near 620 nm (Fig. 4b). In polarized light, there are maxima around 465 and 500 nm and beyond 600 nm (Fig. 4c) which match closely the photokinesis and phototaxis action spectra. Polarized light phototaxis indicates that there could be two light-absorbing pigments, a carotenoid and a flavin, and that energetic transfer could occur between them [19,25]. The phototaxis action spectrum was found to coincide with the action spectrum for oxygen evolution (Fig. 5). In photosynthetic bacteria and algae, the action spectrum for phototaxis coincides with that of photosynthesis, which implies that similar molecules participate in both of these photoprocesses.

To circumvent the difficulties of isolating and extracting the Euglena eyespot with solvents, the absorption spectra can be obtained in situ with the microspectrophotometer. Absorption spectra of the eyespot region show that the peaks lie near 430, 465, and 495 nm, and near 350 nm [19]. In the heat-bleached (HB) mutant that lacks chloroplasts, and hence chlorophyll, the spectrum of the eyespot area shows peaks near 340, 430, 465, and 510 nm. Spectra closer to the base of the flagellum and near the paraflagellar body show absorption peaks at about 440 and 490 nm and two smaller peaks near 558 and 590 nm. When these light-grown Euglena are dark-adapted for 1 hour at 5° C and irradiated with strong white light from 1 to 5 minutes, the absorption peak around 490 nm bleaches, accompanied by an increase in the absorption peak at 440 nm [21].

If a flavin is the photoreceptor molecule, then an analysis of Euglena flavins would give us some idea of its concentration. The total flavins extracted from both light and dark-grown Euglena are of the order of 10^8 molecules per cell. Pagni et al. [14] isolated the eyespot granules and calculated from fluorometric analysis that there was 5 x 10^{-4} μg flavin/ml. Interpretation of these spectra to establish the identity of the photoreceptor molecule is extremely difficult. However, phototactic action spectra and isolation of the pigment from the eyespot area suggests a flavin or flavoprotein to be involved in the photoreceptor process. How the photoexcitation takes place from the photoreceptor, the paraflagellar body, to its locomotor, the flagellum, is unknown.

THE CHLOROPLAST AND PHOTOSYNTHESIS

The photoreceptor for photosynthesis is the chloroplast, a highly ordered structure of membranes (lamellae). The membrane bilayers of lipids and proteins are of the order of 200 Å. The number of chlorophyll molecules per chloroplast, from photosynthetic bacteria to algae to higher plants, is of the order of 10^9 molecules. The number of chlorophyll molecules is directly related to the number of lamellae and hence to the total surface area available on the membranes. Our experimental data indicate that the chlorophyll molecules are spread as monolayers on the surfaces of the membranes and that the cross-sectional area of the chlorophyll molecule is 225 $Å^2$ [23]. This maximizes the surface area of each chlorophyll molecule for light absorption and for energy transfer.

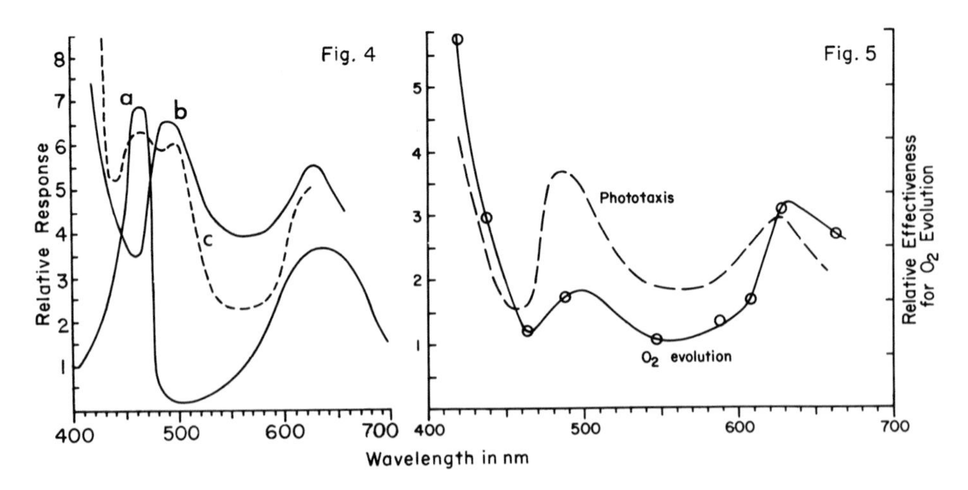

Figure 4. Action spectrum for (a) photokinesis, (b) phototaxis, (c) phototaxis in polarized light. Euglena gracilis.
Figure 5. Action obtained for phototaxis compared to action spectrum for oxygen evolution. Euglena gracilis.

Photosynthesis

In photosynthesis, the light reaction is the photolysis of water and the dark reaction is the reduction of carbon dioxide to carbohydrates. The initial photophysical event is the absorption of light by chlorophyll a in the chloroplast to produce an excited chlorophyll molecule in which an electron is raised from its normal energy level to a higher energy level. Such excited electrons are transferred from chlorophyll to ferredoxin and to cytochromes via flavins and quinones. During this cyclic flow of electrons, the energy which the electrons initially acquired is transferred through oxidation-reduction reactions. The oxidation-reduction reactions are driven by two photochemical pigment systems, Photosystem I and Photosystem II. The reduced electron acceptor donors of Photosystem II produce oxygen by the oxidation of water. These two photopigment systems provide the high energy phosphates (reduced NADP and ATP) needed for the synthesis of carbohydrates and proteins from CO_2 and water.

Chloroplastin

Attempts to isolate photoactive fractions from the chloroplast and to separate the fractions that contain Photosystems I and II have been actively pursued. The chloroplast can be dissociated by detergents (e.g. digitonin, sodium dodecyl sulfate (SDS), etc.) which solubilize the chloroplast. The non-ionic surfactant, digitonin, whose structure is similar to cholesterol, has a strong attraction for dye molecules. In the solubilization process, the pigment-lipid-protein complex of the chloroplast lamellae is opened and chlorophyll micelles are formed. These chlorophyll-containing micelles are referred to as chloroplastin [18,19]. The chloroplastin absorption spectrum is similar to that of the chloroplast. In the chloroplastin micelles, one chlorophyll molecule is associated with one protein molecule with an estimated molecular weight of 40,000 [19]. The chloroplastin micelles are colloidal systems that form aggregates which are stabilized by the chlorophyll molecules. Such micelles can be regarded as liquid crystals and their molecular structure in the micelle is much like that of the chloroplast. Chloroplastin not fixed or stained and viewed in the electron microscope shows particles which range in diameter from 100Å to 1000Å, which compares to the particles in the chloroplast lamellae. Assuming that the chloroplastin micelle is 200Å in diameter and that three such micelles aggregate, they could contain about 225 chlorophyll molecules, 55 carotenoid molecules, one cytochrome and one ferredoxin molecule. This estimate approaches the number of molecules necessary for a functional photosynthetic unit.

Photochemical Activity

Chloroplastin exhibits photochemical activity which shows some similarities to the photosynthetic process; for example, the photoreduction of the dye, 2,6-dichlorobenzenoneindophenol (DBIP), at 600 nm, exhibits a primary photochemical reaction. During photosynthesis, water-splitting reduction provides the chemical reducing power which is trapped by the dye, and the dye is reduced to a colorless form. The molecular turnover of a typical preparation indicated that 7 x 10^3 molecules of dye was photoreduced per molecule of chlorophyll/minute. Chloroplastin which actively photoreduced the dye also caused photolysis, or evolution of oxygen. Photolysis was measured manometrically in completely anaerobic Warburg vessels made oxygen-free to permit a qualitative identification of oxygen with yields of 20 to 30 μl of O_2 in two minutes [6]. A light-catalyzed conversion of inorganic phosphate into labile phosphate (ATP) was also observed over a 1-hour period in a similar anaerobic system containing 2 ml of chloroplastin, 10^{-5} M, 20 μM of Mg^{++}, 30 μM of alpha ketogluterate, 0.3 μM of riboflavin-5-phosphate, 0.6 μM of menadione (vitamin K_3), 2 μM of ascorbate, 5 μg of cytochrome _c_, 55 μM of adenosine monophosphate (AMP) and 4 μg of inorganic phosphate.

Attempts were then made to isolate the two photochemical systems from chloroplastin by the methods of sucrose density gradient ultracentrifugation, chromatography, and electrophoresis. _Euglena_ chloroplastin, when subjected to analytical ultracentrifugation, shows a rapidly sedimenting component, 13.5 S and a slower sedimenting component, 7.8 S. Similarly, spinach chloroplastin shows two components, a fast component, 18.8 S, and a slow component, 7.7 S. These can be considered to contain Complex I and Complex II. The molecular weights estimated from the sedimentation S give about 33,000 (Complex I) and 21,900 (Complex II) for the spinach chloroplastin [16,17]. These values compare favorably to that calculated for _Euglena_ chloroplastin of 21,000 and 40,000 [23]. Chloroplastin subjected to polyacrylamide gel electrophoresis also shows two green chlorophyll-containing bands, and in addition, one protein band. It remains to be determined whether these two chlorophyll have the photochemical Systems I and II.

THE VISUAL PIGMENT - RHODOPSIN

The photoreceptors for visual excitation are the retinal rods and cones of vertebrate eyes. In compound eyes of arthropods and mollusc eyes, they are retinula cells whose rhabdomeres are the photoreceptors. All are membraneous processes of those photoreceptor cells, of packed microtubules as in the invertebrates and lamellae or discs for the vertebrate, which show a striking similarity in structure to the chloroplasts.

All visual pigments thus far isolated from invertebrate and vertebrate eyes are rhodopsins which contain $retinal_1$ or $retinal_2$ (derivable from vitamin A_1 or vitamin A_2) as the chromophore. Rhodopsins in the retinal rods and cones are retinal-protein complexes. The 11-*cis* geometric isomer of retinal is the functional chromophore molecule complexing with opsin to form rhodopsin, through a Schiff base linkage. The visual protein, opsin, is species specific and determines the absorption maxima and molecular weight of each rhodopsin. There are from 10^6 to 10^9 rhodopsin molecules that reside in or on the membranes of the photoreceptors.

The light bleaching of rhodopsin to release retinal from opsin does not proceed directly, but in a series of intermediate steps. In the process, the 11-*cis*-retinal is isomerized to all-*trans*-retinal to a more stable form. The *cis* to *trans* isomerization seems to be the process which best accounts for the experimental observations bearing on the initial photochemical event in the photochemistry of visual pigments [9]. Therefore, the only apparent action of the absorption of light by rhodopsin is the isomerization of the 11-*cis*-retinal to the all-*trans*-retinal.

Bacteriorhodopsin

Let us now turn to our investigations of the extremely halophilic bacteria that grow in a marine environment with an unusually high salt (NaCl) concentration of 25%, that of the Dead Sea and the Great Salt Lake, Utah. The bacterium *Halobacterium halobium* grows at high temperatures near 44° C and in continuous direct sunlight. The adaptability to such an environment indicates a unique biochemistry. These bacteria synthesize large quantities of carotenoids which give the culture a pink to red to purple color. Absorption spectra of the bacteria by microspectrophotometry show a broad absorption with maxima around 420, 468, 505, and 540 nm. The bacteria extracted with 60% acetone show absorption near 465, 496, and 532 nm. These absorption spectra are typical of C_{50}-bacterioruberins and the C_{40}-carotenoids which are the major pigments.

A finding of considerable interest for these bacteria was that their ruptured cell membrane, the "purple membrane," possesses the visual pigment rhodopsin [11,12]. The purple membranes can be isolated from the bacteria and their absorption spectrum resembles the vertebrate retinal cone pigment, iodopsin, with peaks at 560 nm and at 280 nm (Fig. 6a). This pigment was named bacteriorhodopsin. The molecular weight of the complex was found to be 26,000 in a molar ratio of 1:1 retinal to protein.

Bacteriorhodopsin does not bleach in the light. In the presence of sodium borohydride ($NaBH_4$), the 560 nm peak is bleached by light and a new absorption peak at 330 nm appears. The 560 nm peak can also be bleached in the presence of hydroxylamine (NH_2OH) to form the retinaldehyde oxime. Monoethanolamine, $NH_2CH_2CH_2OH$, which is more hydrophobic than hydroxylamine, also bleaches the bacteriorhodopsin. Bacteriorhodopsin can be bleached without light by cetyltrimethylammonium bromide (CTAB) which shifts the 560 nm peak to 369 nm. A reaction with sodium cyanoborohydride, $NaCNBH_4$ (which is more hydrophobic than $NaBH_4$) also bleaches the 560 nm peak. In the photobleaching process of bacteriorhodopsin, it was suggested that an intermediate of a few milliseconds lifetime occurs with an absorption peak at 415 nm, and that a very fast recovery reaction to the original 560 nm state takes place [12], which probably requires another photon.

It was of interest to identify retinal as the chromophore, for flavoproteins can mimic the absorption characteristics of rhodopsin. If the bacterial chromophore was retinal, it would then be of interest to see whether it would complex with opsins from animal retinas to form rhodopsins. Retinal was extracted from the bacterial membranes and purified by chromatography. The retinal eluted from the chromatograms with chloroform showed an absorption peak at 376 nm [26]. From its absorbance, the retinal concentration was found to be 0.64 μg/g of wet weight bacteria. Opsin was then prepared from cattle retinas, since bovine opsin is one of the more stable opsins [8]. The bacterial retinal (in the 1.8% digitonin solution buffered with M/15 phosphate buffer, pH 6.3) was incubated with an excess of the cattle opsin for 3 hours in the dark at room temperature (Fig. 6b). The absorption

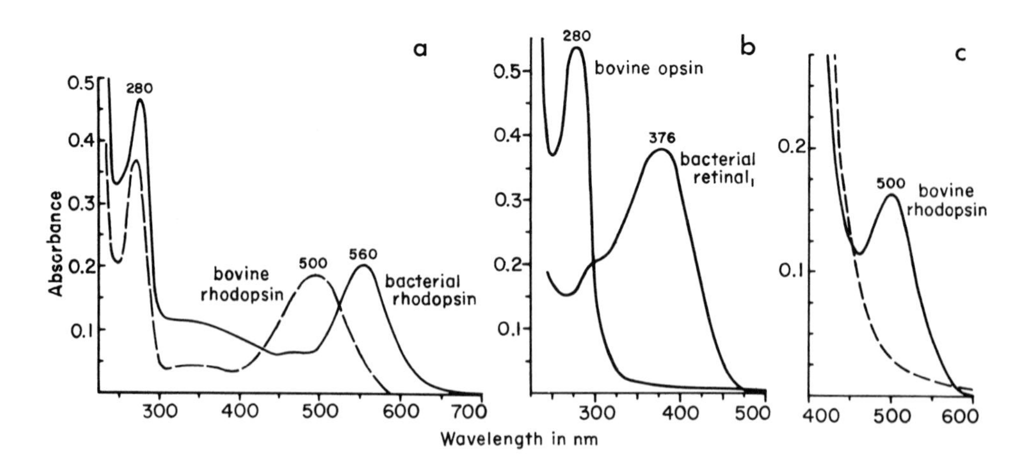

Figure 6. (a) Absorption spectra of bacterial rhodopsin and bovine rhodopsin. (b) Absorption spectra of retinal and bovine opsin. (c) Bovine rhodopsin formed upon incubation of bacterial retinal and opsin, and after irradiation (---) bleaches.

spectrum was shifted to near 500 nm to that of bovine rhodopsin which was photosensitive (Fig. 6c). We have tried to form rhodopsins using the bacterial retinal with opsins prepared from a variety of animal retinas, with globular proteins and with tryptophan. These complexes are difficult to do in the laboratory, but when successful, they shift the absorption peak from around 376 nm to around 500 nm, but are unstable.

The finding of retinal and rhodopsin in *Halobacterium halobium* raises the question of its function. It was found that under anaerobic conditions in the dark, the ATP of the bacterial cells decreases significantly, and light is required to store the original ATP concentration [4]. In the absence of light, however, oxygen is required. Thus, these bacteria have adapted to their environment by incorporating rhodopsin in their cell membrane which converts light energy to chemical energy, a process analogous to photosynthesis.

CONCLUDING REMARKS

Photoreceptors can be considered the cell's short-term "memory" system. Photoreceptor molecules in the photoreceptor membranes are molecularly structured to receive and process information. There are 10^6 to 10^9 receptor molecules in the photoreceptor structure, each of which can receive one photon or a single bit on an area < 100 Å in diameter. Information to the system is received by a light signal of a particular wavelength, and it is stored, or "memorized," until a signal of another wavelength erases the information and restores the molecule to its original memory state. Such a memory system functions in *photoperiodism* in plants, embodied in the red and far-red absorbing forms of phytochrome, the photoreceptor molecule. Continuous red light at 660 nm is effective in altering the plant's response, that is, in preventing flowering. But, if a flash of red light is followed immediately by a short interval of far-red light at 730 nm, the effect of red light is cancelled out. The plant then acts as if its night time had never been interrupted, and it flowers. Other examples could be protochlorophyll to chlorophyll, the *cis-trans* isomerization of retinal in rhodopsin, and the oxidized to the reduced forms of flavins. A relatively simple model can be conceived in which the photoreceptor molecule resides in the membrane in a non-sensory state A until it receives a light signal and enters into a state B. The result is a conformational change that may be very fast, of the order of nanoseconds, or slow, with lifetimes of the order of hours, and the cycle $A \rightleftarrows B$ must be reversible and reproducible.

REFERENCES

1. Berns, D. S. and J. R. Vaughn (1970). Biochem. Biophys. Res. Commun. 39, 1094.
2. Carlile, M. J. (1965). Ann. Rev. Plant Physiol. 16, 175.
3. Curry, G. M. and K. V. Thimann (1961). In: Progress in Photobiology (B. G. Christensen and B. Buchmann, eds.). Elsevier Publishing Co., New York, 127-134.
4. Danon, A. and W. Stoekenius (1974). Proc. Natl. Acad. Sci. (U.S.A.) 71, 1234.
5. Delbrück, M. and W. Shropshire, Jr. (1960). Plant Physiol. 35, 194-204.
6. Eversole, R. A. and J. J. Wolken (1958). Science 127, 1287.
7. Hatchard, C. G. and C. A. Parker (1956). Proc. R. Soc. (London) A235, 518-536.
8. Hubbard, R., P. K. Brown, and D. Bownds (1971). In: Methods in Enzymology (D. B. McCormick and L. D. Wright, eds.) Vol. 18 C, Academic Press, New York, 615-653.
9. Kropf, A. (1976). Nature (London) 264, 92-94.
10. Nathanson, B., M. Brody, S. Brody, and S. B. Broyde (1967). Photochem. Photobiol. 6, 177-187.
11. Oesterhelt, D. and W. Stoeckenius (1971). Nature (London) New Biol. 233, 149-152.
12. Oesterhelt, D. and W. Stoeckenius (1973). Proc. Natl. Acad. Sci. (U.S.A.) 70, 2853-2857.
13. Ootaki, T. and J. J. Wolken (1973). J. Cell. Biol. 57, 278-288.
14. Pagni, P. G., P. Walne, and E. L. Wehry (1976). Photochem. Photobiol. 24, 373-375.
15. Thimann, K. V. and G. M. Curry (1960). In: Comparative Biochemistry, Vol. I (M. Florkin and H. J. Mason, eds.) Academic Press, New York, 243-309.
16. Thornber, J. P., R. P. V. Gregory, C. A. Smith, and J. L. Bailey (1967a). Biochemistry 6, 391-396.
17. Thornber, J. P., J. C. Stewart, M. W. C. Hatton, and J. L. Bailey (1967b). Biochemistry 6, 2006-2014.
18. Wolken, J. J. (1961). Int. Rev. Cytol. 11, 195-218.
19. Wolken, J. J. (1967). Euglena, 2nd ed. Appleton-Century-Crofts, New York.
20. Wolken, J. J. (1969). J. Cell. Biol. 43, 354-360.
21. Wolken, J. J. (1971). Invertebrate Photoreceptors. Academic Press, New York.
22. Wolken, J. J. (1972). Int. J. Neurosci. 3, 135-146.
23. Wolken, J. J. (1973). In: Phytochemistry, Vol. 1 (L. P. Miller, ed.) Van Nostrand Reinhold, New York, 15-37.
24. Wolken, J. J. (1975). Photoprocesses, Photoreceptors, and Evolution. Academic Press, New York.
25. Wolken, J. J. and E. Shin (1958). J. Protozool. 5, 39-46.
26. Wolken, J. J. and C. S. Nakagawa (1973). Biochem. Biophys. Res. Commun. 54, 1262-1266.

ELECTRON-ELECTRON INTERACTIONS AND RESONANT OPTICAL SPECTRAL SHIFTS IN PHOTORECEPTOR MOLECULES

Lawrence J. Dunne

Department of Mathematics,
Chelsea College, University of London,
Manresa Road,
London SW3, England.

The purpose of this paper is to discuss the optical excitation spectrum and electron-electron interactions in a Rhodopsin-like molecule. The task of the theory here is to provide some initial insight into the microscopic behaviour of these photoreceptor molecules and must, considering the complexity of the system involved, be semiempirical and somewhat speculative. Such a preliminary approach is usually required in understanding physical systems. A full microscopic theory of a phenomenon rarely comes out of the blue, but is almost always preceded by a period of interplay between approximate and semiempirical theory and experiment. It is in such a spirit that this paper is presented. This work has benefited from and is related to the work of W.A. Little (1, 2, 3) on the possible occurrence of a superconducting state in macromolecules. Little's theory (1, 2, 3) suggested that certain special types of macromolecules and solids should exhibit superconductivity at room temperature. Unfortunately, as far as can be checked, no one has ever synthesised a structure which has seriously tested the predictions of the theory. Little's proposal, then, is neither substantiated nor is it discredited.

This paper will examine some features of the Little proposal and will argue that the Rhodopsin molecule should be studied in the light of this theory (4). Little (1, 2, 3) examined the possible occurrence of a superconducting state in macromolecules using the criteria for the existence of a superconducting state proposed by Bardeen, Cooper and Schrieffer (BCS), (5). According to BCS, the prime requirement for the occurrence of superconductivity is that an <u>attractive</u> interaction must exist between pairs of electrons in the material and that this

B. Pullman and N. Goldblum (eds.), Excited States in Organic Chemistry and Biochemistry, 187-198.

attraction must be strong enough to overcome the normal screened coulomb repulsion between the two electrons. In other words, bound electron pairs may be formed.

We will concentrate here on the mechanism for the attractive interaction and then examine how such an attraction may be occurring in the Rhodopsin molecule. The existence of such bound electron pairs in the Rhodopsin molecule may be the key to understanding the mechanism of vision even in the absence of the type of long range order usually invoked to describe superconductivity. Quite dramatic changes occur in the physical properties of matter in the presence of attractive pairing correlations. For example, metals become superconducting; liquid ^{3}He becomes superfluid below 3mK (6) and such pairing correlations between nucleons have been invoked to explain in part the collective excitations of nuclear matter. (7). A further common feature of all these systems is that the attractive force is carried by some other field and is an indirect coupling.

Thus, in some solids, the indirect coupling arises from electrons exchanging virtual phonons. In the nucleus, nucleons are thought to exchange virtual mesons and in liquid ^{3}He, the pairing correlation may involve a magnetic mechanism (6). Little's proposal is based on an attractive pairing correlation between pairs of electrons exchanging virtual excitons or electronic excitations. The exciton field carries the force between the two electrons and has some boson properties.

Little proposed that it would be worth looking at structures such as 1.

1.

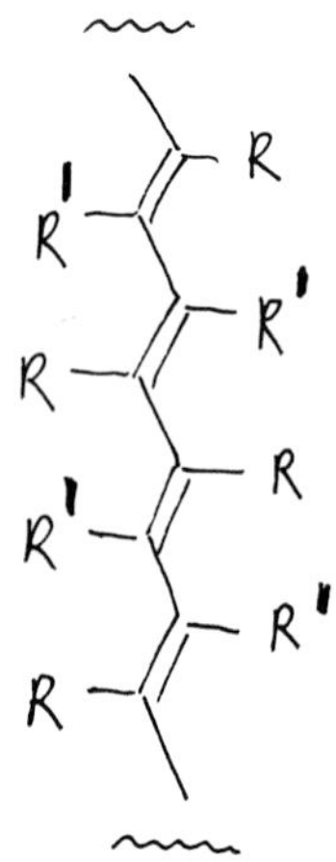

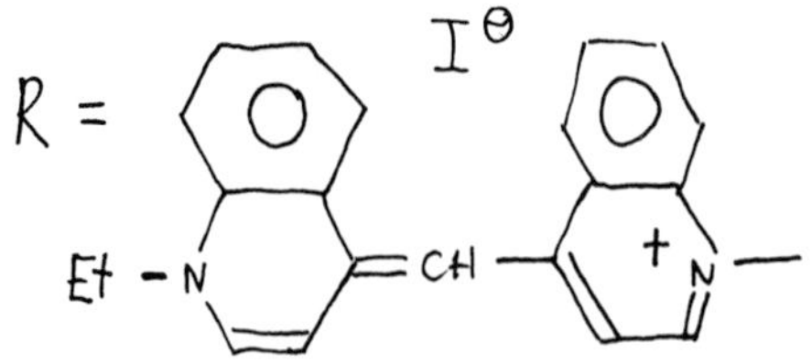

R' = H or R.

The virtual oscillation of charge in the dye R, gives rise to an attractive coupling between the electrons in the polyene, provided the electronic energies of both the polyene and the dye meet some special requirements. In fact, it will be shown that we are looking for a resonance condition when the perturbation due to the dye becomes most efficacious.

To gain some orientation, we will follow a discussion due to Little (2). Fig. 2 shows a pair of electrons at r_1, r_2 fluctuating across a gap of energy $\hbar\omega$ in the presence of a dye whose absorption maximum occurs at 2eV. The field at r_2 resulting from the charge at r_1 in the presence of the dye is the sum of the fields from the charge at r_1 itself and the field due to the induced electric moment in the dye. As the frequency of the fluctuating charges approaches the resonant frequency of the dye, large induced charges appear in the dye. It is believed that under such conditions the potential V (r_1 r_2) would be attractive and bound electron pairs would result. Under such conditions, we conjecture that quite dramatic changes in physical properties of the molecule would result even in the absence of a sharp phase transition.

The Rhodopsin molecule consists of the Schiff base of the polyene 11-cis retinal (possibly in protonated form) which is most probably clathrated into a hydrophobic cleft of the parent opsin molecule. Since the protein opsin is rich in aromatic amino acid residues such as trytophan (λ_{max} $\sim$ 300 nm, 4eV) then it is evident that the Rhodopsin molecule falls into the general class of molecules discussed in Little's paper (1, 2, 3). We will make the qualification here that Rhodopsin may be too small to exhibit the true many body effects expected in an infinite polymer chain.

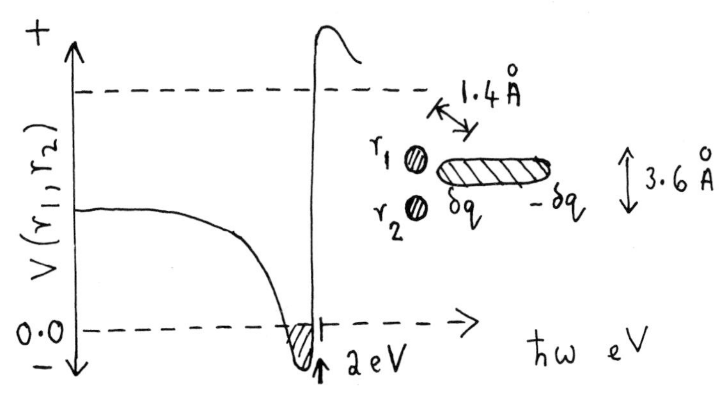

Fig. 2. effective interaction between two fluctuating charges at r_1, r_2 as a function of frequency . The dye has a strong absorption band at 2eV (1200 nm) (after Little (2)).

One of the central phenomenon to be explained in the visual pigments is the spectral shift which occurs when 11-cis retinyl Schiff base (λ max $\sim$ 380 nm, 3.2eV) or its protonated form (λ max $\sim$ 420 nm, 2.7 eV) combines with opsin to form Rhodopsin, varieties of which absorb light almost into the near infra red region of the electro magnetic spectrum. A review of this problem has been given by Pullman and Mantione (8). In this paper we shall discuss a simplified problem in which a long polyene and dielectric medium are intended to mimic the retinyl spine and aromatic side chain residues of the Rhodopsin molecule respectively. Fig. 3 shows a sketch of this hypothetical Rhodopsin-like molecule. Note the common features of this structure with the molecule proposed by Little shown in figure 1.

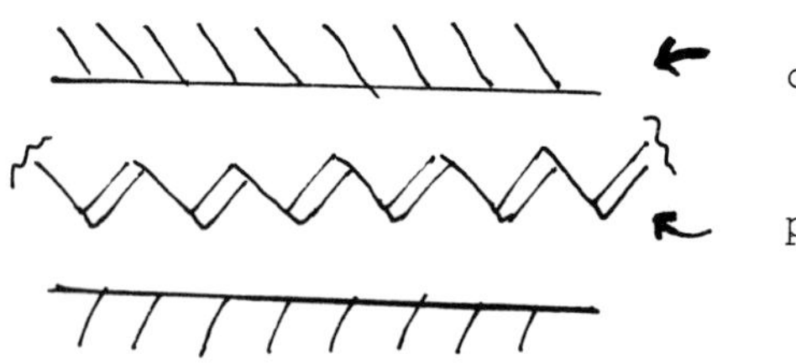

Fig. 3 (see text).

In the next part of this paper we will discuss the effect of the dielectric on the optical absorption spectrum of the polyene and show that a large spectral shift due to the presence of the polarisable dielectric implies an attractive coupling between electrons in the polyene spine. If the spectral shift is due to another mechanism such as charged groups around the chromophore then this theory fails.

THE HAMILTONIAN AND ASSUMPTIONS OF THE MODEL

We define the Hamiltonian for the polyene $\mathcal{H}^0_p$, for the dielectric $\mathcal{H}^0_d$ and their interaction $\mathcal{H}_{pd}$. The total Hamiltonian for the combined system is then given by

$$\mathcal{H}_{total} = \mathcal{H}^0_p + \mathcal{H}^0_d + \mathcal{H}_{pd}$$

$\mathcal{H}_{pd}$ is treated as a perturbation. The Hamiltonian is expressed in the representation of second quantisation. For some general many electron Hamiltonian $\mathcal{H}$ given by $\mathcal{H} = \mathcal{H}^0 + \mathcal{H}'$ where $\mathcal{H}'$ is a particle-particle interaction term, we introduce the Fermion field operators

$$\hat{\psi}(r) = \sum_{\underline{k}} \phi_{\underline{k}}(r) a_{\underline{k}}, \quad \hat{\psi}^*(r) = \sum_{\underline{k}} \phi^*_{\underline{k}}(r) a^+_{\underline{k}}$$

The $a^+_{\underline{k}}$'s and $a_{\underline{k}}$'s are Fermion creation and annihilation operators which satisfy the following anticommutation relations

$$\{a_\ell, a^+_m\} = \delta_{\ell m}, \quad \{a_\ell, a_m\} = \{a^+_\ell, a^+_m\} = 0$$

The $\{\phi_{\underline{k}}(r)\}$ are assumed to be the exact single particle eigenfunctions of $\mathcal{H}^0$ whose eigenvalues are $\{\varepsilon(\underline{k})\}$. These could be obtained <u>in principle</u> from a Hartree-Fock or other self consistent-field procedure.

Letting the k's represent the polyene states and a set of q's the dielectric states, the interaction term $\mathcal{H}_{pd}$ adopts a form

$$\mathcal{H}_{pd} = \tfrac{1}{2} \iint \theta^*(r_2)\psi^*(r_1)\hat{V}(r_1,r_2)\psi(r_1)\theta(r_2)\,dr_1\,dr_2$$

$$= \tfrac{1}{2} \sum_{\underline{k}',\underline{q}',\underline{k},\underline{q}} \langle \underline{q}', \underline{k}' | v | \underline{k}, \underline{q} \rangle c^+_{\underline{q}'} a^+_{\underline{k}'} c_{\underline{q}} a_{\underline{k}}$$

where $\theta(r) = \sum_{\underline{q}} \chi_{\underline{q}}(r) c_{\underline{q}}$. $\theta(r)$ is thus a field operator for the dielectric electron states.

The electronic part of the total Hamiltonian for the Rhodopsin-like molecule is then given approximately by

$$\mathcal{H}_{total} = \sum_{\underline{k}} \varepsilon(\underline{k}) a^+_{\underline{k}} a_{\underline{k}} + \sum_{\underline{q}} \varepsilon(\underline{q}) c^+_{\underline{q}} c_{\underline{q}} +$$

$$\tfrac{1}{2} \sum_{\{\underline{\alpha},\underline{\beta},\underline{\gamma},\underline{\delta}\} \subset \{\underline{k}\}} \langle \underline{\alpha}, \underline{\beta} | v | \underline{\gamma}, \underline{\delta} \rangle a^+_{\underline{\alpha}} a^+_{\underline{\beta}} a_{\underline{\delta}} a_{\underline{\gamma}} + \tfrac{1}{2} \sum_{\{\underline{u},\underline{v},\underline{w},\underline{x}\} \subset \{\underline{q}\}} \langle \underline{u}, \underline{v} | v | \underline{w}, \underline{x} \rangle c^+_{\underline{u}} c^+_{\underline{v}} c_{\underline{x}} c_{\underline{w}} +$$

$$\tfrac{1}{2} \sum_{\underline{k}',\underline{q}',\underline{k},\underline{q}} \langle \underline{q}', \underline{k}' | v | \underline{k}, \underline{q} \rangle c^+_{\underline{q}'} a^+_{\underline{k}'} c_{\underline{q}} a_{\underline{k}}$$

V is the usual coulomb interaction operator.

We stress the following assumptions of the theory.

1. The ground state of the polyene has all the single particle states up to some level $\varepsilon_f - \Delta/2$ occupied. An excited electronic state of the polyene is described in terms of the old ground state plus a number of single particle excitations at some level $\geqslant \varepsilon_f + \Delta/2$ above the ground state. Δ , the energy of the first excited state of the polyene is obtainable approximately from spectral data on the unperturbed polyene. In the retinyl Schiff base Δ is likely to be $\sim$ 3.2 - 2.7eV.

2. The ground state of the dielectric medium has all the single particle states up to some level $\varepsilon_f - E/2$ occupied. An excited electronic state of the dielectric is described in terms of the old ground state plus a number of single particle excitations at some level $\geqslant \varepsilon_f + E/2$ above the ground state. E , the energy of the first excited state of the dielectric is obtainable approximately from spectral data on the opsin molecule or aromatic amino acids. Since trytophan residues absorb strongly at 300 nm, E is likely to be $\sim$ 4eV.

3. We assume that $E > \Delta$

We will now consider some of the scattering processes associated with the interaction term $\mathcal{H}_{pd}$ and calculate approximate expressions for wavefunctions and energy shifts. To achieve this, it is helpful to introduce the concept of a self energy correction. The self energy concept was introduced into solid state physics in connection with the polaron problem by Fröhlich et al. (9). Essentially, an electron moving in a polar crystal with wave vector $\underline{k}$ can emit a phonon (a quantised lattice vibration) with momentum $\underline{q}$ which scatters the electron by a recoil process into a state with $\underline{k}-\underline{q}$ (ignoring 'umklapp' terms). The electronic energy change is $\varepsilon(\underline{k}) - \varepsilon(\underline{k}-\underline{q})$ while the phonon energy is $\hbar\omega_q$. Energy need not instantaneously be conserved in accordance with the Heisenberg uncertainty principle. The electron wave is thus strongly coupled to the optical phonon modes since these are associated with strong electrical polarisation of the medium. Fröhlich (9) noticed that this process can yield a strong correction to the electronic energy levels. The Feynmann diagram for this <u>virtual</u> process is given by

$\underline{k}-\underline{q}$ $\underline{k}$ $\underline{k}$ $\underline{q}$

$\sim\!\sim$ represents occupied virtual phonon mode q.

The phonon excitations here are analagous to the exciton modes of the dielectric in our model. Thus, in a similar way, from the viewpoint of quantum field theory, an electron excited in the retinyl spine of Rhodopsin has a dynamic self energy due to the virtual emission and re-absorption of the excitons (excitation modes) of the surrounding protein. It is helpful to imagine an excited polyene chromophore at some level Δ ($\sim$ 3eV) above the ground state interacting with the surrounding protein which is initially in its ground state. According to Heisenberg's uncertainty principle, the excitation can be virtually transferred to the surrounding protein (E $\sim$ 4eV) provided that the borrowed energy δE is repaid within a time δt subject to the condition $\delta E . \delta t \sim h$. (h is Planck's constant).

The Feynmann diagram considered to be of importance here is shown below.

occupied excited state of polyene → ← ground state of opsin

ground state of polyene → ← occupied excited state of opsin

occupied excited state of polyene → ← ground state of opsin

At this stage it is helpful to simplify the notation to a form which clarifies the physics of the problem. Ziman's formulation of interacting fields is particularly useful here (10).

Let us denote the excited state of the polyene and ground state of the dielectric by $|\underline{k}, \underline{0}_q\rangle$ while the converse virtual state is denoted by $|\underline{k}-\underline{q}, \underline{1}_q\rangle$. This represents the higher energy state with the excitation localised on the dielectric. Standard perturbation theory leads to a modified excited state wavefunction given by

$$|\underline{k}\rangle' \approx |\underline{k}, \underline{0}_q\rangle + \sum_{\underline{q}} \frac{|\underline{k}-\underline{q}, \underline{1}_q\rangle \langle \underline{k}-\underline{q}, \underline{1}_q | V | \underline{k}, \underline{0}_q\rangle}{\mathcal{E}(\underline{k}) - \mathcal{E}(\underline{k}-\underline{q}) - E_q}$$

where $\mathcal{E}(\underline{k}) - \mathcal{E}(\underline{k}-\underline{q}) \sim \Delta$

Using this approximation wavefunction, a self energy correction and thus spectral shift, can be estimated and is given by

$$\mathcal{E}(\underline{k}) - \mathcal{E}'(\underline{k}) \approx \sum_{\underline{q}} \frac{|\langle \underline{k}, \underline{0}_q | V | \underline{1}_q, \underline{k}-\underline{q} \rangle|^2}{E_q - \{\mathcal{E}(\underline{k}) - \mathcal{E}(\underline{k}-\underline{q})\}}$$

$\langle \underline{k}, \underline{0}_q | V | 1_q, \underline{k}-\underline{q} \rangle$ is thus an excitation transfer matrix element.

Thus, provided $E > \Delta$, which is almost certainly the case in Rhodopsin, then a significant lowering of the energy of the excited state can occur. $E - \Delta$ is a resonant denominator. Of course when $E \to \Delta$ the theory blows up, because we have neglected the damping, but this simple theory does predict that for equal couplings then a dielectric composed of benzene rings ($E \sim$ 4.5eV) would not produce as large an energy shift as, say, a system of indole rings ($E \sim$ 4eV). Taking $E - \Delta$ to be ~ 1eV and $|\langle \underline{k}, \underline{0}_q | V | \underline{1}_q, \underline{k}-\underline{q} \rangle| \sim 1$eV then the expected energy shift would be ~ 1eV causing the absorption maximum to shift from around 400 nm to about 600 nm, roughly that observed experimentally. This gives us the order of magnitude for $|\langle \underline{k}, \underline{0}_q | V | \underline{1}_q, \underline{k}-\underline{q} \rangle|$. For this rough estimate we have neglected the first term in the sum because of the large denominator involved. Ordinary perturbation theory is suspect here because of the large energy shifts involved, however it is nevertheless useful to enable one to gain some insight into the problem.

We will now look at the Hamiltonian of the polyene in another way which implicitly includes the effect of the dielectric medium on the modified energies of electron states of the polyene. This is achieved by a renormalisation procedure which shows that such a spectral shift implies an attractive electron-electron interaction between electrons in the polyene spine. Opposed to this attractive coupling is the usual repulsive coulomb interaction between the electrons, coulomb which is very strongly screened by the other electrons. To show that such an attractive coupling should exist, we need to transform the second quantised form of the operator $\mathcal{H}_{pd}$ and then to feed in our order of magnitude estimate for $|\langle \underline{k}, \underline{0}_q | V | \underline{1}_q, \underline{k}-\underline{q} \rangle|$ into the transformed equations.

ELECTRON-ELECTRON INTERACTIONS IN RHODOPSIN

The part of the hamiltonian describing the polyene-dielectric coupling $\mathcal{H}_{pd}$, is given by

$$\mathcal{H}_{pd} = \frac{1}{2}\sum_{\underline{k}',\underline{q}',\underline{k},\underline{q}} \langle \underline{q}',\underline{k}'|V|\underline{k},\underline{q}\rangle\, c^{+}_{\underline{q}'}\, a^{+}_{\underline{k}'}\, c_{\underline{q}}\, a_{\underline{k}}$$

This can be rearranged by use of a unitary transformation equivalent to a change of the representation of the basis functions which approximately renormalises the interaction in such a way that allows concentration on the properties of the polyene electron states in the presence of the dielectric. The polyene excitations can thus be described as "dressed." A simple proof of this type of transformation, which Little used in his paper, is given in detail in Ziman's book (11). For reasons of space, we will merely quote the result that $\mathcal{H}_{pd}$ can be transformed into the form

$$\tilde{\mathcal{H}}_{pd} \approx \frac{1}{2}\sum_{\underline{k}',\underline{k},\underline{q}} \frac{2\,|\langle \underline{k},0_{\underline{q}}|V|1_{\underline{q}},\underline{k}-\underline{q}\rangle|^{2}\, E_{q}}{\{\mathcal{E}(\underline{k})-\mathcal{E}(\underline{k}-\underline{q})\}^{2}-E_{q}^{2}}\; a^{+}_{\underline{k}'+\underline{q}}\, a^{+}_{\underline{k}-\underline{q}}\, a_{\underline{k}}\, a_{\underline{k}'}$$

where $\underline{k}' = \underline{k} - \underline{q}$

(see refs. 1, 2).

This effectively decouples the polyene and dielectric excitations. The interaction may be represented by the Feynman diagram

$\underline{k}-\underline{q}$ $\underline{q}$ $\underline{k}'+\underline{q}$ $\underline{k}$ $\underline{k}'$

The wavy line joining the vertices of the graph signifies that the interaction is not the bare interaction but the indirect coupling arising from electrons exchanging virtual excitons.

Because of the structure of the operators $a^{+}a^{+}aa$ this additional term to the Hamiltonian corresponds to an electron-electron scattering process in the polyene. We observe that if $E_{q} > \{\mathcal{E}(\underline{k}) - \mathcal{E}(\underline{k}-\underline{q})\}$ then this extra term corresponds to an attractive coupling between pairs of electrons in the polyene. Because of the form of the equation for the dynamic self energy correction which could possibly account for the spectral shift in Rhodopsin, we deduce that this side chain induced electron-electron attraction in Rhodopsin could be quite large.

This becomes clearer if we write the spectral shift approximately as

$$\frac{|V_{sc}|^2}{E - \Delta}$$

and the net electron-electron matrix element in the polyene in the form

$$V = V_C - \frac{2|V_{sc}|^2 E}{E^2 - \Delta^2} \quad \text{(see refs. 1, 2)}$$

V_C is the usual screened coulomb interaction. $|V_{sc}|$ is an averaged matrix element. For the theory not to give nonsensical results, we require $E \neq \Delta$. The side chain induced interaction could thus be several eV. We thus make the very important deduction that the electron-electron interaction in Rhodopsin could be attractive provided the side chain induced electron-electron attraction due to the "Little" mechanism exceeds the usual repulsive coulomb interaction between the electrons V_C. Little and Kampas (11, 12) have also discussed an instance of a spectral shift in a porphyrin as a consequence of the type of excitonic mechanism discussed here. If this does occur, then the formation of bound electron pairs could drastically alter the physical properties of the molecule. This would be associated with the Boson like properties of the electron pairs.

It can be shown (7) that a bound state exists for a pair of particles with total kinetic energy $\varepsilon(p)$ in the presence of an attractive pairing interaction V if the following equation can be satisfied

$$V \int \frac{dp}{\Delta_{ee} + \varepsilon(p)} = 1$$

The integral is over a centre of mass momentum space and $-\Delta_{ee}$ is the binding energy of the pair. Following Belyaev (7), who discussed the problem in the context of bound states in nuclear matter, we conclude that even in our essentially one dimensional system, a solution exists when $V \gtrsim 0$ (very weak attraction) if the integral

$$\int dp\, \varepsilon^{-1}(p) = 2M \int dp\, p^{-2}$$

is divergent. Clearly this can be so when $p \approx 0$. An appropriate choice of Δ_{ee} can always make the integral converge and large enough to compensate for the small value of V. M is an effective mass for the pair. This result for the one dimensional case is similar to, but nevertheless distinct from Cooper's well known proof concerning bound states between electron pairs in a degenerate Fermi gas. (13).

Finally, before very briefly discussing some possible experiments, some qualitative comments can be made on the relationship between this theory and the 'sudden polarisation effect'of Salem and Bruckmann (14). In a qualitative way, one expects that the very strongly polarised excited states calculated by these workers for the retinyl chromophore would have a very strong electrostatic influence on the surrounding protein. The excitation transfer matrix element V_{SC} is roughly related to the product of the transition dipoles for the excitations of the retinyl chromophore and also of the aromatic side chains of opsin, (15), and thus the very large changes in electric moment which seem possible on excitation of the polyene are of great interest here. Quite remarkably, one also expects a significant lowering of the coulomb repulsion V_C in the protonated Schiff base due to the presence of the extra positive charge.

The task of the experimentalist in attempting to test this theory is to search for the existence of bound electron pair states in the Rhodopsin molecule.associated with this phenomenon we would expect a very high electronic conductivity. What is needed in the first instance is a systematic series of measurements of the dielectric constant, diamagnetic susceptibility, N.M.R. spectra, infra red spectra and measurements also of electron tunnelling (16) on Rhodopsins in solution in both the illuminated and dark states. With such a set of data we can then begin to discuss on a firmer basis the intriguing possibility that bound electron pairs are involved in vision.

ACKNOWLEDGMENTS

I am grateful to Professor B. Pullmann for inviting me to participate in this Symposium and for providing hospitality. My thanks must also go to the Royal Society of London for a travel grant. I also thank Professors G.M. Bell, H. Fröhlich FRS, W.A. Little, D.H. Whiffen FRS and S.J. Wyard for help and encouragement over the past few years and Dr. N.G. Parsonage for introducing me to molecular superconductivity.

REFERENCES

1. Little W.A., Phys. Rev. A134, 1416, (1964).
2. Little W.A., J. Poly. Sci. C29, 17, (1970).
3. Little W.A., Scientific American 212, No. 2, 21, (1965).
4. Dunne L.J., Physics Letters, 48A, 13, (1974).
5. Bardeen J., Cooper L.N., Schrieffer J.R., Phys. Rev. 108, 1175, (1957).
6. "The Helium Liquids". NATO Advanced Study Institute. Eds. Armitage J.G.M., Farquhar I.E. Academic Press, 1975.
7. "Collective Excitations in Nuclei". (Documents on Modern Physics). S.T. Beylaev, Gordon and Breach, (1968).
8. Mantione M.J., Pullman B. Int. J. Quant. Chem., 5, 349, (1971).
9. For a review with references to earlier work see Fröhlich H., Adv. in Physics 3, 325 (1954). See also "Statistical Mechanics". R.P. Feynmann, W.A. Benjamin, Inc. (1972).
10. Ziman J.M., "Elements of Advanced Quantum Theory". Cambridge University Press, (1969).
11. Kampas F.J., J. Poly. Sci. 29C, 81, (1970).
12. W.A. Little cited in reference 11.
13. Cooper L.N., Phys. Rev. 104, 1189, (1956).
14. Salem L., Bruckmann P., Nature 258, 526, (1975).
15. Davydov A.S., "Theory of Molecular Excitons". Plenum Press, New York, (1971).
16. Dunne L.J., Clark A.D., Physics Letters, 60A, (1977).

CLASSICAL AND NON-CLASSICAL DECAY PATHS OF ELECTRONICALLY EXCITED CONJUGATED DIENES.

S. Boué, D. Rondelez and P. Vanderlinden

Department of Organic Chemistry ; Physical-Organic Chemistry Grouping ; Faculty of Sciences.
Free University of Brussels
50 F.D. Roosevelt Ave., 1050 Brussels, Belgium.

INTRODUCTION

In 1969 Oosterhoff and van der Lugt[1] investigated the butadiene → cyclobutene interconversion and pointed out that a reaction would take place only if there exist such states from which a driving force can derive ; this was an elegant way of showing on theoretical grounds that a molecule electronically excited in its Franck-Condon nuclear configuration will spontaneously distort itself (relax) if the new nuclear configuration corresponds to an energy minimum which correlates with the initial state along a coordinate devoid of significant barriers. If two electronic states do not correlate with each other (regardless ΔH_0) or if they do through an exceedingly high barrier or in a continuous highly endothermic way, no adiabatic interconversion would occur (interconversion does however take place diabatically at avoided surface crossings). According to and within the restrictions of the above statements a molecule, whatever in the ground or in an excited state, is entitled to move on its hypersurface. These motions (changes of internuclear angles and distances) from energy well to energy well are determined by the shape of the potential surfaces ; if one were to make any prediction on chemical (or physical) transformations one should obviously have some knowledge of the potential surfaces. In a remarkable series of papers Michl[2] has shown how qualitative MO arguments can provide an insight in the location of wells, funnels and maximums on the excited states and ground state hypersurfaces and in the connections between them (i.e. the decay from excited states to ground state); the problem of surface touchings and crossings has also been actively investigated by Salem[3] and Devaquet[4].

B. Pullman and N. Goldblum (eds.), Excited States in Organic Chemistry and Biochemistry, 199-207.

Thus it now sounds legitimate to question the fate of a molecule following Franck-Condon photoexcitation ; for highly rigid (cyclic) entities the freedom of motion (distortion) will obviously be limited (but not negligible, e.g. the benzene channel III case[5]) and one expects the F.C. state to lie in an energy well from where it will either fluoresce or intersystem cross or still internally convert to S_0. For flexible molecules, like the acyclic conjugated dienes which will be dealt with in this paper, the field of possibilities is much more widely open ; thus far essentially two types of dienes vibrational relaxation, which we will call "classical", have been analyzed by the theoreticians, namely the butadiene (s-*cis*) → cyclobutene and butadiene → methylene-allyl interconversions. Since the molecular motions on the excited hypersurface depend on the accepting vibrational modes available as well as on the barriers, we have decided to investigate experimentally the effects of alkyl substitution, of isotope substitution and of initial energy content (exciting wavelength) on the behavior of the buta-1,3-diene framework in the condensed phase.

EXPERIMENTAL AND RESULTS

All dienes, either purchased or prepared by conventional methods, were better than 99% pure ; n-pentane (pro analysi ; UCB) was transparent down to 210 nm and cyclohexane (pro analysi ; Merck) had an OD/1 cm of 0.1 at 228.8 nm, i.e. its contribution to light absorption was negligible in our conditions (Dienes solutions OD/1 cm > 3). All photolyses were performed at 20° on calibrated (0.1 M) n-pentane or cyclohexane solutions of dienes, according to a procedure described at length elsewhere[6]. Quantitative progress of the reactions was followed by g.l.c. analysis on an Intersmat IGC 16 machine equipped with a TCD detector.

The table gives the initial quantum yields measured for most unimolecular processes observed on direct photolysis of a wide series of dienes (dimers were not determined except for penta-1,3-dienes-*vide infra*); pertinent to comparison are the following additional data on cyclobutenes formation from buta-1,3-dienes in solution :

$$\text{2-methylbutadiene} \xrightarrow{253.7\ \text{nm}} \text{1-methylcyclobutene} \quad \Phi = 0.09$$

$$\text{2,3-dimethylbutadiene} \xrightarrow{253.7\ \text{nm}} \text{1,2-dimethylcyclobutene} \quad \Phi = 0.12$$

$$\textit{trans}\ \text{hexa-2,4-diene} \xrightarrow{253.7\ \text{nm}} \textit{cis}\ \text{3,4-dimethylcyclobutene} \quad \Phi = 0.02$$

2,4-dimethylpentadiene $\xrightarrow{253.7\ nm}$ 1,3,3-trimethylcyclobutene
Φ = 0.00 (in gas φ, Φ_{max} = 0.05 at 1400 Torr of ethane[7])

cycloocta-1,3-diene $\longrightarrow$ bicyclo[4,2,0]oct-7-ene
Φ = 0.15 at 253.7 nm ; Φ = 0.015 at 228.8 nm.

The penta-1,3-dienes were further studied in somewhat more details and it was observed that the decrease of exciting λ from 253.7 nm down to 228.8 nm not only reduced Φ of chemical processes but also gave rise to an unexpected concentration dependent behavior of the excited state ; this is shown in the figure. The photostationary composition was found to be 41% *cis*-59% *trans* at 0.1 M and 22% *cis*-78% *trans* in the plateau, in reasonable agreement with the classical relationship $[trans]_s/[cis]_s = \Phi(c\rightarrow t).\ \varepsilon(c)/\Phi(t\rightarrow c).\varepsilon(t)$ derived from a simple kinetic scheme (at 228.8 nm, $\varepsilon(c)$ = 17,900 $M^{-1}cm^{-1}$ and $\varepsilon(t)$ = 15,100 $M^{-1}cm^{-1}$)[8]. Worth a mention is the total absence (g.l.c.) of dimers in the 228.8 nm excitation[9] whereas dimerization was a major contribution at 253.7 nm[10].

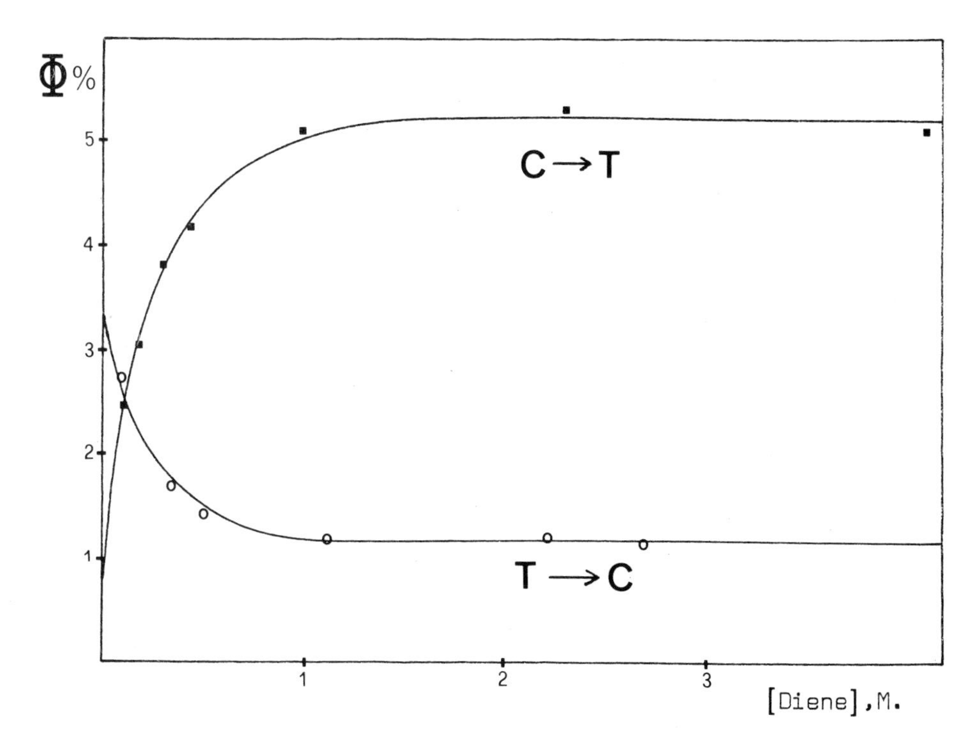

FIGURE. Concentration dependence of the quantum yields for C ⇌ T isomerizations of the penta-1,3-dienes excited with 228.8 nm light in cyclohexane at 20°.

TABLE. Condensed phase photolysis of substituted buta-1,3-dienes.

Starting diene [a]	Photo-products	Quantum yield x 10^2			
		All ^1H dienes		^2H labelled dienes * = CD_3	
		Exciting wavelength, nm			
		253.7	228.8	253.7	228.8
	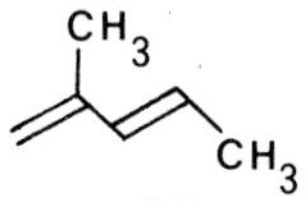 CH_3, CH_3	16	5.5	9	6.4
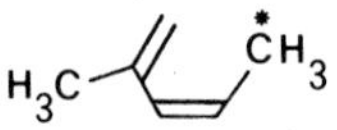 H_3C, $\overset{*}{C}H_3$	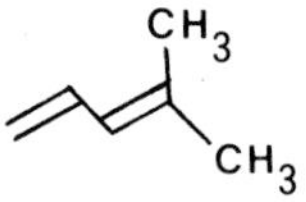 CH_3, CH_3	17	1.7	6	1.2
	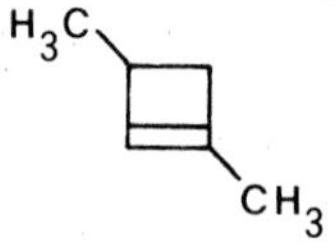 H_3C, CH_3	6	0.9	4	1.1
	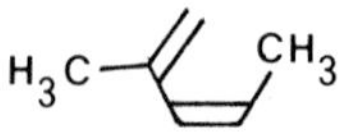 H_3C, CH_3	1.2	1.6	0.6	1.4
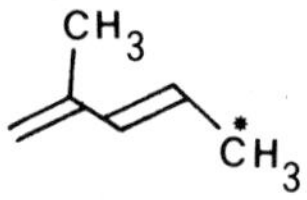 CH_3, $\overset{*}{C}H_3$	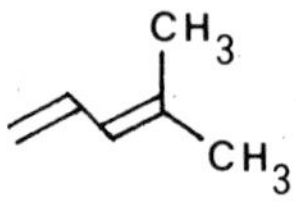 CH_3, CH_3	0.12	0.04	0.00	0.00
	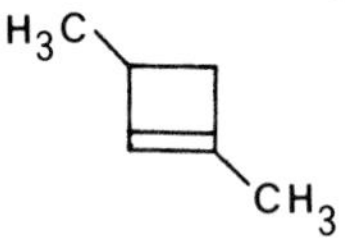 H_3C, CH_3	4.4	0.09	1.6	0.00
	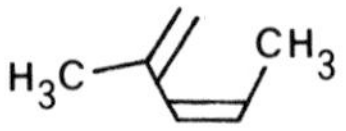 H_3C, CH_3	0.6	0.00	0.5	–
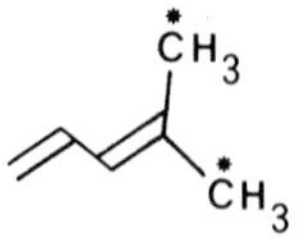 $\overset{*}{C}H_3$, $\overset{*}{C}H_3$	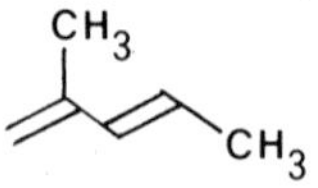 CH_3, CH_3	0.4	0.00	0.00	–
	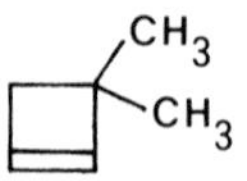 CH_3, CH_3	0.00	0.00	0.00	–
	Others [b]	1.0	0.00	0.00	–

a. Thermodynamic s-*cis*/s-*trans* composition at 20°; ~0.1M solution.
b. See reference 6.

TABLE. Continued.

Starting diene [a]	Photo-products	Quantum yield x 10^2: All ^1H dienes, Exciting wavelength, nm: 253.7	228.8	^2H labelled dienes, * = CD_3: 253.7	228.8
c 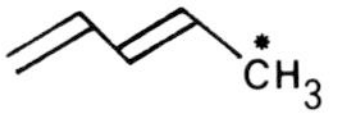	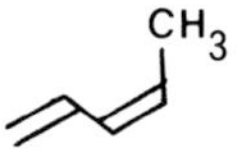	8	2.2	10	1.4
	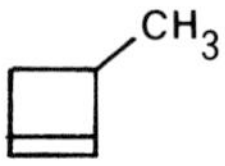	3	0.00	4.4	0.00
c 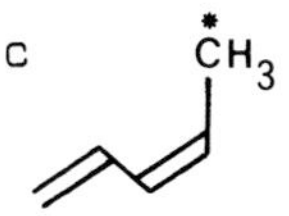	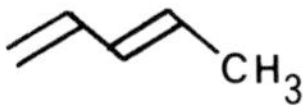	10	2.3	10	4
	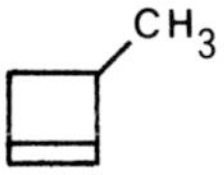	0.3	0.00	0.8	0.00
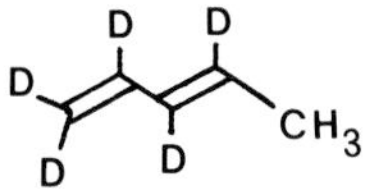	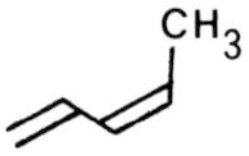	–	–	4	2 [d]
	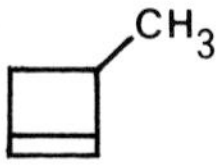	–	–	not determined	
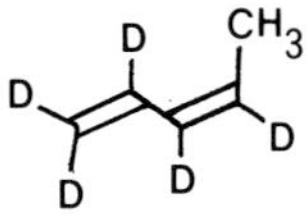	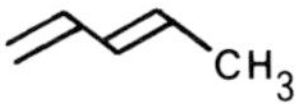	–	–	8	2.5 [d]
	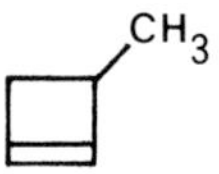	–	–	not determined	
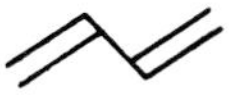	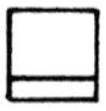	4	0.2	–	–
	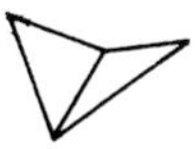	0.34	0.38	–	–

c. 1,3-dimethylcyclopropene is also formed as a minor product at 253.7 nm[11] ; the 1,5 shift of deuterium from CD_3 is undetectable (NMR). d. Determined on a 2×10^{-2}M solution.

DISCUSSION

A. Influence of methyl substitution.

Inclusion of methyl group(s) has at least a threefold consequence which shows up in the photochemistry of buta-1,3-dienes.

(i) When substituted at sterically congested positions, CH_3 groups will modify the s-*cis*/s-*trans* composition and/or draw the ground state diene skeleton aside from planarity.

(ii) Apart from the change in the s-*cis*/s-*trans* ratio, the presence of methyl groups at terminal sp^2 positions decreases Φ (cyclobutene) whereas it increases that yield at the 2 and 3 positions, with respect to the figure obtained with the unsubstituted parent hydrocarbon ; the effects seem to be additive. Since however Φ(cyclobutene) = k(cyclobutene)/Σk_i, it cannot be decided whether the changes in Φ values are accounted for by a change in the energy profile along that coordinate or rather by intrusion of a competitive process k_i but there certainly are factors other than conformational as shown by the comparison between *cis* penta-1,3-diene and 4-methylpenta-1,3-diene and between *cis* 2-methylpenta-1,3-diene and 2,4-dimethyl-penta-1,3-diene.

(iii) Inclusion of a terminal CH_3 group induces a new decay path, the 1,5-sigmatropic shift of a hydrogen, which competes efficiently with the 1,4 cyclomerization and which would sterically require the twist of the unsubstituted end p atomic orbital ; the occurrence of abnormal 1,5 shifts has already been discussed[6] in terms of a nuclear configuration involving the twist of π orbitals about more than one bond.

The low Φ for *cis* $\rightleftharpoons$ *trans* isomerization of the 4-methylbuta-1,3-dienes excited in the singlet state suggest either that relaxation to the methylene-allyl configuration occurs preferentially by a twist at the unsubstituted end of the π system or that relaxation does not correspond to a twist by as much as 90° ; the previously reported[11] formation of 1,3-dimethylcyclopropene (but not of ethylcyclopropene) in the direct photolysis of the penta-1,3-dienes would be consistent with the specific twist at C1 and the ring closure of a methylallyl moiety, although this does not fit into the model proposed by Salem[12].

B. ^2H isotope effect.

Substitution of ^1H by ^2H has a definite quantitative influence on the behavior of singlet excited buta-1,3-dienes (see table) but only the primary effect on 1,5 sigmatropic shifts of hydrogen is clearly understandable. Although the secondary effects are quite important in some cases, no significant trend shows up as to the factors which come into play such as particular vibronic couplings, change in natural lifetime or by-pass of specific pathways to the benefit of others. Therefore the involvement of C-H modes in vibrational relaxation (e.g. for explaining the wavelength effects - *vide infra*) might be not as important as was proposed before[8].

C. Wavelength effect.

The existence of thermally equilibrated s-*cis* and s-*trans* conformers, each of which contributes to light absorption at every wavelength with its proper $\varepsilon(\bar{\nu})$, may lead to results which depend on the exciting λ. The whole of our observations however cannot be accounted for by a change in the relative contribution of ground state conformers to physical and chemical processes. In particular *cis* 2-methylpenta-1,3-diene which is an essentially s-*cis* molecule[6] and cycloocta-1,3-diene which has a rigid cisoid conformation provide a strong evidence that the state populated at 228.8 nm does not cascade down to the state produced by 253.7 nm excitation.

The case of the penta-1,3-dienes is perhaps still more striking due to the peculiar concentration dependent behavior exhibited, in addition to the reduced Φ values, in the 228.8 nm excitation (see figure) ; we have here a quenchable state of finite lifetime displaying properties distinctly different from these which obtain at 253.7 nm. The data are consistently accommodated by the kinetic scheme shown below and which implies the formation of an excimer to justify that $\Phi(cis \rightarrow trans)$ increases with the diene concentration whereas $\Phi(trans \rightarrow cis)$ decreases. (The same scheme applies to both *cis* and *trans* isomers, with of course specific α and β values[13] ; thus only one scheme is developped where D stands for pentadiene, either *cis* or *trans*).

$$D \xrightarrow{I_a} D^*$$

$$D^* \xrightarrow{k_1} \alpha\ cis + (1-\alpha)\ trans$$

$$D^* + D \xrightarrow{k_q} D_2^*$$

$$D_2^* \xrightarrow{k_2} \beta\ cis + (2-\beta)\ trans$$

From that scheme one derives that the isomerization quantum yield Φ_i is given by : $\Phi_i = (\alpha k_1 + \beta k_q[D])/(k_1 + k_q[D])$

(Φ_i refers to *trans→cis* with α and β and to *cis→trans* with (1-α) and (2-β) instead). Putting in numerical values one gets for the *trans→cis* isomerization:

$$\alpha = 3.3\times10^{-2};\ \beta = 1.15\times10^{-2} \quad \text{and} \quad k_q/k_1 \sim 15\ \text{l mol}^{-1}$$

and for the *cis→trans* transformation:

$$(1-\alpha) = 8\times10^{-3};\ (2-\beta) = 5.2\times10^{-2} \quad \text{and} \quad k_q/k_1 \sim 8\ \text{l mol}^{-1}$$

If we assume that k_q = k(diffusion) ∿ 10^{10} s^{-1} we set a lowest lifetime limit of ∿ 1 ns for the quenchable D* (*cis* or *trans*) state produced at 228.8 nm., i.e. a time long enough to warrant thermal equilibrium with the solvent.

If the excimer arises from a bonding overlap involving the C2 and C3 atoms as predicted by Salem[14], then one understands that no dimers are formed any longer in the 228.8 nm excitation.

The present wavelength dependence is reminiscent of the "channel III threshold" observed for benzene ; here the decrease of exciting λ has no influence down to 246 nm and is completed at 230 nm[15]. This is most readily explained by the model proposed by Birks[16] for the benzene case: at short λ a barrier is being overcome which offers the molecule a new vibrational mode for relaxation into an energy well of as yet undetermined configuration (a S_X state) and which necessarily connects with the S_0 valley of the starting isomers.

ACKNOWLEDGEMENTS

We thank Professor R.H. Martin for his interest. We gratefully acknowledge financial support from the Fonds National de la Recherche Scientifique. One of us (P.V.) is indebted to the Institut pour l'Encouragement de la Recherche Scientifique dans l'Industrie et l'Agriculture (IRSIA) for the grant of a fellowship.

REFERENCES

1. W.Th.A.M. van der Lugt and L.J. Oosterhoff, J. Amer. Chem. Soc., 1969, 91, 6042.

2. J. Michl, Pure Appl. Chem., 1975, 41, 507 and references cited therein.

3. L. Salem, J. Amer. Chem. Soc., 1974, 96, 3486.

4. A. Devaquet, Pure Appl. Chem., 1975, 41, 455.

5. J.B. Birks, Photophysics of Aromatic Molecules, Wiley-Interscience, 1970, page 247.

6. D. Rondelez and S. Boué, J.C.S. Perkin II, 1976, 647.

7. R. Srinivasan and S. Boué, J. Amer. Chem. Soc., 1971, 93, 550.

8. P. Vanderlinden and S. Boué, J.C.S. Chem. Comm., 1975, 932.

9. This dismisses the erroneous figure based on a mass balance, which appeared in a preliminary communication[8].

10. M. Bigwood and S. Boué, Tetrahedron Letters, 1973, 44, 4311.

11. S. Boué and R. Srinivasan, J. Amer. Chem. Soc., 1970, 92, 3226.

12. P. Bruckmann and L. Salem, J. Amer. Chem. Soc., 1976, 98, 5037.

13. The same scheme would hold as well if two states were formed, one of which decaying exclusively to its initial ground state and the other giving rise to the excimer. This would only change the α and β values but be of no importance to our reasoning ; the lowest excimer channel contribution compatible with the numerical figures would involve $\sim$ 3.7% of the excited molecules.

14. L. Salem, J. Amer. Chem. Soc., 1968, 90, 553.

15. D.F. Eaton, personal communication (PhD thesis, Caltech 1972).

16. J.B. Birks, Organic Molecular Photophysics, J.B. Birks Ed., Wiley-Interscience, 1973, volume 1, page 48.

Far Ultraviolet Circular Dichroism of Oligosaccharides

C. Allen Bush

Department of Chemistry. Illinois Institute of Technology. Chicago, Il. 60616 U.S.A.

I. Introduction

Although many biological polymers possess both structural and information carrying functions, molecules such as DNA and messenger RNA are usually associated with the latter function while collagen and bacterial cell wall polysaccharides are thought of as mainly structural. That such a division should not be made along strictly chemical lines is shown by the example of ribosomal RNA, whose role is structural, and by the protein hormones which carry information between cells of higher organisms. Although a purely structural and metabolic function has generally been envisioned for the carbohydrate biopolymers, it is instructive to consider what, if any, informational role might be played by the sugars.

Certain data in the recent biochemical literature suggest that the complex carbohydrate chains of glycolipids and glycoproteins may indeed serve in coding and transmitting intercellular information. Morell et al. (1971) showed that ceruloplasmin, in common with most serum glycoproteins, was cleared very rapidly from the circulation when a carbohydrate residue, N-acetyl neuraminic acid (NANA) was removed enzymatically from its carbohydrate chain. Kawasaka and Ashwell (1976) have recently isolated a glycoprotein from liver cell membranes which they identify as the binding site of the asialo glycoproteins. This hepatic binding protein which binds specifically to carbohydrate chains terminating in galactose has many properties in common with the plant lectins. The mitogenesis of lymphocytes induced by lectin binding to the cell surface apparently results from the affinity of these carbohydrate binding proteins for the carbohydrate chains

B. Pullman and N. Goldblum (eds.), Excited States in Organic Chemistry and Biochemistry, 209-220.

of membrane glycoproteins. Similarly the binding of cholera toxin to the surface carbohydrates of cells causes profound metabolic effects on the transport properties of intestinal cells. The membrane surface carbohydrate chain identified as the receptor for cholera toxin binding is the ganglioside, GM_1, a glycolipid having a complex carbohydrate chain, (Fischman and Brady, 1976).

Although the assignment of an important organizational function for the complex carbohydrate chains of glycolipids and glycoproteins is emerging in the biochemical literature, rather little is known about the conformation of these oligosaccharide chains and even less research has been directed at their interactions with other biopolymers. Of the covalent chemical structure of glycoprotein carbohydrates certain general facts are known; they are generally composed of the neutral sugars galactose, mannose and fucose plus the 2-acetamido sugars, N-acetyl glucosamine (GlcNAc) and N-acetyl galactosamine (GalNAc) in addition to sialic acid (NANA). The oligosaccharide chains contain from 2 to 12 residues covalently connected by an N-glycosyl linkage to asparagine in the protein chain or by an O-glycosidic linkage to serine or threonine. The detailed sequence, intersaccharide linkage positions and branching points are known for a few common glycoprotein carbohydrate chains.

II. Circular Dichroism

Speculating that application of some of the methods of biophysical chemistry to the study of complex oligosaccharides might lead to some useful structural information, several laboratories including our own have begun looking at uv CD of some monosaccharides and the simplier oligosaccharides. The absence of chromophores absorbing at wavelengths longer than 190 nm in the neutral hexoses makes necessary the use of vacuum uv spectrometers to study such compounds as galactose, glucose and mannose. Johnson and coworkers have studied the CD of several neutral hexoses and their methyl glycosides in aqueous solution finding bands in the 185 to 165 nm region, (Nelson and Johnson, 1976). Spectra of dry films of polysaccharides have been extended to shorter wavelengths, about 155 nm by Pysh and coworkers (Balcerski et al, 1975).

Studies in our laboratory have concentrated on those carbohydrates having chromophores which may be studied with a conventional CD apparatus. For example the amide chromophore of the 2-acetamido hexoses has optical properties not unlike those of the well studied peptide chromophore. While the CD data of Fig. 1 for three common 2-acetamido hexoses were measured with a vacuum instrument, it is obvious that the band near 210 nm can be easily measured with a conventional apparatus. This band is as-

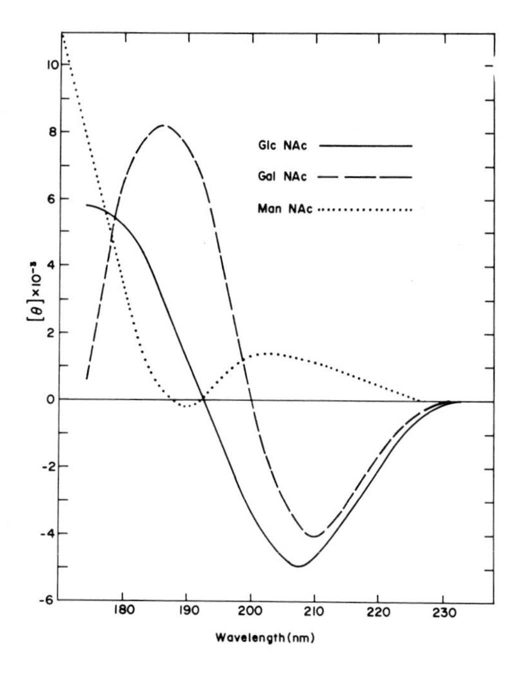

Fig. 1 Circular dichroism of three common 2-acetamido sugars in water.

signed to the n-π* transition of the amide and is thus analogous to the n-π* band seen in α helical polypeptides at 220 nm. Its blue shift in the amido sugars may result from more extensive solvation of the chromophore in the latter case. While the n-π* band is negative in both GlcNAc and GalNAc it is positive in ManNAc as a result of the opposite stereochemistry at C2, (Yeh and Bush, 1974).

Since galactose shows no large CD band at wavelengths longer than 185 nm, the band at 189 nm in GalNAc is presumably due to the amide chromophore, and more specifically to the amide π-π* transition at that wavelength. The weak CD of GlcNAc at 190 nm apparently results from fortuitous cancellations since the CD's of the α and β O-methyl glycosides show strong bands (Fig. 2). In both α methyl GlcNAc and α methyl GalNAc the π-π* band is large and occurs at 185 nm, a wavelength substantially shorter than that for the corresponding β O-methyl glycosides. The sensitivity to stereochemistry seen for this short wavelength band is not observed for the n-π* band. The CD at 210 nm has been shown to be -3 to -5,000 for both the α and β anomers of GlcNAc and GalNAc as well as for their O-methyl glycosides (Fig. 2 and Coduti et al, 1977).

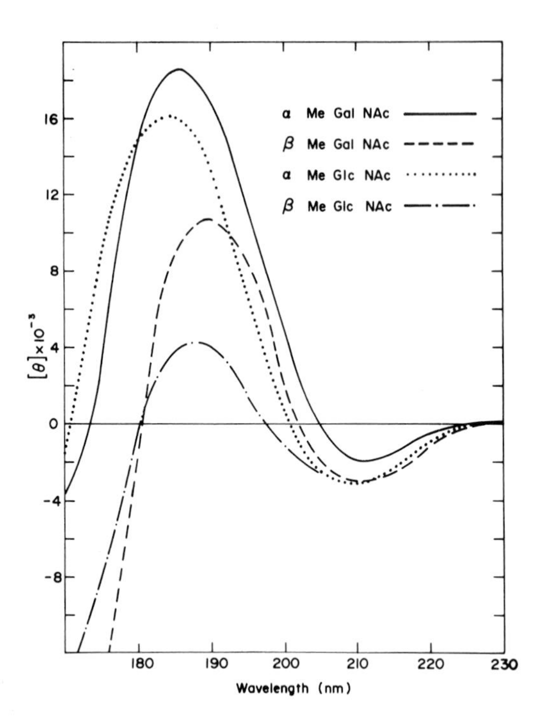

Fig. 2 Circular dichroism of the methyl pyranosides of 2-acetamido-glucose and galactose in water.

The CD spectra in water of a disaccharide and a trisaccharide composed only of GlcNAc (Fig. 3) show negative bands at 210 nm illustrating again the insensitivity of the n-π* band to stereochemistry. The amide π-π* bands are at the longer wavelengths characteristic of β linked residues. (See Figs. 2 and 3). It has been shown in the case of the series of chitin oligosaccharides (β1-4 linked) that the π-π* band shifts to longer wavelengths with an increasing magnitude of ellipticity per residue as a function of chain length (Coduti et al, 1977). It is difficult to ascribe this red shift and increased magnitude with chain length to an exciton mechanism such as that seen in oligopeptides; any reasonable model of the oligosaccharide conformation places the amide chromophores too far apart for any strong coupling effect. A more plausible explanation for this trend is a modification of the pyranoside ring conformation which in turn alters slightly the surroundings of the amide chromophore.

While to date we have been unsuccessful in our attempts at a quantitative theoretical treatment of the π-π* band in acetamido sugars, the n-π* band may be understood within the framework of one electron optical activity theory of Stigter and Schellman (1969). Our calculation of the CD due to the n-π* transition of

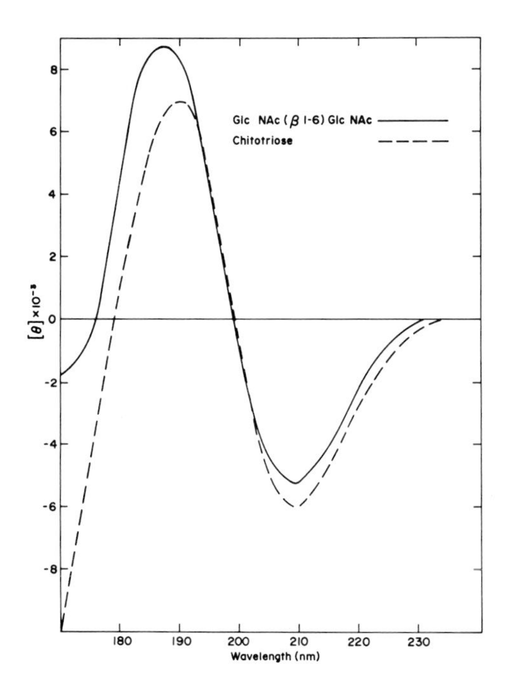

Fig. 3 Circular dichroism of two oligosaccharides composed of 2-acetamido glucose in water.

2-acetamido sugars shows that the electrostatic perturbations of the polar groups near the amide gives rise to the observed CD, (Yeh and Bush, 1974). These calculations rationalize the difference in sign of the CD of GlcNAc and ManNAc and also the relative insensitivity of the n-π* CD to the anomeric configuration of the sugar. The dominant electrostatic perturbation is shown to be that of the hydroxyl group at C3, directly adjacent to the amide chromophore. Changes in the orientation of this group as reflected by changes in the dihedral angle about the C3-O3 bond cause profound changes in the n-π* CD including reversal of the sign, (Yeh and Bush, 1974).

This last observation suggested to us that perhaps a change from aqueous solvents, which act as both donors and acceptors of hydrogen bonds, to a solvent such as hexofluoro 2-propanol (HFIP) which is exclusively a hydrogen bond donor might modify the dihedral angle about the C3-O3 bond and hence the CD in the 210 nm region. Fig. 4 compares the CD of GalNAc in water and in HFIP showing a reversal of the sign of the CD at 210 nm to a positive value in the latter solvent. We have obtained data similar to those of Fig. 4 for GlcNAc and also for both the α and β methyl

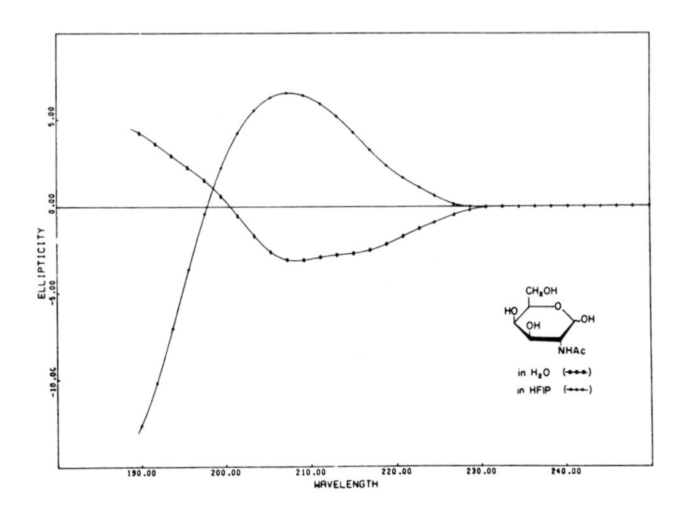

Fig. 4 Comparison of the circular dichroism of 2-acetamido galactose in water and in hexafluoro 2-propanol.

glycosides of GlcNAc and of GalNAc, (Dickinson, et al, 1977). In contrast, 3-O-methyl, β methyl GlcNAc in which the methoxyl group connected to C3 can act only as a hydrogen bond acceptor showed negative CD in both water and in HFIP. In contrast to the situation in aqueous solution, the n-π* CD in HFIP seems quite sensitive to stereochemistry and to the position of substituents, suggesting that the latter solvent might be an interesting one in which to study the CD of oligosaccharides. The CD of GlcNAc (β1-6) GlcNAc in HFIP is found to be simply the average of that for GlcNAc and its β methyl glycoside while that for the β1-4 isomeric disaccharide (chitobiose) shows negative CD at 210 nm. We have ascribed this difference in the CD at 210 nm of these two linkage isomers to the formation of a hydrogen bond in chitobiose between the ring oxygen of the non-reducing terminal residue to the C3 hydroxyl hydrogen of the reducing terminal residue (Dickinson et al, 1977). Since the CD in HFIP of disaccharides containing amido sugars depends on their linkage, it should be possible to detect the linkage in longer oligomers containing 2-acetamido sugars in only one type of linkage. For example, the CD of the tetrasaccharide lacto N tetraose ought to be sensitive to the linkage to its single residue of GlcNAc. Moreover it may be possible to carry this technique for linkage determination over into glycolipids containing a single amido sugar and no other chromophores absorbing in the 210 nm region.

III. Sialic Acid

Sialic acids such as NANA, which are rather common in both glycoproteins and glycolipids possess in addition to an amide chromophore, a carboxylic acid function. In order to discriminate between the contributions of these two chromophores to the

Table I: Gaussian Decomposition of CD

Compound	Electronic Transition	Wave-length (m)	Peak Height [θ] x 10^{-3}	Half Width at Half Maximum (nm)	Rotational Strength (cgs-esu x 10 40)
β-methyl-(methyl-D neuraminid)ate	ester - n-π*	221	2.8	13.3	8.30
β-methoxy-neuraminic acid	carboxy-late n-π*	217	-5.0	11.3	-12.8
	carboxy-late π-π*	189	10.1	8.0	21.1
NANA	carboxy-late n-π*	216	0.4	4.7	.38
	amide n-π*	210	0.5	4.0	0.48
	? π-π*	198	8.1	12.4	24.62
2-3(NAN-LAC)	carboxy-late n-π*	220	-3.0	9.0	-6.03
	amide n-π*	207	-2.2	4.7	-2.43
	? π-π*	199	10.9	10.9	29.3
2-6(NAN-LAC)	carboxy-late n-π*	226	-0.5	8.1	-0.9
	amide n-π*	211	0.8	5.9	1.0
	?	199	3.8	7.0	6.6

CD of NANA, we have synthesized a de-N-acetylated methyl glycoside, methyl ester by methanolysis of NANA (Dickinson and Bush, 1975). The weak absorption band and positive CD at 221 nm for the methyl ester may be assigned to the ester n-π* transition, (Table I). Saponification of the methyl ester afforded the β methyl glycoside of neuraminic acid, a compound having only the carboxylic acid chromophore. The low pK_a of NANA combined with the facile hydrolysis of its glycosidic bonds makes difficult the study of the CD of NANA glycosides in their protonated form, but at neutral pH the carboxylate form shows a negative CD band at 217 nm, a wavelength region of low absorbance, implying that this band must be due to the carboxylate n-π* transition. Although we observe positive CD in the 190 nm region, which is apparently due to a carboxylate π-π* transition, we do not yet have any vacuum uv data to firmly establish the wavelength of its maximum.

With the carboxylate CD bands thus assigned, one can better interpret the CD of NANA and of its oligosaccharides. Decomposition of the CD of NANA and that of two milk trisaccharides, NANA (2-3) lactose and NANA (2-6) lactose into Gaussian bands shows the carboxylate n-π* band to occur between 216 and 226 nm depend-

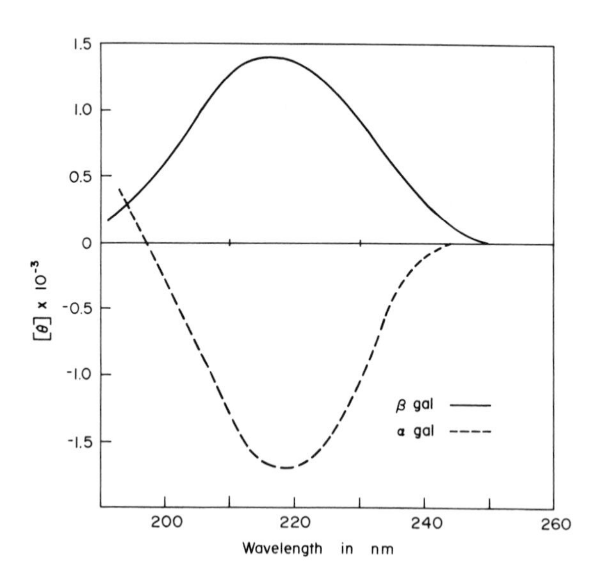

Fig. 5 Circular dichroism of the galactopyranoside pentaacetates in acetonitrile.

ing on the nature of the aglycon. (See Table I.) This analysis also reveals the amide n-π* band to be near 210 nm, the same wavelength region in which it is found in 2-acetamido hexoses. A large positive CD band found in NANA as well as in both trisaccharides near 200 nm is not so readily assigned. Both the amide and the carboxylate π-π* bands are generally found at wavelengths shorter than 192 nm, (Figs. 1-3, Table I). The rather remote possibility of an exciton type coupling mechanism might be tested by vacuum uv CD experiments.

IV. Peracetylated Carbohydrates

Since fully acetylated carbohydrates do not occur in nature, study of their conformational properties appears to have stimulated little interest among biophysical chemists. On the other hand, attachment of this chemically convenient chromophore might make accessible some covalent structural information from CD. A relationship between stereochemistry and the CD of peracetylated sugars is illustrated in Fig. 5, which compares the CD of α and β penta O-acetyl galactose in acetonitrile. Similar curves are

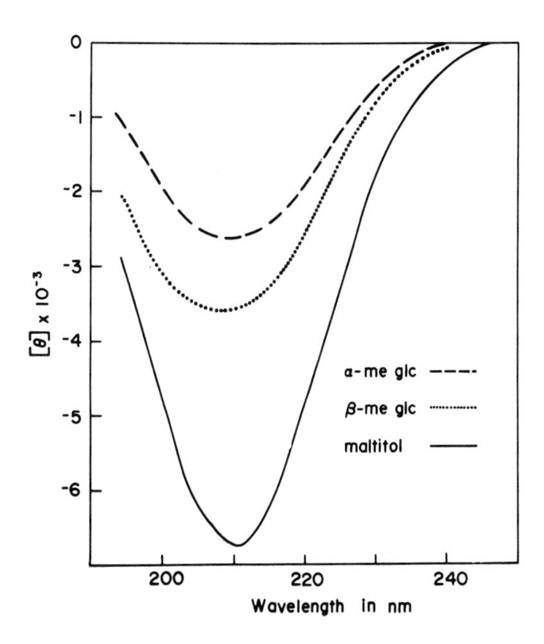

Fig. 6 Circular dichroism of the methyl pyranosides of glucose tetraacetate and of maltitol nonaacetate in trifluoroethanol.

found in trifluoroethanol (TFE) a solvent which along with acetonitrile combines good solvent properties for carbohydrate acetates with excellent uv transparence. The band at 216 nm, which may be assigned to the ester n-π* transition, is the only band seen in the wavelength range studied; the π-π* transition for esters lies below 190 nm. The α and β anomeric galactosyl acetates have oppositely signed CD bands. Similar results for the α and β anomeric glucosyl peracetates (Table II) suggests that the sign of the n-π* CD of these compounds is sensitive not to the stereochemistry of the pyranoside ring but rather to that of the anomeric acetate. That this interpretation is correct and that the sign of the CD is not strongly influenced simply by the presence of axial or equatorial substituents at Cl is shown by the data of Fig. 6 on the α and β O-methyl glycosides of tetra O-acetyl glucose.

Table II. CD Spectral Data for Peracetylated Sugars

	$[\theta]^{a)}$	λ_{max}	reference
α galactose	-1,700	218	d)
β galactose	1,300	216	d)
α glucose	3,100	210	c) and d)
β glucose	1,200	220	c) and d)
β xylose	3,100	208	c)
α methyl glucose	-2,500	210	d)
β methyl glucose	-3,500	210	d)
α cello biose	-4,800	208	c)
β cello biose	-1,600	210	c)
β gentiobiose	-2,400	208	c)
β isomaltose	-3,200	210	c)
β maltose	600	220	c) and d)
mannitol	5,610	215	b) and d)
glucitol	1,000	218	b) and d)
maltitol	-6,800	210	d)

a) [θ] for the entire molecule

b) Bebault et al. (1973)

c) Mukherjee et al. (1972)

d) This work

These observations suggests a simple additivity theory in which the CD of the n-π* band is taken to be the sum of a dominant contribution depending on the configuration of the anomeric acetate and a lesser contribution from the pyranose ring characteristic of the stereochemistry of the sugar. The CD data for seven monosaccharide acetates in Table II can be fit to such a theory with an accuracy of ±500. Since this is well outside experimental error, the theory may be considered an approximate one for monosaccharides. For disaccharides the predictions of this simple theory are less exact. For example this theory predicts β maltose, β cellobiose, β gentiobiose and β isomaltose should all have the same CD, a prediction not well satisfied by the data of Table II. Clearly there is some dependence of the CD of disaccharide peracetates on stereochemistry in addition to that of the component rings and anomeric acetates.

Since alditols obtained from aldoses by borohydride reduction of the aldehyde to an alcohol, have no anomeric carbon atom the CD of their peracetates might be expected to show greater dependence on stereochemistry than do the parent sugars. Bebault et al (1973) have reported the CD at 210 nm of several partially methylated alditol acetates, compounds of some analytical interests because of their importance in the methylation analysis of polysaccharide structures. Galacitol is inactive and glucitol and mannitol have quite different CD's (Table II). Disaccharide alditols, the borohydride reduction products of disaccharides, also have no anomeric acetate and thus might be expected to show a strong dependence of CD on stereochemistry. The CD of such a compound, maltitol nona-acetate is shown in Fig. 6 to be quite large compared to that of β maltose.

Acknowledgement

I wish to thank Dr. S. Brahms of the University of Paris for assistance with the vacuum uv CD measurements. The data on peracetylated carbohydrates were those of Mr. Jeff Schultz. This research was supported by NIH Grant AI-11014 and by NSF Grant CHE-76-16783.

References

J.S. Balcerski, E.S. Pysh, G. Chen and J.T. Yang (1975) J. Am. Chem. Soc. 97, 6274.

G.M. Bebault, J.M. Berry, Y.M. Choy, G.C.S. Dutton, N. Funnell, L.D. Hayward and A.M. Stephen (1973) Canad. J. Chem. 51, 324.

P.L. Coduti, E.C. Gordon, and C.A. Bush (1977) Anal. Biochem. 77,

H.R. Dickinson and C.A. Bush (1975) Biochemistry 14, 2299.

H.R. Dickinson, P.L. Coduti and C.A. Bush (1977) Carbohydrate Res. (in press).

P.H. Fischman and R.O. Brady (1976) Science 194, 906.
T. Kawasaka and G. Ashwell (1976) J. Biol. Chem. 251, 5292.
A.G. Morell, G. Gregioriadis, I.H. Schernberg, J. Hickman and G. Ashwell (1971) J. Biol. Chem. 246, 1461.
S. Mukherjee, R.H. Marchessault and A. Sarko (1972) Biopolymers 11, 291.
R.G. Nelson and W.C. Johnson (1976) J. Am. Chem. Soc. 98, 4296.
D. Stigter and J.A. Schellman (1969) J. Chem. Phys. 51, 3397.
C.Y. Yeh and C.A. Bush (1974) J. Phys. Chem. 78, 1829.

DISCUSSION

PYSH :

Could you estimate the signal/noise ratio in your data in th region from 175-185 nm ?

BUSH :

Typical ratios are 10 : 1 to 20 : 1 when a time constant of 10 seconds is used.

ADIABATIC PHOTOREACTIONS IN ACIDIFIED SOLUTIONS OF 4-METHYLUMBELLIFERONE

Thomas Kindt and Ernst Lippert

Stranski-Institut für Physikalische und Theoretische Chemie der Technischen Universität Berlin, Berlin, Germany (FRG)

4-Methylumbelliferone fluoresces in the spectral range from the near ultraviolet to the red region if dissolved in acidified mixtures of alcohol and water /1/. In all this spectral region even the Laser treshold can easily be exceeded so that a useful dye Laser has been achieved the strong and polarized radiation of which can be tuned by a swivel-mounted grating /2/. There are at least 5 Laser active species in those solutions.

Some of the species which possibly fluoresce are shown in fig 1. In this diagram the solvent is represented by three water molecules, the pH-value is assumed to increase from the bottom (cation c) to the top (anion a), and in the ground state the system solute solvent becomes energetically less favourable when proceeding from left to right; but if the solute is in S_1 the left hand side species might be the more favourable ones.

We failed in verifying that a photochemical ringopening reaction occurs from the cation c*2 or the anion a*2 by addition of an ^{18}O-water molecule hoping to find an ^{18}O-atom in the mass spectrum of the solute molecule after irradiation of the solution. But this failure, of course, is not an antiproof against the occurance of a photochemical ringopening reaction since the back reaction can be very fast.

B. Pullman and N. Goldblum (eds.), Excited States in Organic Chemistry and Biochemistry, 221-231.

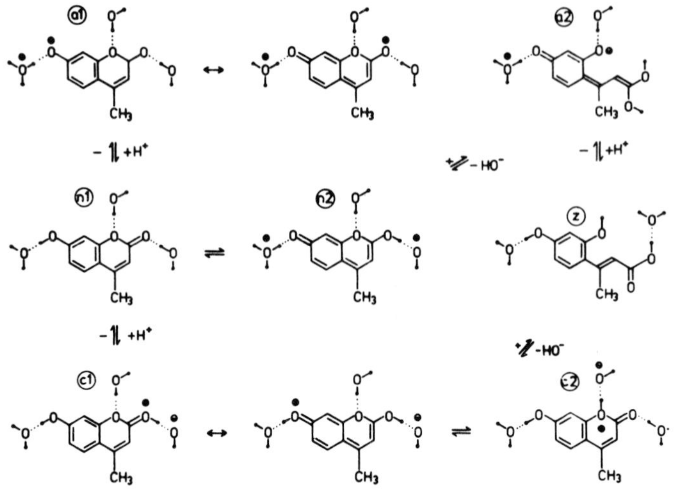

fig. 1: 4 MU in some possible solvated states, schematically. The bridging protons are indicated by heavy points /3/.

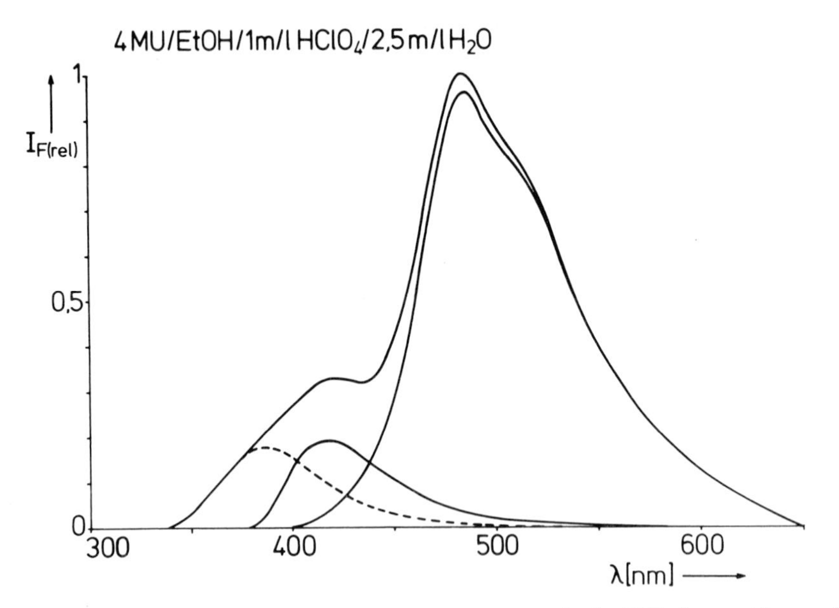

fig. 2: Fluorescence spectrum of 4 MU in ethanolic-solution at high acid concentration and low water concentration. Upper curve: Total fluorescence spectrum. Lower curves: Components of the spectrum. The spectrum was normalized to unity. Excitation wavelength 347 nm. The spectrum was recorded at room temperature.

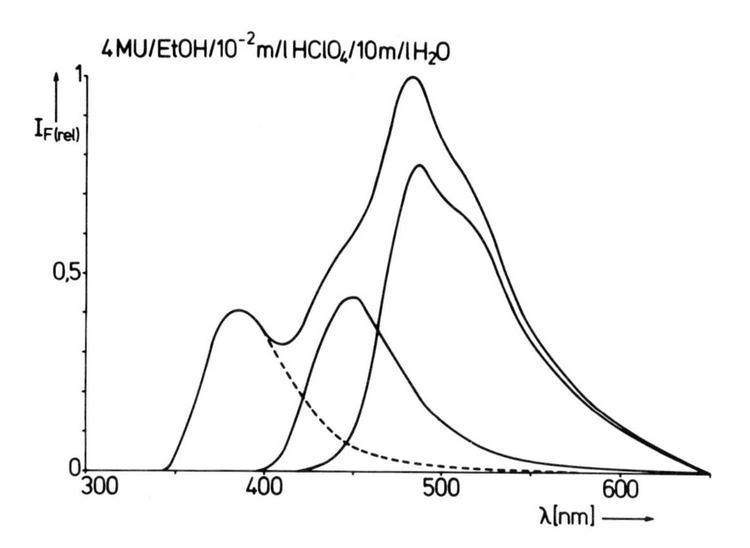

fig. 3: Fluorescence spectrum of 4 MU in ethanolic solution at low acid concentration and high water concentration. Upper curve: Total fluorescence spectrum. Lower curves: Components of the spectrum. The spectrum was normalized to unity. Excitation wavelength: 347 nm; the spectrum was recorded at room temperature.

But we could verify that in appropriate solvent mixtures all fluorescing species can be adiabatic photoreaction products from the initially excited neutral species n_1* since the intensity ratios between the different fluorescence bands do not change with excitation intensity below sufficient small intensities.

In fig. 2 and fig. 3 the fluorescence spectra of more or less acidified solutions are shown and analysed as the super-position of at least four fluorescence bands in each case, namely from n_1*, a_1^*, n_2^*. $n_2'^*$ and n_1^*, c_1^*, n_2^* $n_2'^*$, respectively. The fluorescence band maxima occur at 385, 418, 450, 485 and 530 nm from n_1^*, c_1^*, a_1^*, n_2^* and $n_2'^*$, respectively, in accordance with the results of other authors /4-8/; the bands at 450, 485 and 530 nm do not occur with the corresponding 7-ethoxy compounds. The fluorescence spectra are independent of the wavelength of excitation, but it should be mentioned that the spectra shown in fig. 2 and 3 have been obtained by

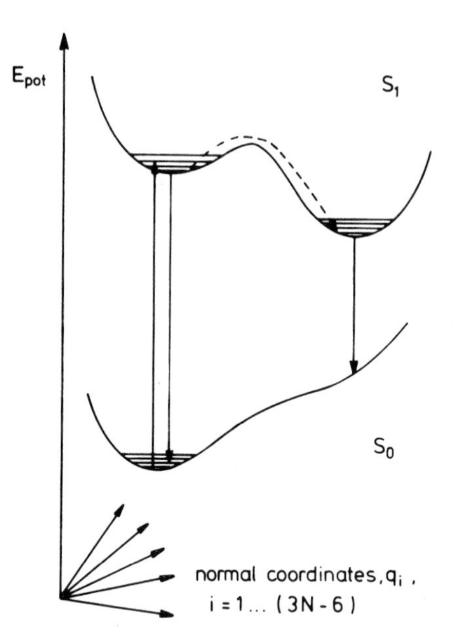

fig 4: Two-dimensional schematic diagram of an adiabatic photoreaction in a multidimensional space.

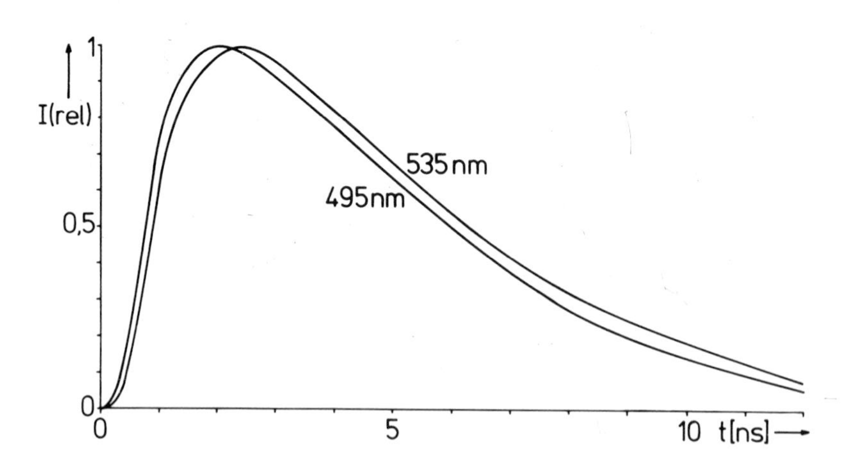

fig. 5: Normalized fluorescence decay at different observation wavelengths of 4 MU in EtOH/0,1 m/l $HClO_4$ /0.5 m/l H_2O obtained by excitation with a mode-locked frequency doubled ruby-laser

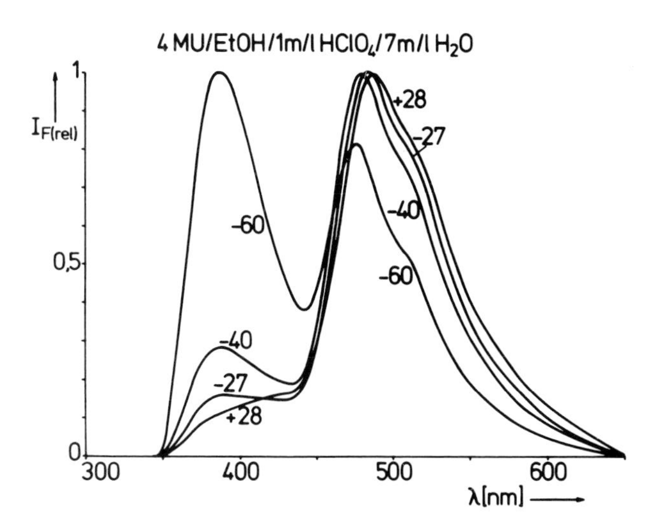

fig. 6: Temperature dependence of the fluorescence of acidified solutions of 4 MU. The spectra are normalized to unity. Excitation wavelength 347 nm. The parameter is the temperature in °C.

excitation at 347 nm, the wavelength of the output of a frequency doubled ruby laser.

In a proceeding paper /9/ we had already mentioned that the long-wavelength fluorescence is quenched if the treshold of the Laser activity of the species n_1* is reached, i.e. if the lifetime of the states n_1* and n_2* are shortened so much that the reaction $n_2^* \rightarrow n_2'^*$ cannot compete anymore with the induced emission $n_2^* + h\nu \rightarrow n_2 + 2\ h\nu$.

One should keep in mind that the spectral position of the fluorescence band of an adiabatic photoreaction product is always situated at the long-wavelength side of the fluorescence band of the initial species (fig. 4), and the product fluorescence occurs at a later time than the initial fluorescence /10/. This effect is even due for the longest-wavelength band at 535 nm which rises and decays later than the 495 nm fluorescence (fig. 5).

There are at least 4 adiabatic photoreactions from n_1* involved in the systems under investigations, namely $n_1^* \rightarrow c_1^*$, $n_1^* \rightarrow a_1^*$, $n_1^* \rightarrow n_2^*$ and $n_2^* \rightarrow n_2'^*$, i.e. a tautomere to n_1* is neither formed

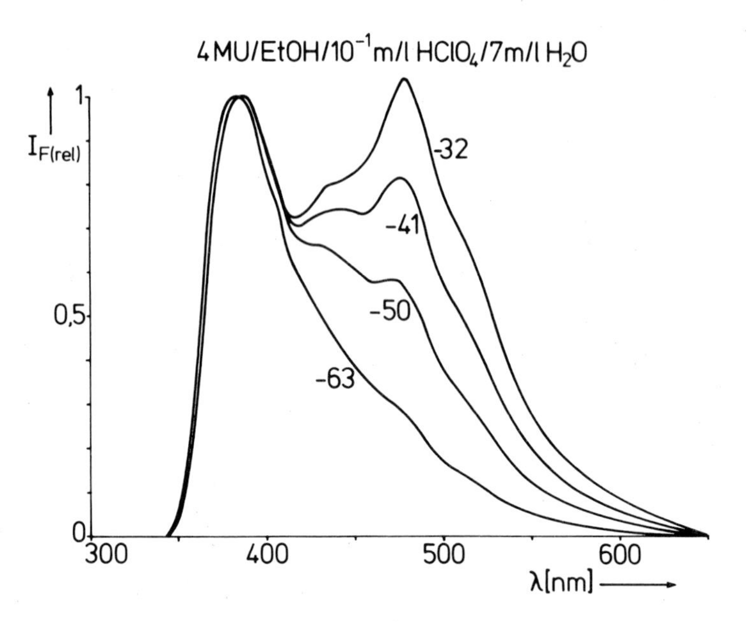

fig. 7: Temperature dependence of the fluorescence of acidified solutions of 4 MU. The spectra are normalized to unity. Excitation wavelength 347 nm. The parameter is the temperature in °C.

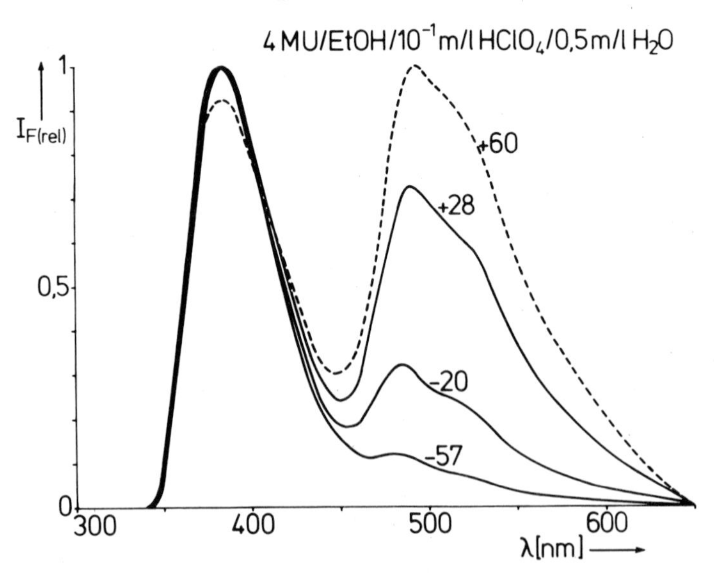

fig. 8: Temperature dependence of the fluorescence of acidified solutions of 4 MU. The spectra are normalized to unity. Excitation wavelength 347 nm. The parameter is the temperature in °C.

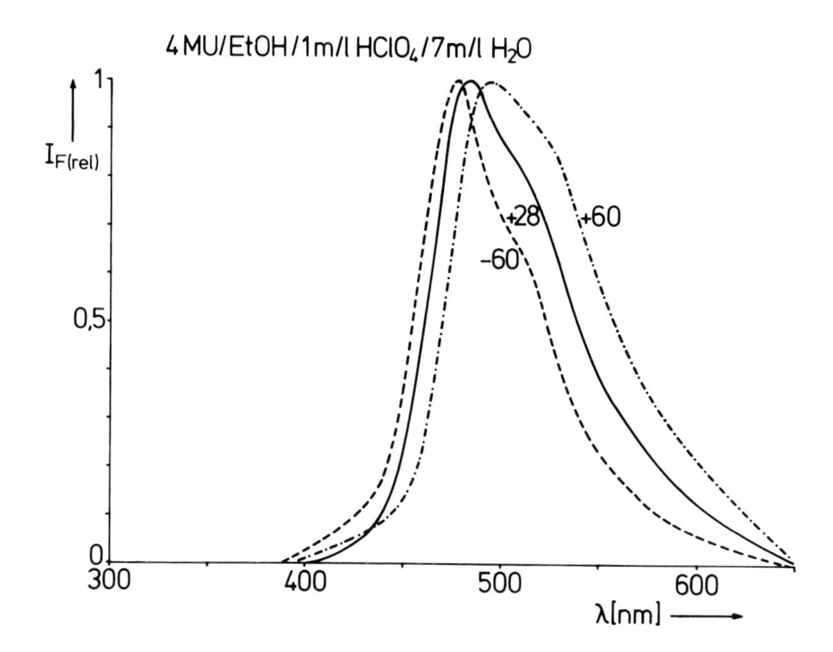

fig. 9: Temperature dependence of the long wavelength components of the fluorescence of an acidified solution of 4 MU. The spectra were normalized to unity. Excitation wavelength: 347 nm. The parameter is the temperature in °C.

via a_1* nor c_1*. At high water concentrations in ethanolic solutions containing no acid the anion a_1 exists also in its ground state, and when excited at sufficient long wavelength only the fluorescence from a_1* at 450 nm occurs but not from n_2* at 485 nm. On the other side in acidified ethanolic solutions (~ 1 mol/l $HClO_4$) the relative cation fluorescence quantum yield increases rapidly if the water concentration becomes less than a certain value (i.e. of about 3 mol/l). In pure acetic acid which contains no water the fluorescence from n_2* also does not occur but only the fluorescence from n_1*. Hence the tautomerization reaction needs just the presence of water molecules.

The temperature-dependence of the rate constants of protonation and deprotonation is not the same (fig. 6,7); at high temperatures protonation gains but at low temperature <u>de</u>protonation gains, and the temperature-dependence of tautomerization is

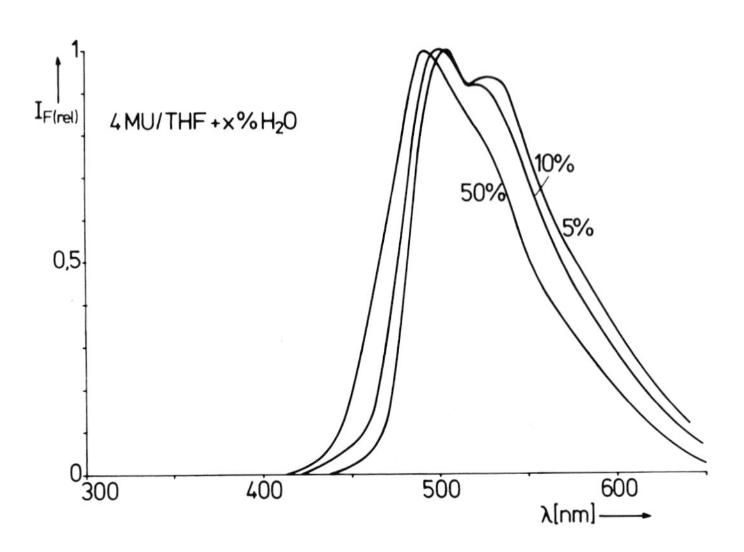

fig. 10: Dependence of the long wavelength fluorescence components of 4 MU on the water content of the solution: Spectra of $5 \cdot 10^{-5}$ m/l 4 MU in THF, 5%, 10%, 50% by volume of water added to the solution. Excitation wavelength 325 nm. The spectra were recorded at room temperature and were normalized to unity.

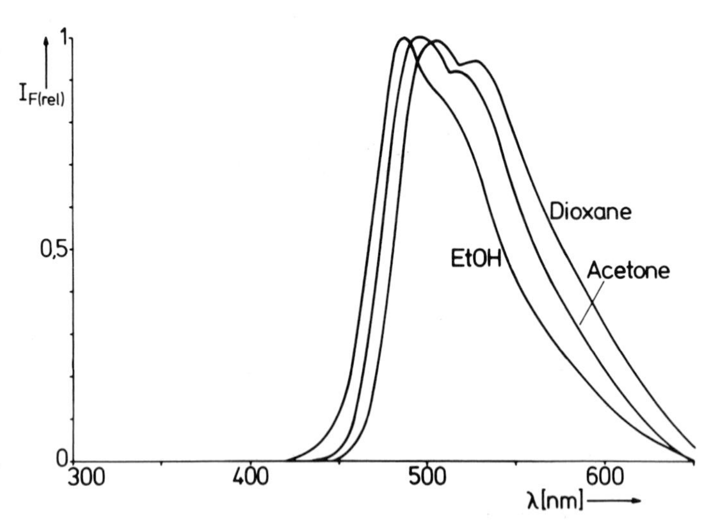

fig. 11: Dependence of the long wavelength fluorescence components of 4 MU on the proton acceptor strength of the solvent in binary solvent mixtures containing the same mole fraction of water (~ 0.2). Excitation wavelength 325 nm. The spectra were recorded at room temperature and were normalized to unity.

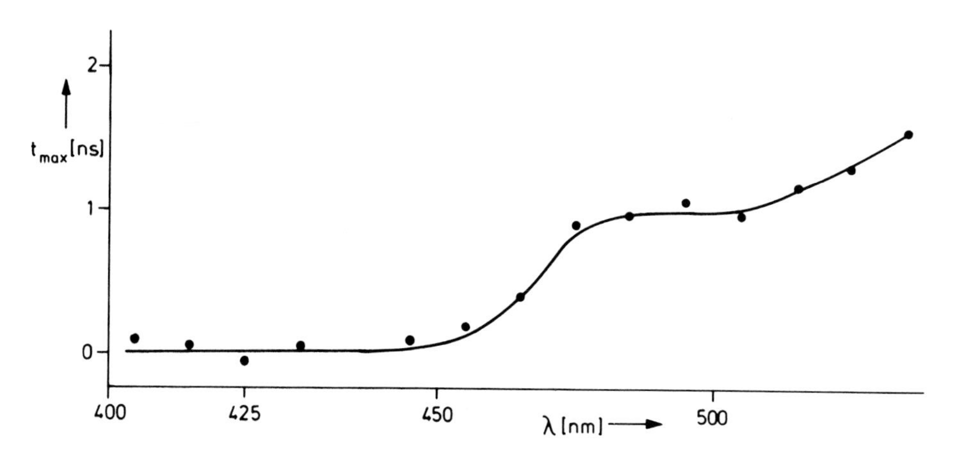

fig. 12: Time of the maximum of the fluorescence decay curve at different observation wavelengths for an acidified ethanolic solution of 4 MU (4 MU/EtOH/10^{-1} m/l $HClO_4$/0.5 m/l H_2O)

somewhere in between, but in all cases the n_2*concentration increases rapidly with temperature (fig. 8).

But there exist still further adiabatic photoreactions concerning the hydrogen bonding state of the species mentioned so far as can easilv be seen from the fluorescence spectra of the tautomer n_2^* (fig. 9, 10, 11) which show an equilibrium between two states named n_2^* and $n_2'^*$. This equilibrium is shifted to lower concentrations of the state $n_2'^*$ by decreasing the temperature, by increasing the water concentration, or by decreasing the proton acceptor strength of the solvent. In dioxane of instance the n_2^* fluorescence maximum is to be seen at 510 nm.

In any case the equilibrium $n_2^* \rightleftarrows n_2'^*$ is reached rather fast as compared with the lifetime: both species have the same decay function (see fig. 5).

The longer the wavelength of the fluorescence observed the longer is the time after which the maximum of the fluorescence decay function appears as it should be expected for such a system of adiabatic photoreactions (fig. 12).

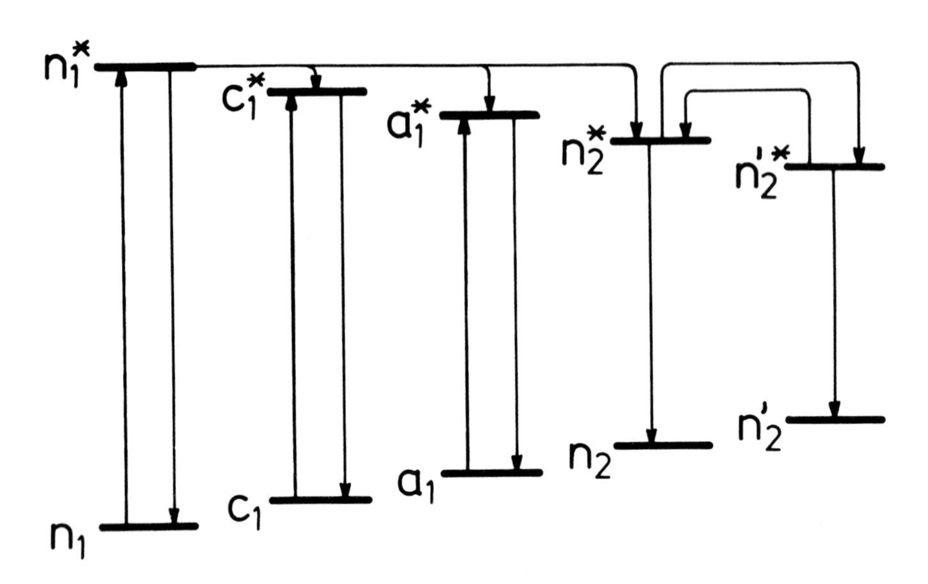

fig. 13: Adiabatic photoreaction pathways of 4 MU in acidified ethanolic solutions.

The schematic diagram fig. 13 summarizes the observed results.

References:

/1/ Fink, D.W. and Koehler, V.R., Anal.Chem. 1970, 42, 990-993

/2/ Takakusa, M. and Itoh, U., Opt.Comm. 1973, 8-10

/3/ Lippert, E., in "The hydrogen bond", Eds. P.Schuster et.al., North-Holland Publ.Co., Amsterdam 1976

/4/ Nakashima, M., Sousa, J.A. and Clapp, R C , Nature 1972, 235, 16-18

/5/ Yakatan, G.J., Juneau. R.J. and Schulman, S.G., Anal.Chem. 1972, 44, 1044-1046

/6/ Bergmann, A. and Jortner, J., J.Luminescence 1973, 6, 390-403

/7/ Dienes, A., Shank, C.V. and Kohn, R.L., IEEE J.Quantum electron., 1973, QE-9, 833-843

/8/ Groves, M.R., Haydon, S.C., and Williams, O.M., Opt.Comm. 1973, 9, 42-47

/9/ Kindt, T., Lippert, E. and Rapp, W., Z.Naturforschung, 1972, 27a, 1371-1373

/10/ Lippert, E. in "Organic Mol.Photophysics, Vol. 2, Ed. J.B.Birks, Wiley and Sons, London 1975

SPECTRAL AND PHOTOCHEMICAL ASSESSMENTS OF INTERACTIONS OF THE FLAVIN RING SYSTEM WITH AMINO ACID RESIDUES

Donald B. McCormick

Section of Biochemistry, Molecular and Cell Biology, and the Division of Nutritional Sciences, Cornell University, Savage Hall, Ithaca, New York 14853 (USA)

INTRODUCTION

The light-reactivity of flavins has been of interest since the recognition that the vitamin B_2 activity extracted from natural materials was attributable to yellow, fluorescent compounds, which were subsequently found to function as oxidation-reduction coenzymes bound to appropriate proteins (1).

As shown below, there is a modest diversity of substituent variations in the isoalloxazine portion of natural flavins. The majority of flavoproteins operate catalytically by utilizing the vitaminic riboflavin portion ($R_1 = H, R_2 = CH_3, R_3 = H$) within the coenzymes, flavin mononucleotide (FMN, $R_3 = PO_3H_2$) and the predominant flavin-adenine dinucleotide (FAD, $R_3 = PO_3$-AMP) (2). A fraction of FAD is covalently linked by 8α-attachment to N or S within histidyl and cysteinyl residues, respectively, in certain flavoproteins (3,4).

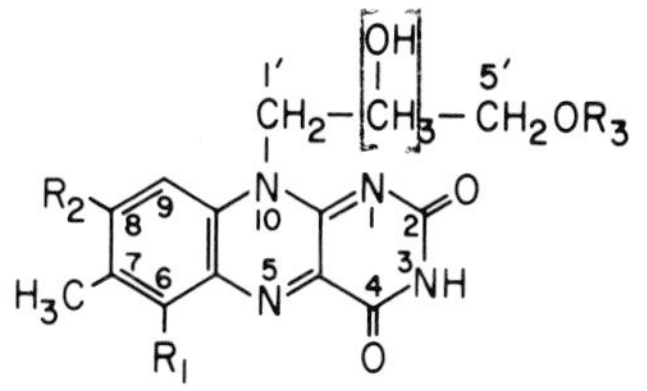

R_1 = H, OH

R_2 = CH_3, CH(H or OH)-peptide, OH, $N(CH_3)_2$

R_3 = H, $(CHO)_n$, PO_3H_2, PO_3-AMP

The differences in redox properties and specificities of flavin-dependent enzyme are largely the result of the particular environment of amino acid residues constituting the flavocoenzyme-binding site. To understand the functional versatility of flavoproteins, then, it is necessary to delineate those residues involved in binding of flavocoenzymes and to elucidate the nature of those intermolecular associations that affect flavin functionality.

Since alterations in spectral and photochemical properties of flavins caused by interaction of their isoalloxazine ring system with certain amino acid residues help reveal the nature of such associations, considerable study has been made of inter- and intramolecular models, as well as enzymes (5).

B. Pullman and N. Goldblum (eds.), Excited States in Organic Chemistry and Biochemistry, 233-245.

NATURE OF ASSOCIATIONS

Tryptophanyl and tyrosyl residues appear to be the best candidates for most commonly providing sites within enzymes for complexing the isoalloxazine moiety of flavins, which have been found to associate intermolecularly with numerous indolic and phenolic compounds (6). Indoles are somewhat more avid complexers than phenols in aqueous solutions, although stability constants for 1:1 complexes range from 10 to 1000 M^{-1}. Both heats of formation and entropy changes are negative. Hydrophobic interaction, i.e. solvent ordering, appears important, and the operation of dispersion forces may be facilitated between the stacked, coplanar ring systems. Polarizability and general electron donor-acceptor relationships are important in some cases where crystallographic examination reveals a proximity of stacking compatible with charge transfer to the flavin. Hydrogen bonding with the pyrimidinoid edge of the flavin becomes important in relatively nonpolar, aprotic media.

An important aspect of complexing concerns the redox/acid-base equilibria of flavins. Of the total of nine species that can be chemically generated in acidic, neutral, and basic solutions of oxidized (quinone), half-reduced (semiquinone), and fully reduced (hydroquinone) flavins, the pKa values mandate that only five are of likely significance within a pH range that is biologically reasonable (7). These are the three neutral redox forms and the anionic semi- and hydroquinones. At physiological pH, tryptophan and especially tyrosine have a considerable affinity for the semiquinone, as well as quinone, but complex least well with the hydroquinone flavin; disulfides, such as cystine, also have a rather strong affinity for the flavoquinone (8,9). It has been suggested that the protonated ε-amino of a lysyl residue could conceivably enhance stabilization of the red anionic semiquinone by interaction at N^5, whereas a side-chain carboxyl from a glutamyl or aspartyl residue could stabilize the blue neutral form; a histidyl imidazole could allow a ready shift between the two (2).

SPECTRAL PROPERTIES

Investigations of interactions between the flavin nucleus and amino acids have been facilitated by covalently bridging the two chromaphores with alkyl chains of varying lengths. In this way, the extent of interactions, as reflected by changes in spectroscopic properties, is markedly enhanced. Most often, such models have been synthesized by amide-linking the α-amino group of the amino acid or its ester to ω-carboxyalkyl chains of 7,8-dimethylisoalloxazines substituted at positions 10 or 3 (10):

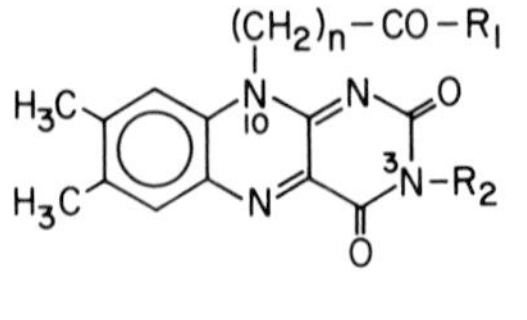

$(CH_2)_n-CO-R_1$; H_3C ; H_3C ; N ; 10 ; N ; O ; 3 ; $N-R_2$; N ; O

R_1 = Trp, Tyr, Phe, His, Met, Ala, etc.

R_2 = H, CH_2CO-R_1

As expected, those amino acids that complex best with flavoquinone, viz. tryptophan and tyrosine, generally cause the greatest alteration in flavin absorption spectra. There is a broadening and bathochromic shift of the visible band with the 1:1 flavin complexes. The locations of maxima and, less significantly, the minima in the flavin difference spectra, shown in Fig. 1 for intramolecular compounds in aqueous solution, are shifted to longer wavelength in the order Trp > Tyr > Phe and

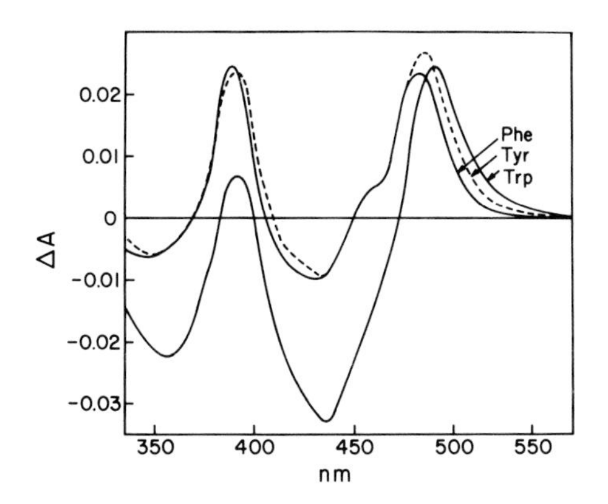

Fig. 1. Difference spectra of 2×10^{-5} M 10-flavinyl amino acid methyl esters versus 10-carboxymethylflavin in 0.05 M sodium phosphate buffer (pH 7).

as the alkyl bridge at 10 is shortened. There is also an increase in hypochromicity near 270 nm, greatest with Trp, as the number of methylene groups in the flavin side chain is decreased (11).

More remarkable changes occur in the intensity of fluorescence. The efficiency decreases markedly in a nonlinear fashion with decreasing chain length between interacting flavin and amino acid methyl ester, as seen in Fig. 2, where only a fraction of the flavin fluorescence, compared to free flavin in different solvents, remains (10,11). The

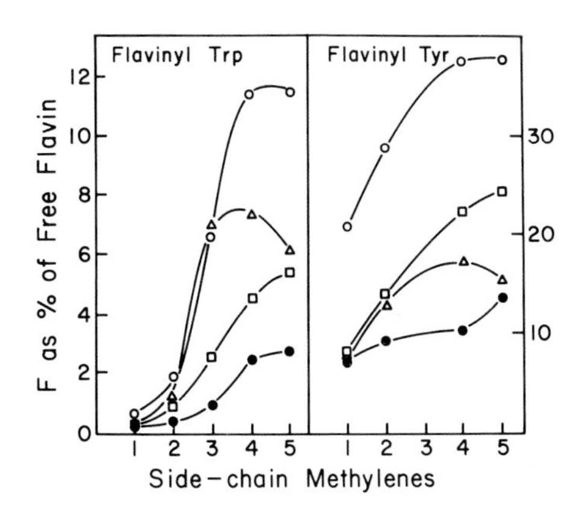

Fig. 2. Fluorescence of 10-flavinyl tryptophans and tyrosines, expressed as % of that of the corresponding ω-carboxyalkylflavins, as a function of the number of methylene groups in the flavin side chain. Flavins were 2×10^{-5} M at 25° C. Symbols: (o), N,N-dimethylformamide; (△), chloroform; (□), commercial absolute ethanol; (●), 0.05 M sodium phosphate buffer, pH 7.

effect of increasing temperature, which leads to disruption of "dark" complexes but increases collisional quenching, is seen in Fig. 3 (11). The fluorescence of the aromatic amino acid is mutually quenched, as

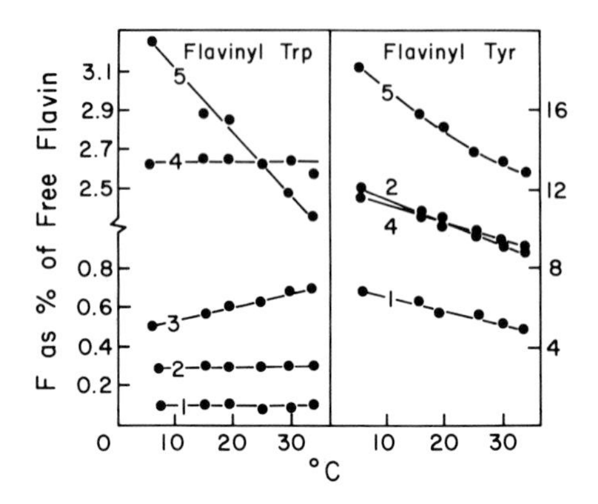

Fig. 3. Fluorescence of 10-flavinyl tryptophans and tyrosines, expressed as % of that of the corresponding ω-carboxyalkylflavins, as a function of temperature. Flavins were 2×10^{-5} M at 25° C. Numbers (n) refer to the methylene groups in the flavin side chain.

is shown in Fig. 4 for flavinyl tryptophans in aqueous solutions (12). Within 10-flavinyl histidines (13), the extent of quenching of flavin

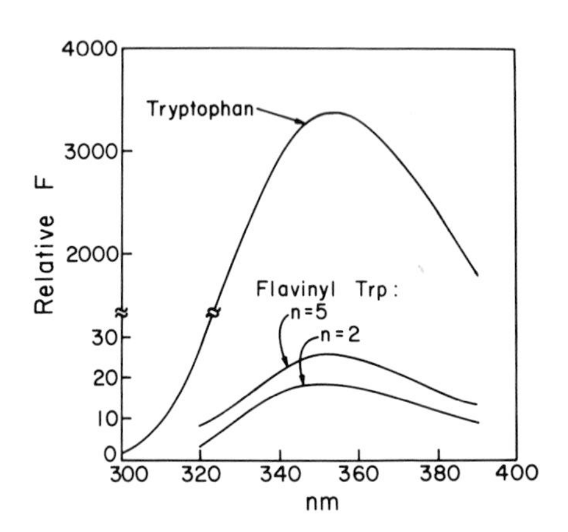

Fig. 4. Fluorescence spectra of tryptophan and 10-flavinyl tryptophans in 0.05 M sodium phosphate buffer (pH 7) at 25° C. Numbers (n) refer to the methylene groups in the flavin side chain.

fluorescence is less than with the Trp, or even Tyr compounds, where complexing is a more significant factor. Moreover, as seen in Fig. 5, the pH-fluorescence profile reflects that only the nonprotonated imidazole ring is responsible for the quenching. It is obvious here, as in other cases, that the more effective quenching derives from nonbonding electron pairs on an electronegative atom. This is not only found when close contact is favored with the ring N of an indole (tryptophan) or imidazole (histidine) and the hydroxyl O of a phenol (tyrosine), but also with an S, such as in the thioether function of methionine (14). Since methionine, like histidine, does not tend to form intermolecular complexes with flavins in aqueous solutions, very little quenching of flavin fluorescence is measured, even when the amino acid is in considerable molar excess, and essentially none when the sulfur

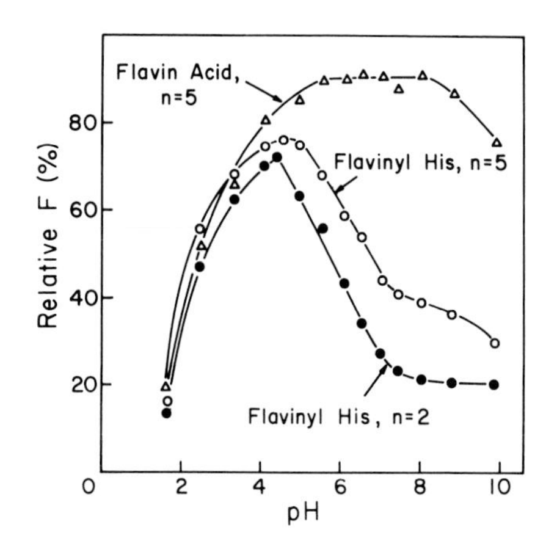

Fig. 5. Effect of pH on the flavin fluorescence of 10-carboxypentyl-flavin and 10-flavinyl histidines. Aqueous solutions contained 10^{-5} M flavin in various buffers at 25° C. Numbers (n) refer to the methylene groups in the flavin side chain.

is oxidized to the level of sulfoxide or sulfone. However, when constrained to an intramolecular amide, where frequency of interactions are favored in water, there is a significant quenching effect that is diminished by an organic solvent, as shown by the data in Table 1. These and other studies on the effects of solvents and temperature on

Table 1. Fluorescence of 2×10^{-5} M 10-Flavinyl Met (n = 5) Relative to Riboflavin in Ethanol/0.01 M Sodium Phosphate (pH 7) at 25° C.

Ethanol (%,v/v)/Buffer	F (%) Relative to Riboflavin
0	40
20	48
50	66
80	71

the changes in intensity and polarization of flavin fluorescence indicate that quenching in these 10-flavinyl amino acids is due to ground-state complex formation, more important with Trp or Tyr residues, as well as kinetic dissipation of excited-state energy through collisions, which may account for most of the quenching with proximal His or Met residues. Differences in fluorescence properties between the 10- and 3-substituted flavinyl amino acids indicate that orientation of the complexing moieties influences the extent and type of quenching (11).

Proton magnetic resonance spectroscopy has been applied to an investigation of the solution conformers of the 10-flavinyl amino acids (13, 15). The flavinyl amides of Trp, Tyr, or Phe with two to five methylene groups in the flavin side chain all associate intramolecularly in water in a stacked manner such that the aromatic portions are in a planar orientation. The conformations are opened at higher temperatures or by dimethyl sulfoxide. In this organic solvent, the flavinyl amides of the methyl esters of the aromatic amino acids are more unfolded, but the amino acid moiety still somewhat shields the benzenoid portion of the flavin that is more proximal to the side chain. Generally similar behavior is characteristic of the flavinyl histidines, except that associations are weaker and more subject to disruption by a lowering of pH from near neutrality.

From infrared spectroscopic changes in the flavin carbonyl stretching frequencies seen upon interactions in relatively nonpolar solvents, e.g. chloroform, it is clear that hydrogen bonding from a phenolic OH or indolic NH may become significant in an environment where protic solvents, such as water, are excluded. For example, flavinyl O-methyltyrosine ($n = 1$), but not the flavinyl tyrosine, exhibits an absorption band in the 4-carbonyl region of the flavin near 1740 cm^{-1} (11).

When an electronegative atom, such as N or S, is covalently attached at the flavin 8α-position, such as occurs naturally within some flavoproteins (3,4), the permanent electronic effects induce spectral changes in the flavin that are even more dramatic. In general, amino acids, viz. histidine and cysteine, that become attached in such a manner cause a hypsochromic shift in the near-UV-visible band due to electron withdrawal from the flavin. Some increase in acidity of the neutral flavin semiquinone results, as indicated by modest decreases in pKa values compared to riboflavin.

The fluorescence of 8α-N-histidyl flavins appears typical in the weakly acid range but is strongly decreased upon deprotonation of the imidazole, which has pKa values of 4.5 for N^3-linkage and 5.0 for N^1-linkage (16). An additional mutual quenching from a neighboring Trp has been found in the N^1-histidyl flavin peptide isolated from β-cyclopiazonate oxidocyclase (17). An 8α-thioether function causes a decrease in fluorescence, which is marked throughout the pH range (~3 to ~9) maximal for flavin fluorescence (14). The yields and mean lifetimes for fluorescence of monoamine oxidase-type S-(N-acetyl)cysteinyl flavins, compared to riboflavin in dilute aqueous buffer, are given in Table 2. The additional quenching contributed by a vicinal C-terminal tyrosine can

Table 2. Fluorescence Yields and Lifetimes of 8α-S-Linked Flavins Relative to Riboflavin in 0.05 M Potassium Phosphate (pH 7) at 25° C.

Flavin	Relative F (%)	τ (x 10^{-9} sec)
RF	100	4.3
AcCysRF	8.9	2.4
AcCys(RF)Tyr	6.6	1.4

also be noted. The most important contribution to diminishing population of excited-state singlet must involve interaction of the pairs of sulfur with the hydrogen-bonding solvent, since fluorescence is largely recovered by shifting to aprotic solvents, as is illustrated in Fig. 6, or when the sulfide is oxidized to sulfoxide or sulfone. The weaker quenching contributed by the Tyr can be essentially abolished by solvent change.

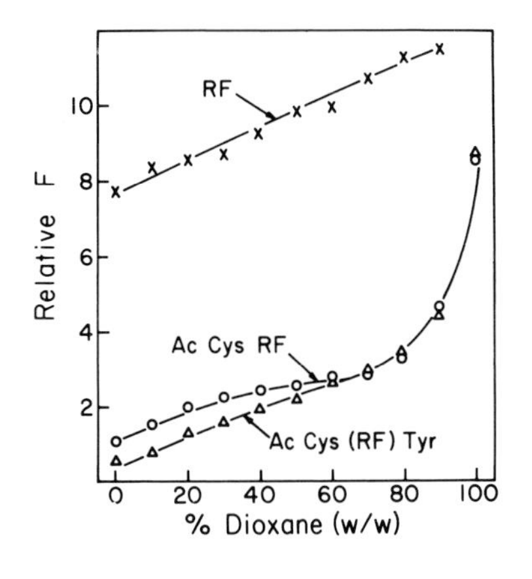

Fig. 6. Fluorescence of 10^{-5} M 8α-S-linked flavins compared to riboflavin in dioxane-water solutions at 25° C.

Circular dichroic properties are also altered by the Tyr proximal to the S-linked flavin within the active-site peptide of mitochondrial monoamine oxidase (18). Such changes as arise are known to involve the disposition of the chiral centers of the ribityl side chain, as well as Tyr, both of which are motion-restricted when complexing of the latter occurs with the optically inactive isoalloxazine nucleus. Strong CD changes also arise from Tyr residues that have been found critical for stability of the thiohemiacetal linkage of the 8α-S-FAD in a Chromatium cytochrome c_{552} flavoprotein (19,20). Here it is probable that a sandwiching of flavin between both N- and C-terminal tyrosyls occurs within the active site Tyr-Thr-Cys(flavin)-Tyr isolated.

PHOTOCHEMICAL REACTIVITIES

The photochemical reactivity of flavins has been used to probe the active sites of flavoproteins (21,22) and has provided additional insight into the characteristics of complexing with aromatic amino acids. Light-excited flavins can sensitize aerobic photooxidations of amino acids and proteins (23,24). It is known that some of the singlet species in the lowest excited state of the flavin can undergo intersystem crossover to the triplet level (25-27). The reactive triplet flavin can abstract hydrogen from its own side chain or a photooxidizable compound to form flavin radical (semiquinone). Some of the flavin radical can undergo reverse electron transfer and also disproportionate to flavoquinone and dihydroflavin, but most is rapidly reoxidized to the original ground-state flavoquinone under aerobic conditions. Studies have been done on the photoreduction of flavins by several amino acids (28, 29). The mechanisms by which aerobic solutions of amino acids are photooxidized by FMN have been analyzed (30).

While both tyrosine and, especially, tryptophan are susceptible to such photooxidation, their complexing propensities for flavin significantly affect the rates involved. The fact that these amino acids in aqueous solutions partially quench the fluorescence of flavins reflects the formation of complexes, wherein radiationless decay from excited singlet states of flavin can occur. Although such complexing decreases the population of excited-state flavin singlet and, therefore, the fraction that can undergo intersystem crossover to photochemically reactive triplet, the effect is too small in dilute, aqueous solutions to account for the large decreases in photoreduction rates (31). For example, under conditions where photobleaching of flavin is inhibited by approximately 90%, tyrosine only quenches the flavin fluorescence by 3% and tryptophan by 12%. As an increase in temperature decreases the inhibition of photoreduction, collisional quenching of flavin triplet by the aromatic amino acid is ruled out as a primary mechanism.

From flash photolysis studies (32-34), it has been shown that phenolic and indolic compounds are good quenchers of flavin triplet, during which process the quencher is oxidized as the flavin is reduced to the radical. Tyrosine and its N-acetyl and ethyl ester derivatives serve nearly equally well as photoreductants, whereas 4-methoxyphenylalanine is less effective (35). Such phenols as 2,6-dimethylphenol (33), tyrosine (34), and the simple tyrosyl derivatives (35) have no appreciable effect on the subsequent aerobic oxidation of the flavin radical, which reacts with O_2 more than ten times faster as the anionic than the neutral form (33). These findings support the contention that the main photochemical reaction involves the formation of phenoxy-type radicals that do not significantly participate in the flavin radical decay in competition with reasonable concentrations of O_2. With tryptophan, though, there is a significant decrease in the rate of O_2 reaction with the flavin radical (34). In this case, the greater extent of complexing and, perhaps, more complete shielding of the flavin may contribute.

In addition to tryptophan and tyrosine, the photooxidation of histidine, methionine, and cysteine in the presence of flavins has been shown to occur rather readily (36). Although one pathway may again follow abstraction of hydrogen from such substrates by the flavin triplet, as is the main mechanism in the photooxidation of aliphatic amines and amino acids (30), reaction of triplet with O_2 to generate singlet oxygen, which then reacts, is certainly important for some of the aromatic amino acids. It has been shown that the rates of reaction of the latter, including tyrosine and tryptophan, increase with oxygen concentration (36). In the flavin-sensitized photooxidation of histidine, it has been demonstrated that a 1,4-cycloaddition of singlet oxygen to the imidazole ring produces a cyclic peroxide that decomposes to give aspartic acid via at least 17 intermediates (37).

The effect that the intramolecular interactions, observed within the synthetic flavinyl amino acids, have on the ability of the light-excited flavin portions to oxidize aromatic amino acids intra- and intermolecularly has been examined by colorimetric determinations of the amino acid remaining (38,39). Results obtained for the Trp, Tyr, and

His series are illustrated in Figs. 7, 8, and 9, respectively. The rates for the photooxidation of amino acid by intermolecularly supplied

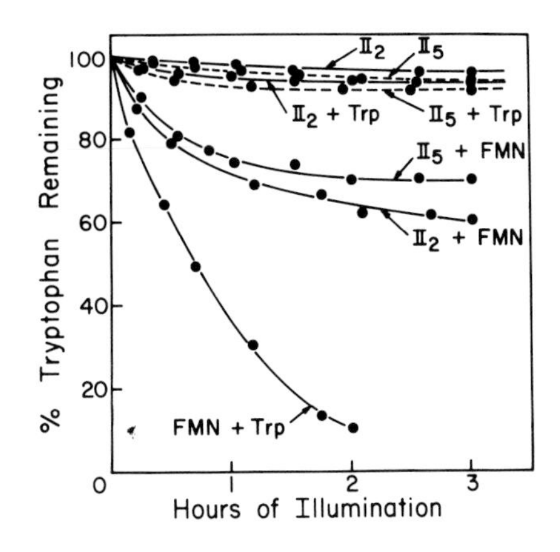

Fig. 7. Flavin-sensitized photooxidations of 10^{-4} M 10-flavinyl tryptophans (II) or FMN and tryptophan in aerobic 0.05 M sodium phosphate buffer (pH 7) at 25° C.

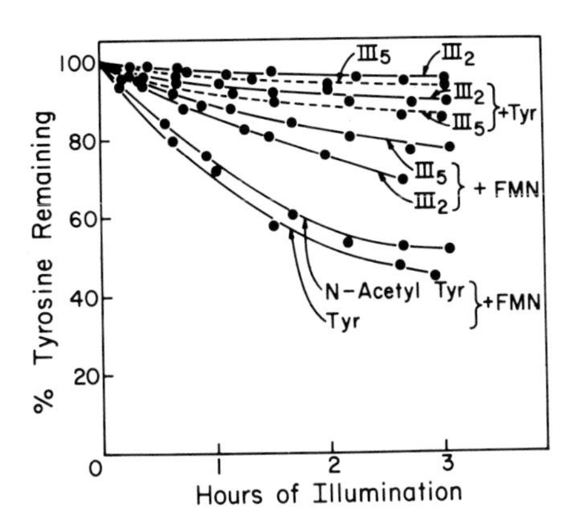

Fig. 8. Flavin-sensitized photooxidations of 10^{-4} M 10-flavinyl tyrosines (III) or FMN and tyrosine in aerobic 0.05 M sodium phosphate buffer (pH 7) at 25° C.

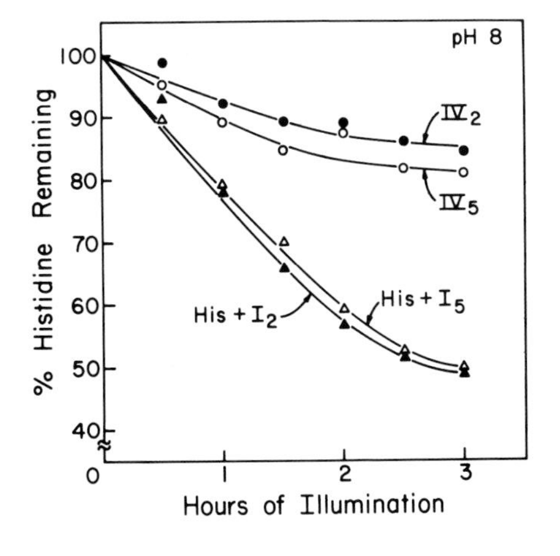

Fig. 9. Flavin-sensitized photooxidations of 3×10^{-4} M 10-flavinyl histidines (IV) or the corresponding ω-carboxyalkylflavins (I) and histidine in aerobic 0.05 M sodium phosphate buffer (pH 8) at 25° C.

flavins are much greater with free flavin than with flavinyl amino acid compounds. The latter, especially those with a shorter methylene bridge between flavin and amino acid, lose but little of the amino acid portion when exposed to light. The rates of the reactions with 1:1 flavinyl amino acid plus amino acid or plus free flavin, as compared to those for the flavinyl amino acid alone, are increased by small and large percents, respectively. Effects of pH, temperature, and solvent on the rates for the Trp and Tyr series are generally similar (38). Up to the optima, higher pH values, higher temperatures, and a dipolar, organic solvent, such as dimethyl sulfoxide in buffer, favor the reactions. Overall, the results indicate that the close association of a flavin to an amino acid residue within the complexes prevents the flavin from effecting oxidation of amino acids. However, the complexed amino acid can be relatively easily photooxidized by intermolecular flavin that can be activated. The mechanisms involved have been clarified, at least for Tyr, by the use of flash photolysis (35). When a Tyr or O-methylTyr is covalently attached to an alkyl side chain at the 10-position of the flavin, the considerable intramolecular complexing that results markedly decreases the formation of flavin triplet (Table 3) and, therefore, the radical yield (Table 4). The rate of triplet decay is not much different than for noninternally complexed flavins, but extensive intramolecular radical decay occurs, and the rate of O_2 oxidation of radical is decreased (Table 5). A shorter alkyl chain is

Table 3. Yields of Flavin Triplet and the Rate Constants for Their Decay in the Presence and Absence of N-Acetyl Tyrosine Ethyl Ester in Anaerobic 2% Dimethylformamide.

Flavin (2×10^{-5} M)	Equimolar AcTyr Et Ester	Triplet Yield (%)	Decay* ($10^4\ s^{-1}$)	Quenching† ($10^9\ mol^{-1}\ s^{-1}$)
Lumiflavin	-	24	1.6	
	+			2.2
Riboflavin	-	22	1.4	
	+			2.0
Tetraacetylriboflavin	-	30	1.2	
	+			2.2
10-Flavinyl phenylalanine methyl ester (n = 1)	-	18	1.1	
	+			2.3
3-(10-Methyl)flavinyl tyrosine methyl ester (n = 1)	-	7.3	1.1	
	+			1.8
10-Flavinyl O-methyltyrosine methyl ester (n = 5)	-	2.4	1.4	
10-Flavinyl tyrosine methyl esters (n = 1,5)	-	<1		

*Pseudo-first-order constants at the concentrations involved.
†Apparent second-order constants, calculated by dividing the pseudo-first-order constant obtained in the presence of quencher by the concentration of quencher.

Table 4. Yields of Flavin Radical and the Second-Order Rate Constants for Their Decay in the Presence and Absence of N-Acetyl Tyrosine Ethyl Ester in Anaerobic 2% Dimethylformamide.

Flavin (2×10^{-5} M)	Equimolar AcTyr Et Ester	Radical Yield (%)	Decay Rate (10^9 mol^{-1} s^{-1})
Lumiflavin	-	5.3	7.3
	+	13	2.8
10-Flavinyl phenylalanine	-	3.0	3.8
methyl ester (n = 1)	+	9.7	1.1
3-(10-Methyl)flavinyl tyro-	-	4.8	5.3
sine methyl ester (n = 1)	+	6.5	2.5
10-Flavinyl O-methyltyrosine	-	1.3	*
methyl ester (n = 5)	+	1.7	3.2

*This decay is by a first-order process with a rate constant of 8×10^3 s^{-1}.

Table 5. Yields of Flavin Radical and the Pseudo-First-Order Rate Constants for Their Oxidation by O_2 in the Presence and Absence of N-Acetyl Tyrosine Ethyl Ester in Air-Saturated 2% Dimethylformamide.

Flavin	M $\times 10^5$	Equimolar AcTyr Et Ester	Radical Yield (%)	Oxidation Rate (10^3 s^{-1})
10-Flavinyl phenylalanine methyl ester (n = 1)	10	+	1.6	1.6
10-Flavinyl O-methyltyrosine methyl ester (n = 5)	3.9	+	0.3	0.7
3-(10-Methyl)flavinyl tyrosine methyl ester (n = 1)	1.9	-	0.7	1.4
		+	1.3	0.9
10-Flavinyl tyrosine methyl ester (n = 1)	2.1	-	0.2	0.5
		+	0.3	0.2
10-Flavinyl tyrosine methyl ester (n = 5)	4.7	-	0.1	1.0
		+	0.3	1.1

more effective than a longer one for decreasing triplet production, but the greater proximity of a photooxidizable Tyr to the flavin nucleus within the former allows a slightly higher intramolecular radical yield. Attachment of Tyr by a short chain from the 3-position of the flavins has only a modest effect on the production of flavin triplet and its decay. There is less radical production from internal than from external Tyr, and the rate of O_2 oxidation of the flavin radical generated by such intermolecular photoreductants as N-acetyl tyrosine ethyl ester or EDTA is somewhat decreased.

CONCLUSIONS

The spectral and photochemical properties of flavins have permitted the application of diverse techniques in the continuing examination of flavin interactions with amino acid residues within flavoproteins (5). The present state of knowledge is insufficient to warrant specific conclusions as to how each residue within the flavin region exerts its special influence, but some generalities have emerged. Often, Tyr and Trp profer sites for complexing the isoalloxazine ring system and may alter its redox function by preferentially binding the quinone and, especially, semiquinone forms. Proximity of these aromatic residues generally leads to modest absorbancy changes, which include a slight bathochromic shift of the visible band of the flavin and hypochromicity in the ultraviolet region, where both flavin and amino acid contribute. With tight complexing, there is a marked quenching of fluorescence of both flavin and amino acid and a decreased susceptibility of the latter to photooxidation. The proximity of a His does not so strongly contribute to binding of flavin unless attached by 8α-substitution via N^1 or N^3 of the imidazole, but a flavin fluorescence-pH optimum reflecting ionization of the imidazole can be generated. Close association with other residues, e.g. Met and especially S-linked Cys, can also lead to quenching of flavin fluorescence throughout the pH range. The natural 8α-substituted flavins, which exhibit hypsochromic shifts in the near-UV-visible band, also have a reduced triplet yield. Many of such light-associated properties of flavins, including their participation in bioluminescence (40), provide a means for elucidating both their structure and that of the contiguous environment. More productive research in this area of photochemistry and photobiology can be expected.

ACKNOWLEDGMENT

The author's studies in the work described were supported by Research Grant AM-04585 from the National Institute of Arthritis, Metabolism, and Digestive Diseases, United States Public Health Service.

REFERENCES

1. Beinert, H. (1960). In The Enzymes, Vol. 2A (Boyer, P.D., Lardy, H., and Myrbäck, K., Editors), p. 339. Academic Press, New York.
2. Hemmerich, P., Nagelschneider, G., and Veeger, C. (1970). FEBS Lett. 8: 69.
3. Singer, T.P. and Edmondson, D.E. (1974). FEBS Lett. 42: 1.
4. Edmondson, D.E. and Singer, T.P. (1976). FEBS Lett. 64: 255.
5. McCormick, D.B. (1977). Photochem. Photobiol., in press.
6. Tollin, G. (1968). In Molecular Associations in Biology (Pullman, B., Editor), p. 393. Academic Press, New York.
7. Cerletti, P. (1971). Acta Vitaminol. Enzymol. 25: 169.
8. Draper, R.O. and Ingraham, L.L. (1970). Arch. Biochem. Biophys. 139: 265.
9. Yeh, L.-S.L. and Ingraham, L.L. (1976). In Flavins and Flavopro-

teins (Singer, T.P., Editor), p. 765. Elsevier, Amsterdam.
10. Föry, W., MacKenzie, R.E., and McCormick, D.B. (1968). J. Heterocyclic Chem. 5: 625.
11. MacKenzie, R.E., Föry, W., and McCormick, D.B. (1969). Biochemistry 8: 1839.
12. Wu, F.Y.-H. and McCormick, D.B. (1971). Biochim. Biophys. Acta 229: 440.
13. Johnson, P.G. and McCormick, D.B. (1973). Biochemistry 12: 3359.
14. Falk, M.C. and McCormick, D.B. (1976). Biochemistry 15: 646.
15. Föry, W., MacKenzie, R.E., Wu, F.Y.-H., and McCormick, D.B. (1970). Biochemistry 9: 515.
16. Edmondson, D.E. and Singer, T.P. (1973). J. Biol. Chem. 248: 8144.
17. Kenney, W.C., Edmondson, D.E., Singer, T.P., Steenkamp, D.J., and Schabort, J.C. (1976). Biochemistry 15: 4931.
18. Falk, M.C., Johnson, P.G., and McCormick, D.B. (1976). Biochemistry 15: 639.
19. Walker, W.H., Kenney, W.C., Edmondson, D.E., Singer, T.P., Cronin, J.R., and Hendriks, R. (1974). Eur. J. Biochem. 48: 439.
20. Kenney, W.C., Edmondson, D.E., and Singer, T.P. (1974). Eur. J. Biochem. 48: 449.
21. McCormick, D.B. (1970). Experientia 26: 243.
22. Tu, S.-C. and McCormick, D.B. (1973). J. Biol. Chem. 248: 6339.
23. Galston, A.W. (1950). Science 111: 619.
24. Frisell, W.R., Chung, C.W., and MacKenzie, C.G. (1959). J. Biol. Chem. 234: 1297.
25. Tether, L.R. and Turnbull, J.H. (1962). Biochem. J. 85: 517.
26. Holström, B. (1964). Arkiv Kemi 22: 329.
27. Radda, G.K. and Calvin, M. (1964). Biochemistry 3: 384.
28. Byrom, P. and Turnbull, J.H. (1967). Photochem. Photobiol. 6: 125.
29. Penzer, G.R. and Radda, G.K. (1968). Biochem. J. 109: 259.
30. Penzer, G.R. (1970). Biochem. J. 116: 733.
31. Radda, G.K. (1966). Biochim. Biophys. Acta 112: 448.
32. Vaish, S.P. and Tollin, G. (1970). Bioenergetics 1: 181.
33. Vaish, S.P. and Tollin, G. (1971). Bioenergetics 2: 61.
34. Gillard, J.M. and Tollin, G. (1974). Biochem. Biophys. Res. Commun. 58: 328.
35. McCormick, D.B., Falk, M.C., Rizzuto, F., and Tollin, G. (1975). Photochem. Photobiol. 22: 175.
36. Taylor, M.B. and Radda, G.K. (1971). Methods Enzymol. 18B: 496.
37. Tomita, M., Irie, M., and Ukita, T. (1969). Biochemistry 8: 5149.
38. Wu, F.Y.-H. and McCormick, D.B. (1971). Biochim. Biophys. Acta 236: 479.
39. Johnson, P.G., Bell, A.P., and McCormick, D.B. (1975). Photochem. Photobiol. 21: 205.
40. Hastings, J.W. and Wilson, T. (1976). Photochem. Photobiol. 23: 461.

PHOTOELECTRON SPECTROSCOPY OF CARBONYLS.
BIOLOGICAL CONSIDERATIONS[1]

S. P. McGlynn, D. Dougherty, T. Mathers and S. Abdulner

Departments of Chemistry
LSU(Baton Rouge) and UNO(New Orleans)
Louisiana, USA

INTRODUCTION

Ultraviolet photoelectron spectroscopy (UPS) has had enormous impact on chemistry. It is assured that it will have comparable influence on molecular biology. The reasons are straightforward: First, the UPS measurement technique[2] is simple and direct; second, a great deal of interpretive ease is guaranteed by Koopmans' theorem[3]; and, third, because of Koopmans' theorem, the UPS data relate in a straightforward way to a very basic electronic structure concept, namely the canonical molecular orbital.

Koopmans' theorem consists of two parts[4]. The first part is a simple energy equivalence $I(k) = -\varepsilon(k,SCF)$ which equates the kth ionization limit to the negative of the kth spin-orbital energy. The second, and more important part is a selection rule which states that the only events of significant probability are those which depopulate individual spin-orbitals. Put another way, this rule states that "shake-up" and "shake-off" events are improbable. Consequently, the mapping of UPS events onto the filled valence spin-orbital set is isomorphic. If spin-orbital coupling is negligible (i.e., $\xi < 20$meV), this isomorphism simplifies further and refers now to the MO (and not MSO) set.

The majority of all UPS works are implicitly based on this isomorphism. Usually, such works generate a set of empirical numbers (UPS data) and a set of theoretical numbers (a quantum chemical scheme), and invoke the Koopmans' energy equivalence as the sole exemplar of the isomorphism. Such a mapping tech-

B. Pullman and N. Goldblum (eds.), Excited States in Organic Chemistry and Biochemistry, 247-256.

nique is inadequate. First, the quantum chemical scheme is usually algorithmic and not of HF-SCF quality; second, the Koopmans' energy equivalence, because of correlation and relaxation neglects, is probably accurate only to 1 or 2eV; and, third, the mapping is best vested in a comparison of the empirical cationic charge density (more specifically, the charge density of the positive hole) with that of the canonical HF-SCF MO from which charge has been removed in the ionization event. Indeed, one could wax cynical and assert, with much truth, that our trust in the verity of Koopmans' theorem stems not so much from the existence of a great deal of corroborative data as from a dearth of contrary evidences[4].

Despite this, it must be admitted that considerable UPS experience now exists; that Koopmans' theorem, in both its aspects, appears to be highly valid; that the expertise pertinent to the recognition of non-Koopmans' situations does exist; and that the net result has been a most concrete vindication of the orbital concept and an accretion of information on the electronic structure of molecules. Unfortunately, since the UPS technique involves gas phase measurements, almost all of this experience refers to small, non-polar (or weakly-polar) molecules. Little or no data is available for the large polar molecules which constitute the majority type in the biological regime: These molecules usually decompose at temperatures considerably lower than those required for gasification.

EXPERIMENTAL

The gas pressures required for UPS measurements have a lower limit of $\sim 1\mu (\sim 10^{-3}$mm Hg). It has been found that many biological materials which decompose readily in the ambient atmospheric environment do not do so (or, at least, do not do so quite as readily) in high vacuum. The general procedure which we have followed is based on the availability of the heated sample probe. It consists of a slow heating rate of $\sim 1^{o}$C/minute until the sample pressure is $\sim 1\mu$. The spectrum is then recorded and its constancy in time checked. A number of decomposition indicators exist. These may be as obvious as inspection of the sample for change of color or charring; as definitive as mass spectrometry; or as sensitive as the UPS spectrum itself to slight contamination by a variety of small but rather probable decomposition products. Indeed, the instrumental sensitivity to contaminants is such that little difficulty is experienced in defining either the optimum temperature range or the decomposition products which appear when the upper limits are exceeded. Of course, spectra cannot be obtained for those materials (eg., Vitamin K_1, protoporphyrin IX dimethyl ester, etc.) which decompose before pressures of $\sim 1\mu$ are achieved.

The spectra reported here were obtained on a Perkin-Elmer PS-18 photoelectron spectrometer. The sample area was capable of sustaining temperatures as high as 350°C to within ±2°C. Normal resolution (∿20meV) degrades for T ≳ 140° and, at 300°C, is 75-100meV. Our experience suggests that the great majority of small to medium-sized biological molecules (vide infra) will yield to such measurements.

BEYOND UPS

The restriction to gas phase studies is irksome. If such studies were feasible in the solid state, the decomposition attendant to heating could be avoided and molecular size or complexity would not be a limiting factor. It is our opinion that X-ray fluorescence[4] and Auger spectroscopy[4,5] provide pertinent techniques for this purpose. A schematization of the UPS and X-ray fluorescence experiments is shown in Figure 1. It is clear that both processes contain similar information, that of X-ray fluorescence being greater since it also provides a physical (as opposed to chemical) means of assessing MO shapes and their AO compositions[4]. The Auger process is complementary to X-ray fluorescence; it also contains similar information, and is intense where X-ray fluorescence is minimal (e.g., carbon and oxygen compounds) and weak where X-ray fluorescence is intense (i.e., the heavier elements).

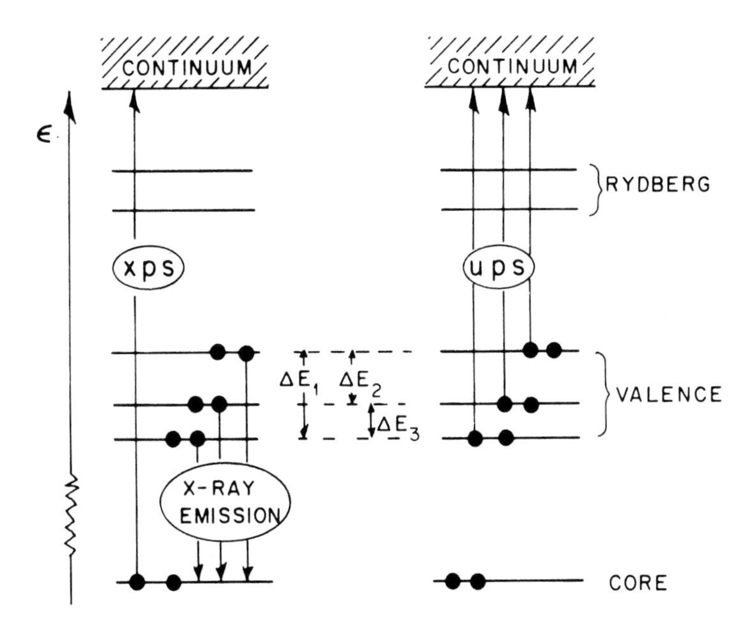

Figure 1. A comparison of UPS and X-ray fluorescence events in an orbital format.

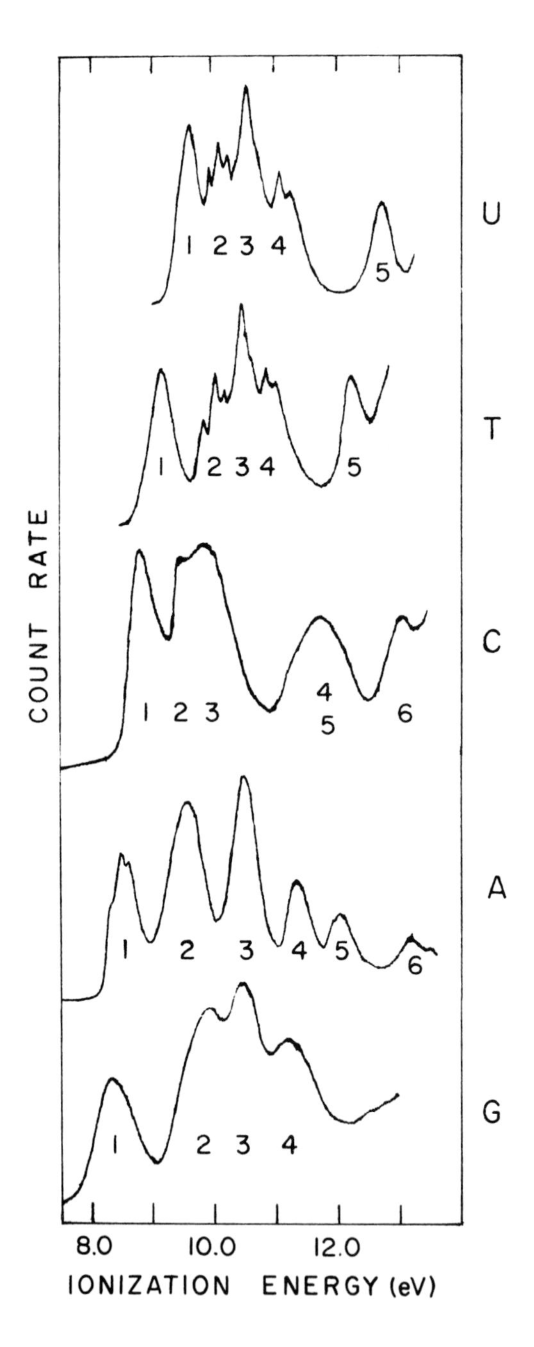

Figure 2. Photoelectron spectra of Uracil (U), thymine (T), cytosine (C), adenine (A) and guanine (G) in the low-energy range. Ionization events are numbered serially in order of increasing energy.

A SMATTERING OF RESULTS

(i) The DNA/RNA Bases

The low-energy UPS region of the five DNA/RNA bases are shown in Figure 2. The spectra exhibit considerable detail: For example, I(2) and I(4) of uracil and thymine possess distinct vibronic structure.

Detailed discussion of these spectra are available elsewhere[6,7,8]. Hence, we will be brief. The nature of the ionization events may be determined by chemical mapping techniques[9,10]. For example, an ionization event which removes charge from a -C=C- unsaturation region will be very sensitive to methyl substitution at this site, but relatively insensitive to such substitution at sites far removed from this region. Similarly, methyl substitution of an amine group will have a large effect on the ionization event which removes a π electron from an orbital with large amplitude on this group, and little or no effect on one which removes an n electron from a carbonyl oxygen. By such means, the ionization event identifications of Figure 3 have been established.

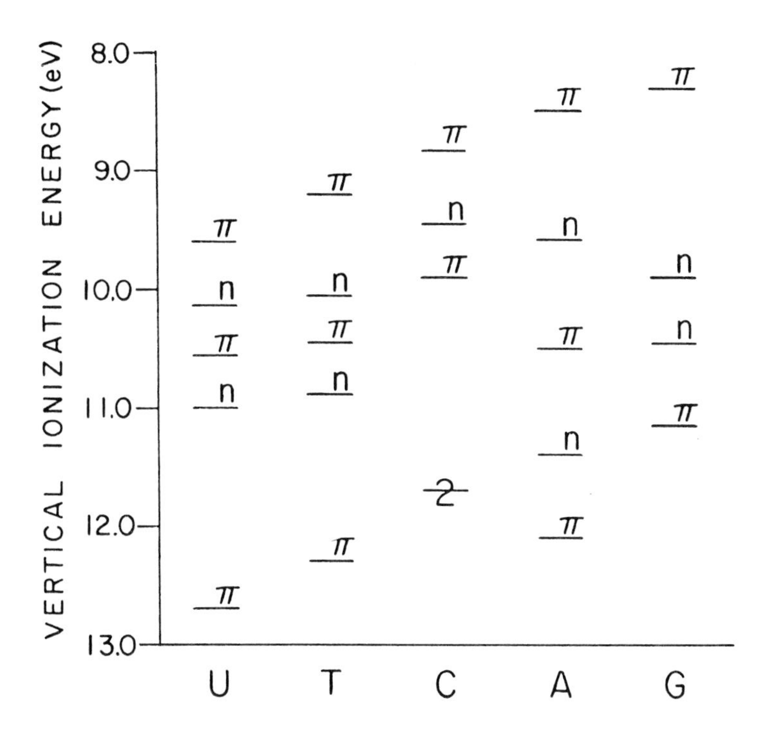

Figure 3. Plot of the vertical ionization energies and MO assignments of the four or five highest occupied MO's for uracil (U), thymine (T), cytosine (C), adenine (A) and guanine (G).

It is well to emphasize that Figure 3 contains more empirical information on the electronic structure of the DNA/RNA bases than the totality of that previously available. And Figure 3 does not exhaust the empirical data base! Indeed, the chemical mapping ploy enables one to distinguish the π-ionization events as $\pi(NH_2)$, $\pi(-C=C-)$, $\pi(-C=O)$, etc. and the n ionization events as n_+ or n_-, where the notation which is either bracketed or subscripted carries information on both the symmetry and spatial distribution of charge in the cationic state (i.e., in the vacated canonical MO). It is precisely this sort of information which is required in order to elicit the isomorphism between the UPS data and SCF-MO results: One simply compares chemically-mapped charge density distributions with MO SCF charge densities and allows adequate variance in the equivalence $I(k) = -\varepsilon(k,SCF)$ to take account of charge reorganization and electron correlation deficiencies.

(ii) The Pullman k-Indexing

The UPS data, via the frontier MO approximation, provide information on a great diversity of biological phenomena. For example, the lowest-energy $I(\pi)$ event contains information on relative π-electron donor efficiencies (e.g., the abilities to interrelate, to complex, or to react) and the lowest-energy $I(n)$ event makes very specific reference to hydrogen-bonding abilities. These topics have been discussed in admirable fashion by Pullman[11]. Unfortunately, not being possessed of any UPS data, Pullman was forced to calculate the quantities of interest. Thus, he used a simple Hückel approach to obtain the Koopmans' equivalent of $I(\pi)$, namely $\varepsilon(\pi)$. According to Pullman, the π donor ability is encapsuled in the coefficient k of $\varepsilon(\pi) = \alpha + k\beta$ where α and β are average Coulomb and exchange integrals for a large range of heteroatomic biologicals. Hence, it is of interest to evaluate the relevance of the Pullman indexing.

A plot of $I(\pi)$ versus the Pullman k coefficient is shown in Figure 4 for a series of biologicals as well as, for reference, a set of simple aromatic hydrocarbons. Several observations are pertinent

---First, despite some scatter, the biomolecule regression line constitutes a remarkable vindication of the Pullman attitudes.

---Second, as expected, the biomolecule regression line lies considerably higher than that of the aromatic hydrocarbons.

---Third, although $I(1)$ is of π-type in most instances, it

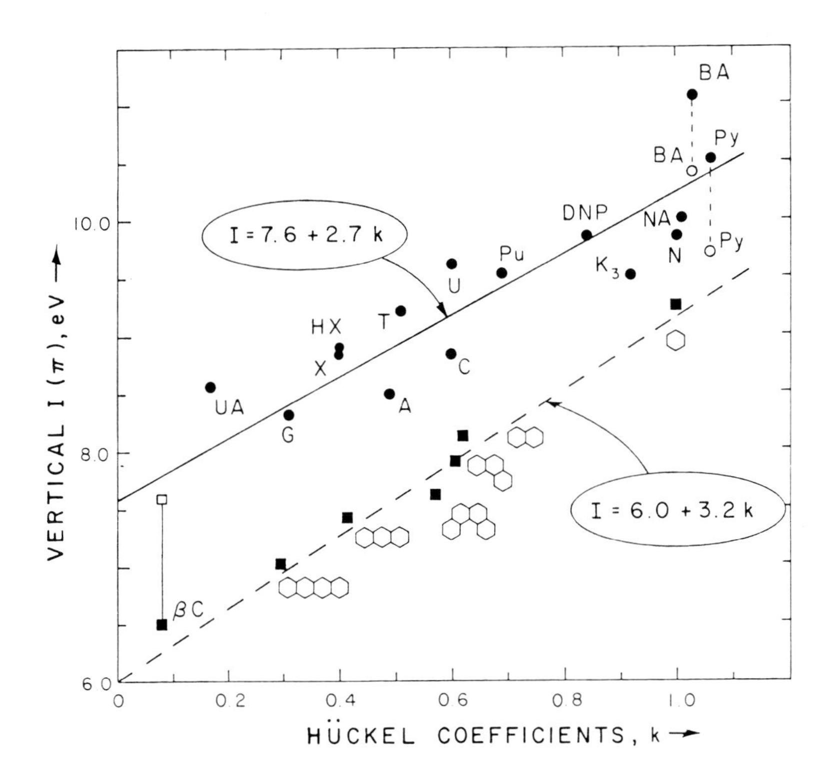

Figure 4. A plot of I(π, vertical) versus the Hückel coefficient, k. The symbolism is βC:β-carotene; UA:uric acid, G: guanine; X-xanthine; HX:hypoxanthine; A:adenine[a]; T;thymine; C;cytosine; U:uracil; Pu:purine[b]; DNP:2,4-dinitrophenol; K_3: vitamin K_3; N:nicotinamide; NA:nicotinic acid; BA:barbituric acid; Py:pyrimidine[c]. The solid circles denote I(π, vertical); the open circles denote I(n, vertical). For βC the open square and the solid square denote I(π, vertical) and I(π, adiabatic), respectively.

[a] S. Peng, A. Padva and P. R. LeBreton, Proc. Natl. Acad. Sci. USA, 73, 2966(1976).

[b] N. S. Hush and A. S. Cheung, Chem. Phys. Letters, 34, 11(1975).

[c] R. Gleiter, E. Heilbronner and V. Hornung, Angew. Chem. Internat. Edit., 9, 901(1970).

is of n-type in barbituric acid (and, possibly, nicotinic acid and nicotinamide). In these instances, frontier MO considerations based on the supposition I(1) = I(π) are wrong.

---Fourth, these data provide direct measures of absolute

(and relative) electron donor abilities. Such knowledge is at the heart of biochemical processes[11].

---Fifth, although UPS data do not directly speak to the subject of electron affinities, these data, in conjunction with N → V spectroscopic information, do lead to facile extraction of electron affinity information.

(iii) Electron Donor Scaling

In order to emphasize the relevance to electron donating abilities, we present in Figure 5 the experimental counterpart of a theoretical scaling first given by Pullman[11]. This scaling is a graphic rendition of the power of the UPS technique.

CONCLUSION

The application of UPS to biology is in its infancy. Further development of the technique, within UPS restrictions, will proceed along two lines: The ability to work at $P < 1\mu$, which will decrease T and, hence, the possibility of decomposition; and the use of substitution devices (such as methylation or trimethylsilylation) which will render many involatiles volatile.

The UPS technique, being limited to gas phase studies, must be superseded by others which are pertinent to the solid state and which can extract similar information. The primary replacement techniques, all presently in development, are:

---X-ray fluorescence spectroscopy (particularly for molecules containing heavier elements)

---Auger spectroscopy (particularly for molecules consisting of C, O, N and H)

---ESCA, when adequate resolution is attainable at high electron kinetic energies[5].

In our opinion, it is these latter techniques, which are not limited by molecule size or volatility, which will provide the most potent probes for the electronic structure of larger biomolecules (e.g., nucleotides, steroids, etc.).

In the meantime, even with UPS, there is much that must be done.

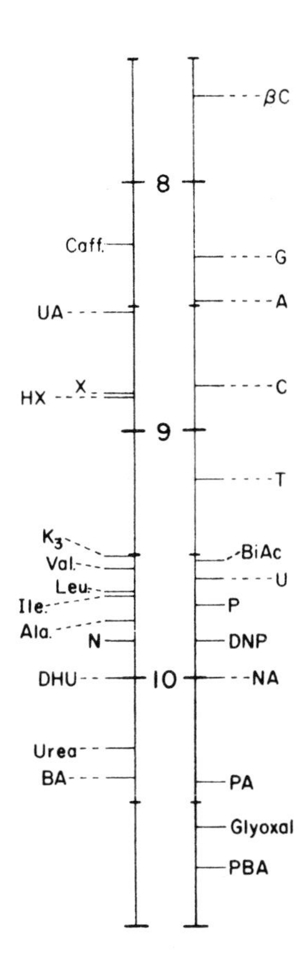

Figure 5. An absolute electron-donor ability (or basicity) scale for a variety of biologicals -- an experimental counterpart to a computational scaling by the Pullmans[11]. The center scale is vertical ionization energy (eV) for the lowest ionization event. The symbolism is that of figure 4 with the additions Caff:caffeine; BiAc:biacetyl[a]; Val:ℓ-valine[b]; Leu:ℓ-leucine[b]; Ile:ℓ-isoleucine[b]; Ala:ℓ-alanine[b]; P:pyruvamide[a]; DHU:dihydrouracil; PA:pyruvic acid[a]; PBA:parabanic acid[c].

[a]S. P. McGlynn and J. L. Meeks, J. Electron Spectrosc. Relat. Phenom., 6, 269(1975).

[b]L. Klasinč, J. Electron. Spectrosc. Relat. Phenom., 8, 161(1976).

[c]J. L. Meeks and S. P. McGlynn, J. Amer. Chem. Soc., 97, 5079 (1975).

REFERENCES

1. This work was supported by the US-ERDA GRANT No. ERDA-(40-1)-3018.
2. D. W. Turner, C. Baker, A. D. Baker and C. R. Brundle, "Molecular Photoelectron Spectroscopy", Wiley-Interscience, London, 1970.
3. T. Koopmans, Physica, 1, 104(1934).
4. K. Wittel and S. P. McGlynn, Chem. Revs., in press.
5. T. A. Carlson, "Photoelectron and Auger Spectroscopy", Plenum Press, New York, 1975.
6. S. Peng, A. Padva and P. R. LeBreton, Proc. Natl. Acad. Sci. USA, 73, 2966(1976).
7. D. Dougherty, K. Wittel, J. Meeks and S. P. McGlynn, J. Amer. Chem. Soc., 98, 3815(1976).
8. D. Dougherty, E. Younathan, R. Voll, S. Abdulner and S. P. McGlynn, Biochem., in press. D. Dougherty and S. P. McGlynn, J. Chem. Phys., in press.
9. D. Dougherty and S. P. McGlynn, J. Amer. Chem. Soc., in press. J. L. Meeks, H. J. Maria, P. Brint and S. P. McGlynn, Chem. Revs., 75, 603(1975).
10. A. Padva, T. J. O'Donnell and P. R. LeBreton, Chem. Phys. Letters, 41, 278(1976).
11. See, for example, B. Pullman and A. Pullman, "Quantum Biochemistry", Wiley-Interscience, London, 1963.

EXCITED STATES OF SATURATED MOLECULES

Mark S. Gordon and James W. Caldwell
Department of Chemistry
North Dakota State University

I. INTRODUCTION

Advances in computational technology and quantum mechanical methodology have brought accurate calculations on molecular structure and electronic spectroscopy within the realm of feasibility for moderate sized molecules. It is noteworthy, however, that in the area of excited states the great bulk of calculations have been carried out on unsaturated molecules. This can be understood in two ways. First, in the early years of quantum chemistry, prior to the availability of fast computers, planar conjugated molecules allowed the investigator to concentrate on the π-electron network, thereby, reducing the computational effort. In recent years, the natural inclination has been to re-evaluate the early results with more accurate techniques. Second, the electronic spectra of saturated molecules, especially those of acyclic alkanes, are difficult to interpret[1] and therefore, provide less comparative information. Consequently, theoretical investigations of excited states of saturated molecules have been few. In particular, with the exception of a preliminary study of methane surfaces using a minimal basis set,[2] *ab initio* calculations on alkane excited states have been limited to the vertical excitations in methane[3,4] and ethane.[5]

The long-range goal of our research program is to gain a better understanding of the spectroscopy and photochemistry of saturated molecules, with particular emphasis on alkanes and alkylsilanes. To attain this goal a necessary first step is the examination of excited state potential energy surfaces. This, in turn, will allow investigations of excited state vibrational manifolds, band oscillator strengths, and ultimately the study of vibronic interactions.

As is always the case in quantum mechanical investigations, a central issue in the study of potential energy surfaces is the choice of basis set. One wants a balance between a basis which is reliable on the one hand and efficient on the other. Efficiency is particularly important if one is to extensively study a number of surfaces. The use of most standardized basis sets of moderate size, such as STO4G[6] or 4-31G[7] is questionable since the exponents for such bases are typically

B. Pullman and N. Goldblum (eds.), Excited States in Organic Chemistry and Biochemistry, 257-270.

determined for the ground state. This problem is complicated further for excited states of saturated molecules. At least for small systems (e.g., water,[8] methane,[3] ethane[5]) it is well known that the spectra are dominated by excitations to diffuse "Rydberg" molecular orbitals. Thus, an adequate representation of the vertical excited states requires the inclusion of diffuse functions in the basis set.

Having said this, it is important to note that the dominant role played by Rydberg functions in low-lying vertical states may significantly decrease as the molecule distorts from its ground state geometry. This has been pointed out by numerous authors, including Mulliken[9] and Flouquet and Horsley.[8] It is important to know whether the essential features of a surface are qualitatively or quantitatively effected by the inclusion of diffuse functions. While the answer to this question is likely to depend on the specific surface, a reasonable possibility exists for gaining some insight into this problem without an exhaustive analysis of each surface.

The present paper is a preliminary attempt to answer the two basic questions raised in the preceding paragraphs. A moderate basis set including diffuse functions is introduced in the following section and is used to predict vertical excitation energies in methane and ethane singlet states in Section III. An analysis of selected surfaces of low-lying states of methane is presented in Section IV, with particular emphasis on the role of Rydberg functions as the molecule distorts.

II. METHODOLOGY

Ground state calculations have been carried out using the usual Hartree-Fock-Roothaan formalism.[10] Excited state wavefunctions have been generated using configuration interaction including all single excitations from the ground state valence MO's.

Four basis sets have been considered. Two of these, STO4G[6] and 4-31G,[7] are well known and require no further explanation. The remaining two, STO4G+ and 4-31G+, are the standard bases augmented by Rydberg functions placed at the molecular center of mass. In 4-31G+ one s and three p primitive gaussians with exponent 0.017 are added to 4-31G. This exponent has been suggested by Buenker and Peyerimhoff,[5] and small increases and decreases from this value raised most calculated excitation energies.

Since STO4G is a minimal basis set, the Rydberg exponents for STO4G+ were determined by minimizing the energy for the lowest state in each irreducible representation. The final exponent was then obtained by averaging over the states considered. The average optimal exponents generated in this way for methane and ethane singlet states are 0.020 and 0.017, respectively. Similar calculations are being carried out for the corresponding triplet states and for propane.

III. VERTICAL EXCITATIONS

A. Methane

The excitation energies for the lowest five singlet states of methane are listed in Table I. The results for the four basis sets discussed above are compared with those obtained by Montagnani and co-workers (MRS)[3] using second order perturbation theory and by Williams and Poppinger[4] (WP) using the equations of motion method. Not surprisingly, the excitation energies predicted by the standard basis sets are considerably larger than the others. In addition, relative to the accurate results of WP, STO4G predicts the order of these states incorrectly. The remaining four calculations are in good agreement. This is particularly important in the case of STO4G+ since calculations using this basis set use considerably less computer time than those using the less accurate 4-31G basis. As expected, all five states are dominated by excitations to Rydberg MO's. Since STO4G is a minimal basis set, a pertinent question is whether the addition of diffuse functions improves the results due to the specific nature of the extra functions or if the improvement is simply due to the variational principle. Partial evidence that the latter is not the case is provided by a comparison of the 4-31G and STO4G+ results, the latter being a larger basis set. Furthermore, the diffuse functions make a negligible contribution to the ground state.

It should be noted that the lower excitation energies predicted by STO4G+, relative to 4-31G+, are not unexpected. The latter provides a better description of the ground state, while the two are probably not too dissimilar in their treatment of the excited states. Similar comments apply to the equations of motion results (WP) since these were obtained with a basis set similar to STO4G+. Thus, quantitative differences among the last four columns in Table I should not be taken too seriously.

B. Ethane

The excitation energies predicted by STO4G+ for ethane singlet states are compared with those of Buenker and Peyerimhoff(BP)[5] in

Table I. Vertical Excitations in Methane Singlet States(ev)

State	STO4G	4-31G	STO4G+	4-31G+	MRS[1]	WP[2]
$1T_2$	24.19	14.07	10.79	11.21	11.18	10.24
$2T_2$	24.78	16.18	12.01	12.47	12.26	11.68
1E	22.67	15.64	11.97	12.47	12.33	11.72
$1T_1$	22.16	15.32	11.99	12.50	12.37	11.77
$2A_1$	28.61	17.76	12.41	12.65	12.61	12.38

[1] R. Montagnani, P. Riani and O. Salvetti, Theoret.Chim.Acta,32, 161 (1973).
[2] G.R.J.Williams and D. Poppinger, Mol. Phys.,30, 1005 (1975).

Table II. Even though the latter authors used a double zeta plus Rydberg basis set and included double excitations in the CI, the qualitative and quantitative agreement is reasonable. This is encouraging since it appears that the smaller basis set will provide a realistic alternative for larger (and thus more expensive) molecules. Once again, all states listed in Table II are dominated by excitations to Rydberg MO's.

IV. EXCITED STATE SURFACES

A. Preliminary Remarks

In general, to fully understand the nature of an excited state surface one must carry out calculations at a great many points. However, as illustrated by Woodward and Hoffman[11] and by Salem and co-workers,[12] one can at least gain some qualitative insight about a surface with much less expense.

With regard to the importance of Rydberg functions as a function of geometry, one can make the following comments:

(1) While vertical states of saturated molecules are commonly referred to as "Rydbergs," it must be recognized that in general the only restriction on the nature of a state is that imposed by symmetry. Within the framework of configuration interaction, all configurations which transform according to a given irreducible representation can mix. Thus, what one really means by a molecular Rydberg state is one in which the dominant configuration <u>throughout the surface</u> involves an excitation to a MO which is clearly Rydberg in character. Such a state will always have mixed with it contributions from valence excitations, even though such contributions are expected to be rather small in the vertical states.

(2) It appears that in small saturated molecules the low-lying

Table II. Vertical Excitations in Ethane Singlet States (ev)

State	STO4G+	BP[1]	Experiment[2]
$1E_g$	9.61	9.16	
$2A_{1g}$	10.04	9.21	
$1A_{2u}$	10.39	9.86	~9.4
$1E_u$	10.41	9.91	
$2E_u$	10.63	9.99	
$2A_{2u}$	10.64	9.99	
$1A_{1u}$	10.50	10.04	

[1] R.J. Buenker and S.D. Peyerimhoff, Chem.Phys., 8, 56 (1975).
[2] D.R. Salahub and C. Sandorfy, Theoret. Chim. Acta, 20, 227 (1971).

vertical states are Rydberg states as defined above. In propane, for example, the lowest 30 excited states may be characterized as Rydbergs.

(3) The photolysis of small alkanes is dominated, at least near the threshold excitation, by elimination of molecular hydrogen. The hydrocarbon fragment (e.g. CH_2, CH_3CH, etc.) may no longer be strictly characterized as saturated; thus, it is more likely that this product will have low-lying valence excited states.

Consider now a low-lying vertical (Rydberg) state of the parent which correlates with a low-lying product valence state. Let us assume for simplicity that each state is the lowest of its symmetry. In order for the parent Rydberg state to dissociate to the product valence state the contribution of the initially dominant Rydberg configuration must decrease as the parent relaxes to products. In view of the high density of states in alkanes and remembering that the vertical parent valence states will lie rather high in energy, what this really means is that an avoided crossing is expected at some point between the initial excitation and dissociation. This is illustrated qualitatively below:

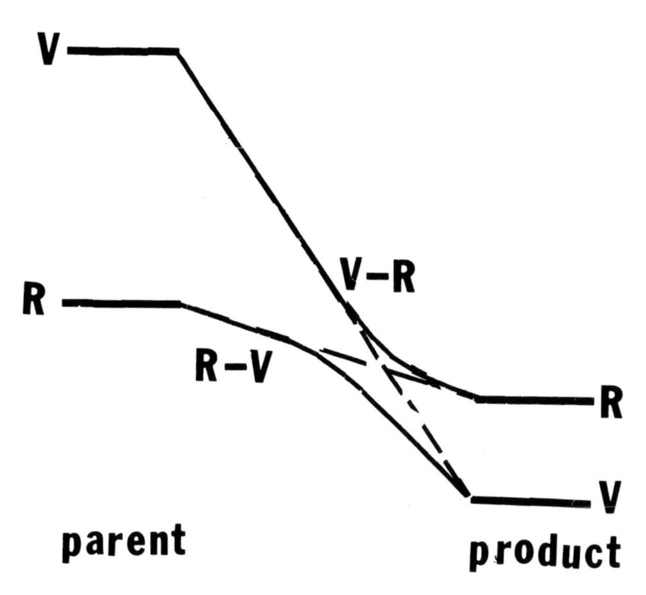

One may refer to the upper and lower curves as V-R and R-V hybrid states, respectively. Such a situation has been characterized by Salem and co-workers[12] as a type D avoided crossing. If the avoided crossing occurs close to the parent, the features of an R-V surface are expected to be characteristic of a valence excited state. In a saturated molecule this is likely to mean that the state will be dissociative. If the crossing occurs close to the product, the features of the R-V surface will be characteristic of a Rydberg state and will probably have a minimum. Of course, the details of the surface will depend on a number of additional factors such as other crossings and the relative energies of the two sets of states. For example, if the product Rydberg state lies above that of the parent, one might expect a local maximum on the R-V surface and a local minimum on the V-R surface:

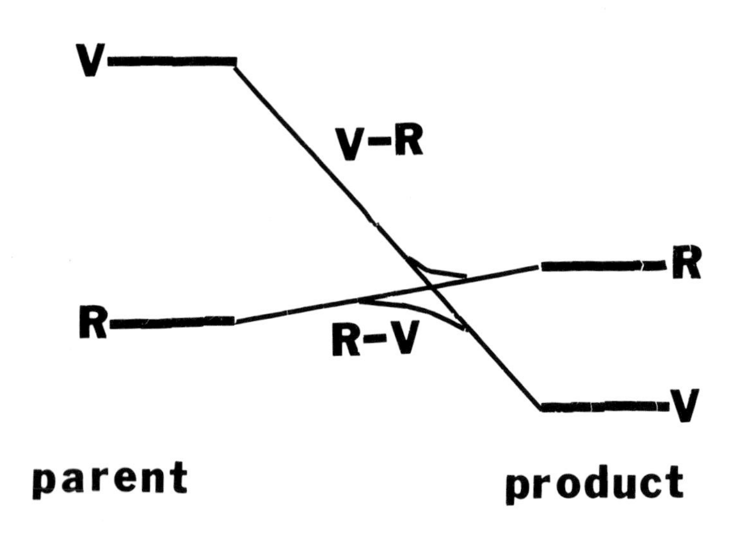

On the other hand, if the lowest product state of a given symmetry is a Rydberg, it is reasonable to expect the surface to be quite similar to that of the corresponding positive ion. It is important to note in this regard that while molecular excited states may or may not be hybrids of one type or another, the ion has no choice. Since the upper electron has been removed, the ion state must behave as a pure Rydberg. Only when the excited molecular state retains Rydberg character throughout the surface is the often used analogy with the corresponding ion state strictly valid.

Finally, it should be noted that qualitative considerations such as those presented above are not expected to be as useful for rationalizing the primary products of saturated molecule photolyses as the Woodward-Hoffman rules are for rationalizing thermal reactions. The difference is that in thermal decompositions the lowest energy (essentially vibrationless) path may be the most likely path for the molecule to follow, whereas an electronically excited molecule typically has considerably more energy than that required to follow the minimum energy route.

With the above comments in mind we now turn to a consideration of a number of excited singlet state surfaces of methane. Unless otherwise specified, the calculations have been carried out using the 4-31G+ basis set. It is to be noted that most excited states of methane are Jahn-Teller states and are therefore expected to distort from tetrahedral symmetry. Most of the discussion below will center around C_{2v} and C_{3v} structures; however, preliminary probes of other possible distortions will also be considered. In all distortions from tetrahedral symmetry the Rydberg functions were kept on the carbon.

B. 1 $^1B_1$ (C_{2v})

Initially, calculations on this state were carried out such that all geometric parameters were allowed to vary, subject to the constraint of C_{2v} symmetry. The resultant behavior is consistent with earlier STO4G[2] and 4-31G[13] calculations on this state in that the molecule dissociates to $^1B_1CH_2$ and $^1A_1H_2$ with no barrier. To investigate this dissociation path in greater detail the geometry was reoptimized at a number of points along the path for fixed values of the long CH bond-length. The results are displayed in Figure 1, with all energies quoted relative to the separated products at their optimal geometries. Included for comparison are the vertical energy and a plot of the ground state energy at the B_1 geometries. Note that the two curves cross so that, to the extent that the ground state calculations are valid (see below), the possibility exists for a radiationless transition from $^1B_1$ to the ground state.

Also shown in Figure 1 is the 4-31G energy for the vertical B_1 state. As noted earlier (Table I) this is nearly 3 ev above the corresponding 4-31G+ energy. In contrast, the energies at dissociation differ by only 1 kcal/mole for the two basis sets. It appears then that the lowest singlet B_1 state (1 T_2 in T_d symmetry) is an R-V hybrid. Further evidence for this conclusion is presented in Figure 2 where the net Mulliken population in the Rydberg basis functions is plotted as a function of the CH bondlength. P_R drops dramatically on distortion from the vertical geometry (R_{CH} = 1.081Å) from 0.9 to less than 0.05 by R_{CH} = 1.5Å, and the lowest $^1B_1$ state of methylene is clearly a valence state.

It must be noted here that a change in the ground state SCF electronic configuration occurs between 1.5 and 2.0Å. At 1.5Å, the configuration is

$$(1a_1)^2(2a_1)^2(3a_1)^2(1b_2)^2(1b_1)^2\ (4a_1)(2b_2)(5a_1)(2b_1)(6a_1),$$

while at 2.0Å it is

$$(1a_1)^2(2a_1)^2(1b_2)^2(3a_1)^2(4a_1)^2\ (1b_1)(5a_1)(2b_2)(6a_1)(2b_1).$$

The dominant configurations before and after the $1b_1 \leftrightarrow 4a_1$ crossing are ($1b_1 \rightarrow 4a_1$), ($1b_1 \rightarrow 6a_1$) and ($4a_1 \rightarrow 1b_1$), ($4a_1 \rightarrow 2b_1$), respectively. In both cases the second configuration listed has a larger CI coefficient. In both cases the first four virtual orbitals are mainly Rydbergs, while $6a_1$(1.5Å) and $2b_1$(2.0Å) are valence MO's; however, greater mixing of Rydberg and valence basis functions occurs at the longer bondlength.

While this result means that a proper description of the ground state requires a CI treatment including both of the above configurations, it is not likely that the major conclusions with regard to the B_1 state will be seriously altered. The crossing occurs after this state has essentially become a valence state. From 1.5Å on, therefore,

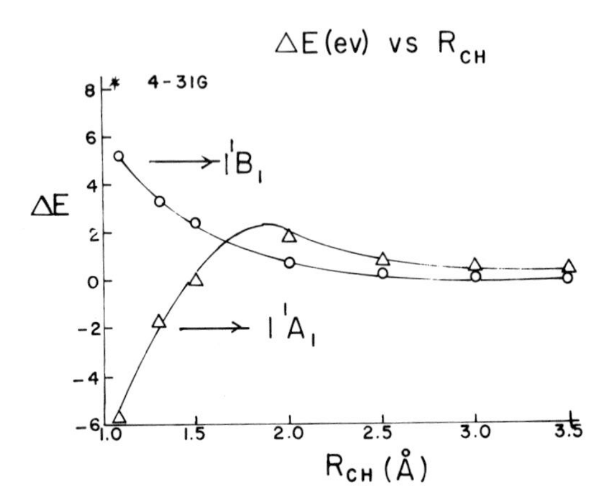

Figure 1

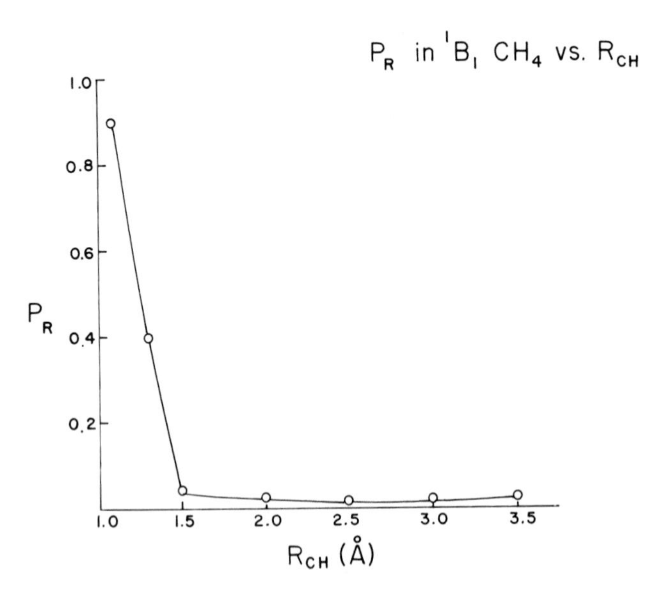

Figure 2

the B_1 curve should be virtually the same as the 4-31G curve. In the latter, $6a_1$ is the lowest virtual orbital, and the crossing is $1b_1 \leftrightarrow 6a_1$. Here, the overwhelming dominant configuration is . . . $(1b_1)^1(6a_1)^1$ at both bondlengths; thus, the excited state is unaffected by the crossing.

C. $1\ ^1B_2\ (C_{2v})$

The lowest $^1B_2$ state of methylene at its optimal geometry (R=1.072Å, angle = 147.1°) is clearly a Rydberg state with P_R = 0.9503. Based on the earlier discussion, it is likely that the corresponding parent state is also a Rydberg with a barrier to dissociation. In fact, there are a number of possible C_{2v} methane states which could lead to $^1B_2$ CH_2, depending on the symmetry of the H_2 co-fragment. One candidate is $^1B_2$ methane (+ $^1A_1H_2$); however, as noted earlier,[2] this state is difficult to isolate. Both $1\ ^1B_1$ and $1\ ^1B_2$ arise from $1\ ^1T_2$ in T_d symmetry. Since $^1B_1$ is lower in energy, attempts to optimize the $^1B_2$ geometry at small CH_2-H_2 bondlengths results in the two pairs of hydrogens inverting their positions. As a result, the B_2 state transforms into B_1. This could mean that the two states are vibronically coupled and that molecules initially in $1\ T_2$ are most likely to dissociate to $^1B_1$ methylene.

D. Other C_{2v} States

The lowest $^1A_1$ and $^1A_2$ excited states of methane have well-defined minima within C_{2v} symmetry. $^1A_2$ is slightly (0.2 ev) lower in energy, and the two states have very similar optimal geometries using 4-31G+. Both are clearly Rydberg states in C_{2v} with a net Rydberg population of about 1.0.

The geometries predicted for $^1A_2$ by the four basis sets are listed in Table III. Note that the extent of distortion from the vertical geometry is much greater when Rydberg functions are not included. This is particularly true for the minimal basis set. Similar comparisons are found for the A_1 state. It is also noteworthy that STO4G+ predicts

Table III. Optimal Geometries for $1\ ^1A_2\ (C_{2v})$[1]

	STO4G	4-31G	STO4G+	4-31G+
R_{CH}	1.360	1.247	1.193	1.190
$R_{CH'}$	1.157	1.158	1.084	1.062
α(HCH)	92.29	100.84	72.67	63.52
α(HCH')	122.37	121.76	113.62	114.02

[1] Bondlengths in Å, angles in degrees.

$^1A_2$ to be ~0.3 ev below $^1A_1$, in qualitative agreement with 4-31G+, while the other two basis sets place $^1A_1$ nearly 1 ev below $^1A_2$ and predict rather different geometries for the two states. The fact that STO4G+ reproduces the geometries and relative energies predicted by 4-31G+ with reasonable accuracy is an important result. The smaller basis set should be reliable for carrying out initial calculations on excited state surfaces. This will result in a considerable savings with regard to computer time.

It should be recognized that the two states discussed above are not necessarily absolute minima on their respective potential surfaces. In C_{2v} symmetry methane has five nontotally symmetric normal modes, each of which could, in principle, lower the energy. Small distortions along the two B_1 and two B_2 internal symmetry coordinates leads to an increase in energy for both states; however, the A_2 twisting motion (Figure 3), which reduces the symmetry to C_2, lowers the energy of 1 $^1A_2$ (2 ^1A) and has almost no effect on 2 $^1A_1$ (3^1A) for a value of 10° for the twist angle. Interestingly, the 2A state opens both the HCH and H'CH' angles to ~140° and ultimately attains tetragonal D_{2d} symmetry with a bondlength of 1.112Å. The state (1^1E in D_{2d}) retains its Rydberg character, and the final geometry is nearly identical to that found for CH_4^+ ($^2B_2$) by Pople and co-workers.[14] This is not too surprising since the dominant configuration in the neutral Rydberg state arises from an excitation from the $1b_2$ MO in D_{2d} symmetry. Pople, et al. also quote a geometry for CH_4^+ in C_{2v} symmetry with an energy slightly higher than that of the D_{2d} structure. Based on the results presented here, it is likely that this C_{2v} structure is not a true minimum.

Using the 4-31G basis set, this same state (2A in C_2 symmetry) twists all the way to a planar D_{4h} structure. In D_{4h} symmetry the the state transforms as B_{1u}, and the 4-31G energy is 2.7 ev lower than the 4-31G+ D_{2d} structure. Adding the Rydberg functions and re-optimizing the bondlength in square planar symmetry results in a negligible change in the energy and a very small population in the Rydberg functions. Thus, it appears that this surface has two minima, one strongly Rydberg in character and the other essentially a valence state. It must be pointed out, however, that the D_{4h} structure may still be unstable to further distortions.

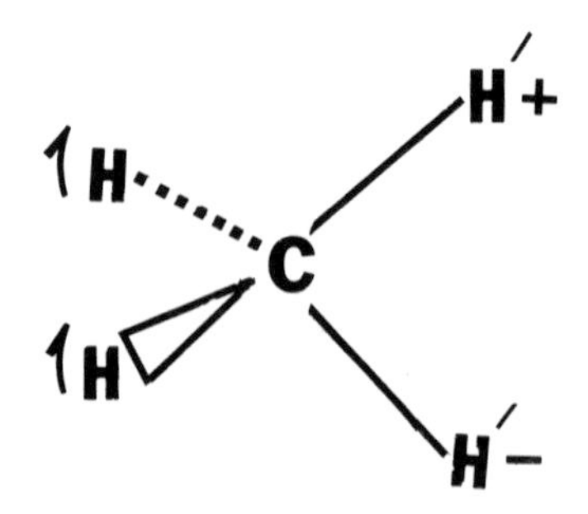

Figure 3. Methane A_2 Symmetry Coordinate.

E. C_{3v} States

Within the restriction of C_{3v} symmetry a minimum is found for the lowest state in each irreducible representation. All three states are dominated by Rydberg excitations at their optimal geometries. The relative energies for each basis set are listed in Table IV, where again the close agreement between STO4G+ and 4-31G+ is evident. Note also that, according to 4-31G+, at the optimal A_2 geometry there are three E states with lower energy.

As shown earlier (Table III) the basis sets with (without) Rydberg functions tend to agree with each other reasonably well with regard to geometry predictions, but the diffuse functions do have a significant effect on the predicted geometry. As an example, the A_1 geometries are compared in Table V, where it may be seen that the long CH bondlength is calculated to be about 0.5Å too long if the diffuse functions are not included.

In C_{3v} symmetry, methane has three E normal modes, each of which distort the molecule to C_s symmetry. Since 1 ^1E is a Jahn-Teller state, such distortions are expected to lower the energy and split the state into an A' and A" component. This is, in fact, the case. The A' component distorts through C_s symmetry, increasing its valence character, and ultimately merges into C_{2v} and dissociates to $^1B_1$ CH_2 and $^1A_1$ H_2. The surface of the A" component has not yet been established.

While the 2 $^1A_1$ state is not a Jahn-Teller state, it too distorts to C_s symmetry. The analysis of this surface has not yet been completed; however, there appears to be a minimum in C_s symmetry with the state being dominated by a Rydberg excitation. The A_2 state is apparently

Table IV. Relative Energies of Methane Singlet C_{3v} States (ev)

	STO4G	4-31G	STO4G+	4-31G+
$2A_1$	0.66	0.0	0.0	0.0
1E	0.0	1.33	0.80	0.98
$1A_2$	1.27	1.66	2.06	2.35

Table V. Optimal Geometries for 2 $^1A_1$ (C_{3v})[1]

	STO4G	4-31G	STO4G+	4-31G+
R_{CH}	1.890	2.085	1.316	1.472
$R_{CH'}$	1.098	1.078	1.100	1.081
α(HCH')	101.53	93.16	94.46	94.25

[1] Bondlengths in Å, angles in degrees. H refers to the odd bondlength.

stable to distortions to C_s symmetry, and investigation of the 2E and 3E surfaces has not yet been initiated.

V. SUMMARY

(1) For the study of excited states of saturated molecules the minimal + Rydberg basis set (STO4G+) appears to be preferable to the standard (and more time consuming) 4-31G basis.

(2) One should be able to gain some insight about an excited state potential energy surface by comparing correlated parent and product states. This is illustrated by the $^1B_1$ surface of methane which originates as a Rydberg state in T_d symmetry and dissociates to valence methylene and H_2 states.

(3) Only those states which retain their Rydberg character throughout the surfaces are characterizable by the corresponding ion surfaces.

(4) Of all the methane excited state surfaces studied to date only one, the relatively high-lying 1 $^1A_2$, retains C_{3v} symmetry. The remaining states either dissociate to $CH_2 + H_2$, distort to a symmetry for which C_{2v} is a subgroup, or possibly $(2A_1$ in $C_{3v})$ distort to a lower symmetry. There are at least two, and perhaps three, channels to the dissociative 1 B_1 state. This is consistent with the observed threshold photochemistry of methane:[15]

$$CH_4 \xrightarrow{h\nu} CH_2 + H_2$$

Acknowledgement:

The computer time for this project was made available by the North Dakota State University Computer Center. The authors have benefited from numerous enlightening discussions with Professor R.D. Koob.

REFERENCES

1. J.W. Raymonda and W.T. Simpson, J.Chem.Phys., 47, 430 (1967).
2. M.S. Gordon, Chem.Phys.Lett., 44, 507 (1976).
3. R. Montagnani, P. Riani and O. Salvetti, Theoret.Chim.Acta, 32, 161 (1973).
4. G.R.J. Williams and D. Poppinger, Mol.Phys., 30, 1005 (1975).
5. R.J. Buenker and S.D. Peyerimhoff, Chem.Phys., 8, 56 (1975).
6. W.J. Hehre, R. Ditchfield, R.F. Stewart and J.A. Pople, J.Chem. Phys., 52, 2769 (1970).
7. R. Ditchfield, W.J. Hehre and J.A. Pople, J.Chem.Phys., 54, 724 (1971).
8. F. Flouquet and J.A. Horsley, J.Chem.Phys., 60, 3767 (1974).
9. R.S. Mulliken, Chem.Phys.Lett., 14, 141 (1972).
10. C.C.J. Roothaan, Revs.Mod.Phys., 23, 69 (1971).
11. R.B. Woodward and R. Hoffmann, The Conservation of Orbital Symmetry, Academic Press, N.Y., 1970.
12. L. Salem, J.Am.Chem.Soc., 96, 3486 (1974); Science, 191, 822 (1976); W.G. Dauben, L. Salem and N.J. Turro, Accts.Chem.Res., 8, 41 (1975).
13. M.S. Gordon and J.W. Caldwell, in preparation.
14. W.A. Lathan, W.J. Hehre, L.A. Curtiss and J.A. Pople, J.Am.Chem. Soc., 93, 6377 (1971).
15. J.R. McNesby and H. Okabe, Adv. Photochem., 3, 157 (1964).

DISCUSSION

SALEM :

As you depart from tetrahedral symmetry, and as you observe an orbital crossing, do you also observe a crossing of two B_1 configurations derived from different (T_1, T_2) vertical states ?

GORDON :

To answer this question properly would require a closer examination of the regions of interest (1.08 -1.3 Å and 1.6 -2.0 Å); however the following can be said. For small distorsions from T_1, $1T_2$, $2T_2$, and $1T_1$ will all have B_1 components and all are expected to decrease in energy. The relative decreases will determine the occurrence of avoided crossings involving the three B_1 states. Such an occurrence is most likely for the B_1 states deriving from $2T_2$ and $1T_1$ since these only differ by 0.03 ev in T (Table I). The corresponding $1T_2$ - $2T_2$ separation is 1.26 ev.

In the region of the orbital crossing (≃ 1.7 Å $1B_1$ is already a valence state (see Figure 2), while $2B_1$ appears to be a Rydberg in this vicinity. An avoided crossing between these two would imply a reverse type D crossing, and there is no evidence for this.

EVIDENCE AND REACTIVITY OF A TWISTED FORM OF MEDIUM SIZE CYCLO-ALKENE RINGS PRESENTING A DOUBLE BOND PAST ORTHOGONALITY

J. JOUSSOT-DUBIEN, R. BONNEAU, P. FORNIER de VIOLET

Université de Bordeaux I, laboratoire de Chimie Physique A, 33 405 TALENCE CEDEX (France)

Cis trans isomerization of olefins has been and still is a topic of large interest to photochemists from several points of view : mechanistic considerations, photostationary state reached after prolonged irradiation and energy transfer to and from the olefins. Furthermore the ethylenic bond being a relatively simple chromophore, its properties can easily be approached by theoretical considerations.

It is well known that upon twisting a carbon-carbon double bond, the ground state leads to an orthogonal singlet diradical whereas the singlet excited states gives rise to an orthogonal zwitterionic form.

A provoking prediction has been made by Salem (1) according to which isomerization of unsymmetrical olefins in polar solvents should lead to a double-well potential due to avoided surface crossing between the ionic excited configuration and the ground covalent one as illustrated in figure 1. Ab initio calculations do confirm these views, however no experimental evidence has yet been obtained.

We thought that this would be an interesting problem to tackle. To put all chances on our side we decided that we would investigate medium size alkene rings in order to avoind rapid cis-trans isomerization which would be the obvious deactivating process of the metastable zwitterionic orthogonal olefins (MZO).

Not knowing what the lifetime of MZO would be, we first planned a laser-flash photolysis experiment of cyclohexene. The disymmetry was introduced by substituting a phenyl ring on one of

B. Pullman and N. Goldblum (eds.), Excited States in Organic Chemistry and Biochemistry, 271-281.

the ethylenic carbon. This substitution also shifted the absorption band of the alkene to the 260 nm region where our quadruple neodymium laser furnishes nanosecond pulses with energy of about 50 mJ.

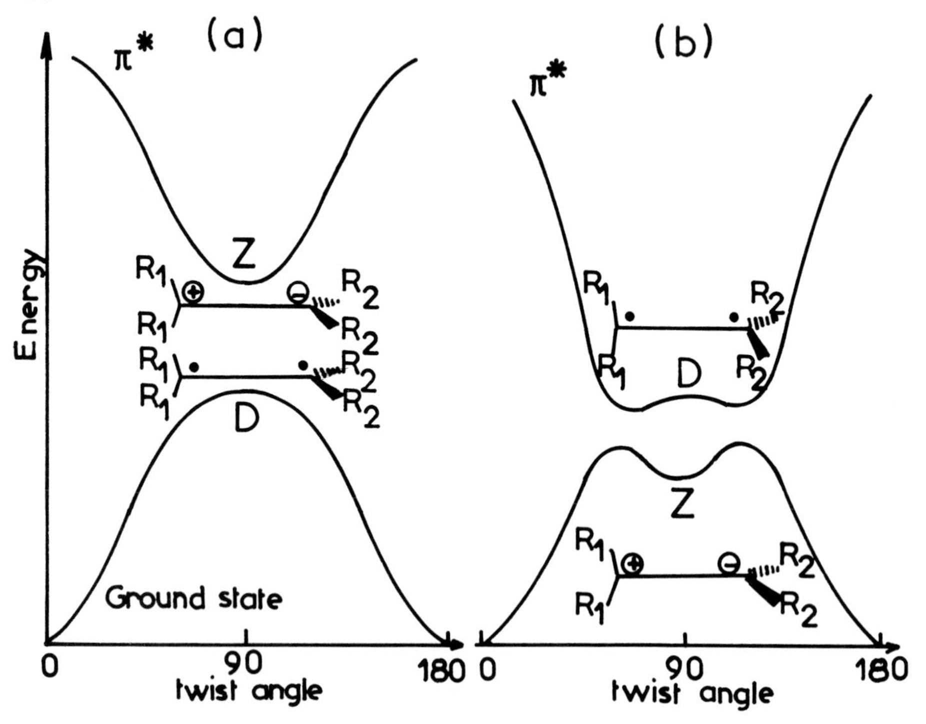

Figure 1 - *Energy of the ground, D, and zwitterionic state Z of an unsymmetrical substitued ethylene in a non polar (a) and in a polar solvent (b) (ref. 1).*

The first experiments were performed in methanolic solution at room temperature and indeed a transient species was observed, absorbing in the 300-430 nm range with a maximum at 380 nm. The lifetime of this species was found to be 9 μs and independent of dissolved oxygen. The next experiments were obviously to change the polarity of the solvent to deteet any effect of stabilization of the species. A similar transient absorption was observed in cyclohexane and acetonitrile with lifetime in the same range as found in methanol, 9 μs and 14 μs respectively (2).

Although these findings were not encouraging for our search of the MZO, we pursued our effort using phenylcycloheptene (3). Upon flashing this compound in cyclohexane, methanol and acetonitrile solutions we observed, between 350 and 280 nm, a transient absorption unaffected by dissolved oxygen and lasting seconds under subdued analytical light but adversly affected by increasing the intensity of the analytical beam.

As matter of fact this species is so long lived that we were able to trap it in an ethanolic glass at 90K when phenyl-cycloheptene is irradiated under continuous illumination with the 254 nm Hg line. Under these conditions a photostationary state is obtained which is in favor of the initial product. Irradiating at 313 nm, bleaches the photoproduct, with recovery of the starting material. Heating also bleaches the absorbance at 300 nm but the initial absorbance in phenylcycloheptene is not recovered.

These results point strongly to the assignment of the transient species to the "trans" phenylcyclohexene and cyclohep-tene respectively in which the double bond is past orthogonality.

Indeed referring to figure II in which approximate potential surfaces of the cycloalkenes are represented as sums of the usual dumb-bell curves for cis-trans isomerization of styrene (4) and steeply rising curves due to ring strains for twist angles past orthogonality, the main features of our findings can be interpreted satisfactorily.

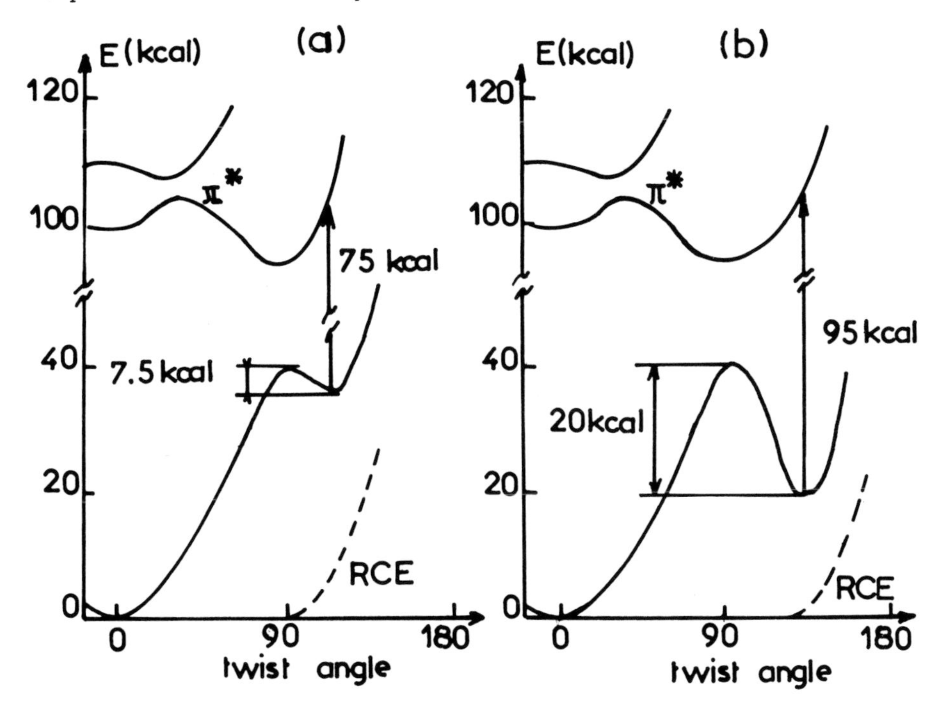

Figure 2 - *Potential surfaces of phenylcyclohexene (a) and phenylcycloheptene (b) as a function of twist angle, taking into account the ring constraint energy (RCE).*

1°) The absorption spectra of the "trans" forms of cycloalkenes are red-shifted with respect to that of the stable cis isomers, the shift being larger for cyclohexene than for cycloheptene.

2°) The photobleaching of the cycloheptene transient in the glassy matrix is in fact the back trans-cis photoisomerization which also accounts for the reduced lifetime of "trans" cycloheptene under intense analytical beam irradiation.

3°) 1-phenylcycloheptene is not recovered when the glass is melted because the 20 $Kcal.M^{-1}$ activation energy barrier is too high to allow thermal re-isomerization to compete with the formation of products. Indeed the rate constant for trans-cis thermal re-isomerization is expected to be of the order of $3.10^{-4}s^{-1}$ whereas the rate of disappearance by chemical means is about 30 times faster, even at room temperature.

The last test that remained to be made concerned the chemical properties of the strained forms, which according to known schemes of reactivity of organic molecules, should be very reactive towards electrophylic reagents, in particular the simplest one : H^+.

Adding known amounts of HCl to methanolic solutions of the cycloalkenes and measuring by flash photolysis the decay of the transient absorption, it is indeed found that H^+ very efficiently quenches the species that give rise to the absorptions. Rate constants of quenching, k_Q, equal to 7.10^6 M^{-1} s^{-1} for "trans" phenylcyclohexene and a much smaller value, 10^3 M^{-1} s^{-1} for "trans" phenylcycloheptene are determined from these measurements.

This is the order of reactivity expected from strain considerations. It is therefore tempting to identify the transient we observed with the intermediate postulated by Kropp (5) on purely photochemical grounds and tentatively assigned by him to the "trans" form of the alkene to account for the acid catalyzed photoaddition of alcohols :

A rigorous proof of the identity of the transient we observe and the intermediate postulated by Kropp is very difficult to carry out, because the conditions of the flash photolysis experimentation - small volume irradiated ~20 ml, dilute solutions used ~ 10^{-3} M, minute quantities of product formed ~ 1 %- are so different from the conditions of continuous irradiations where product analysis is the goal.

To test this coherence, a kinetic scheme must be assumed. Let it be the following in the case of cis-phenylcyclohexene (abbreviated c-PC_6) irradiated in methyl alcohol at room temperature, to which HCl is added ; t-PC_6 stands for the strained "trans" form and no provision is made to know whether the "trans" form is a primary photochemical step or not, i.e. whether there is a triplet step in between c-PC_6 and t-PC_6 :

c-PC_6	$\xrightarrow{h\nu}$	t-PC_6	φ quantum yield
t-PC_6	$\longrightarrow$	c-PC_6	k_{tc} } $k_1 = 1/\tau$
t-PC_6	$\xrightarrow{MeOH}$	carbocation + $MeOH^-$	k'_r } k_r
t-PC_6	$\xrightarrow{MeOH,\ t\text{-}PC_6}$	other products	k''_r
t-PC_6	$\xrightarrow{H^+}$	carbocation	k_Q
carbocation	$\xrightarrow{MeOH}$	ether	α chemical yield

According to this scheme the quantum yield of disappearance of PC_6 is given by the relation :

$$\Phi = \varphi \frac{k_r + k_Q [H^+]}{k_1 + k_Q [H^+]}$$

and the quantum yield of formation of the ether is :

$$\Phi' = \alpha \varphi \frac{k'_r + k_Q [H^+]}{k_1 + k_Q [H^+]}$$

when $[H^+] = 0$, $\Phi_o = \varphi k_r/k_1$ and $\Phi'_o = \alpha\varphi k'_r/k_1$

when $[H^+] \longrightarrow \infty$, $\Phi \longrightarrow \Phi_\infty = \varphi$ and $\Phi'_\infty = \alpha\varphi$

From these relations one can define the ratios :

$$\frac{\Phi - \Phi_o}{\Phi_\infty - \Phi} = \frac{\Phi' - \Phi'_o}{\Phi'_\infty - \Phi'} = \frac{k_Q}{k_1}[H^+] = k_Q \tau [H^+] \qquad I$$

which is a linear function of the H^+ concentration.

We have measured, under continuous irradiations, the quantum yields, on the one hand, of disappearance of cis-phenylcyclohexene followed by UV spectroscopy and, on the other hand, of the formation of the ether followed by gas chromatography, both as a function of pH.

Results expressed as the ratios defined above are given on figure 3. Indeed a good linear plot is found, giving for the slope the value of 70 which is precisely the one that is determined for the ratio k_Q/k_1 from flash photolysis experiments.

This good agreement is the final argument we needed not only to assign definitely the transient we have found to the "trans" form of cyclohexene but also to identify it to the trans intermediate postulated in the photochemistry of this medium sized ring compound.

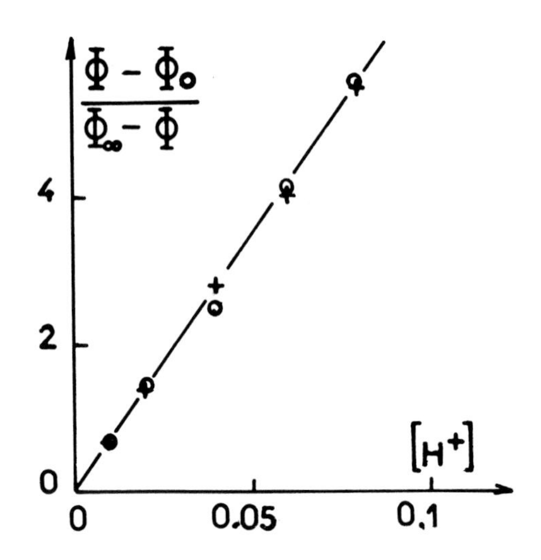

Figure 3 - *Values of the ratio I (see text) as a function of H^+ concentration.*

o — o : phenyl 1-cyclohexene disappearance ;
+ — + : ether formation.

At this point our findings must be compared with those reported for other strained trans forms. A "trans" cycloheptenone was identified several years ago by IR spectroscopy in a rigid glass at low temperature (6, 7) but no UV spectrum was given by the authors. However in a report to a recent meeting a UV spectrum was assigned to a transient "non cis isomer" of cycloheptenone at room temperature by flash photolysis (8).

We were able to confirm this last point without difficulty. Indeed cycloheptenone flashed in various solvents gives rise to a transient absorption peaking at 265 nm which has all the characteristics of a "trans" form (9).

Incidentally this "trans" form has a lifetime that is strongly dependent on the cycloheptenone concentration and the nature of the solvent. This is interpreted in term of two competing reaction channels depleting the "trans" form :

- dimer formation which is the main process occuring in non polar or non protic media
- addition product with alcohols favored by an increasing acid character of the substrate.

Assuming a mechanism in which the "trans" form is a common intermediate in these reactions we have compared the data taken from Nozaki et al (10) obtained under continuous illumination and expressed as a ratio :

$$\frac{[\text{addition product}]}{[\text{add. prod.}] + [\text{dimers}]}$$

with our flash data from which we can calculate the comparable ratio :

$$\frac{k_r\,[\text{ROH}]}{k_r\,[\text{ROH}] + k_{dim}\,[\text{Cycloheptenone}]}$$

in which k_r and k_{dim} are rate constants respectively for addition reaction and dimer formation with the cis isomer :

	MeOH	iPrOH	tButOH
$\frac{k_r[\text{ROH}]}{k_{dim}[\text{C}] + k_r[\text{ROH}]}$	0.97	0.65	.10
$\frac{[\text{add prod}]}{[\text{dimer}] + [\text{add.prod}]}$	1.0	0.67	.05

The good agreement we found between these two ratios shows undoubtedly that the "trans" form is indeed the reactive precursor of the products obtained.

We have made several attempts to record the IR spectrum of the long live "trans" phenylcycloheptene in an inert matrix at low temperature in order to compare it with the one previously given for "trans" cycloheptenone, but so far with no success.

This is understandable if one considers the spectral properties of the two compounds.

In the case of cis-cycloheptenone, photoisomerization can be induced by irradiating the n π transition in the near UV (figure 4). This absorption band being weak, a fairly high concentration of the starting material can be irradiated affording an in-depth formation of the "trans" isomer. Furthermore, the irradiation wavelength is far from the strongly absorbing $\pi\pi$ transition of the trans form which very efficiently reverses the isomerization.

Quite different spectral conditions are presented by phenylcycloheptene. At the wavelength of irradiation the exctinction coefficient of the cis form is high, meaning that under conditions favorable for IR spectroscopy, the UV light hardly penetrates the sample glassy material. This is aggravated by the overlap of the "trans" absorption which builds up under the cis, affording a photostationary state in favor of the starting cis derivative.

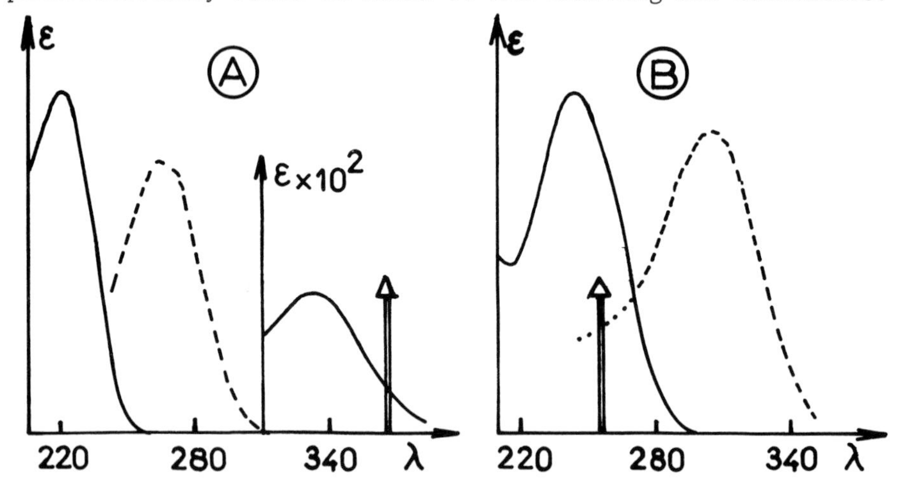

Figure 4 - *Absorption spectra of the cis (——) and trans (---) forms of cycloheptenone (A) and phenylcyloheptene (B). The double arrows indicate the excitation wavelengths in continuous irradiation experiments.*

Finally we have also examined acetyl-1-cyclohexene (11) whose photochemical properties are also consistent with a "trans intermediate (12) and for which a transient species absorbing near 280 nm and having a lifetime of about 20 ns had already been recorded by laser photolysis (8).

We confirm this finding and in addition we observed a second transient absorption whose maximum is at 345 nm and decaying with lifetime of 15 μs in cyclohexane and acetonitrile and 0.35 μs in methanol. This second species can be assigned with confidence to the "trans" form of the acetyl-1-cyclohexene, but the first one is much more difficult to identify. Incidentally cycloheptenone and cyclohexenone also give rise to this short live transient (8).

If, as Goldfarb said in the case of cycloheptenone, and as we can infer from our own experimentation on acetylcyclohexene the species absorbing a 280 nm is the precursor of the "trans" form and not a product of chemical degradation, it can be the excited singlet (S_1) or triplet (T_1) of the cis form or the orthogonal singlet (^1Z) or triplet (^3D) or eventually, as suggested by Goldfarb, a strained "trans" conformer.

Excited singlet S_1 can be ruled out from lifetime consideration. Triplet state T_1 ($E_T \sim$ 70-80 Kcal) would be efficiently quenched by piperylene ($E_T \sim$ 58-60 Kcal) but this is not the case (8). Orthogonal singlet ^1Z could have a lifetime in the ten nanoseconds range if it were stabilized by polar solvent but the transient lifetime is hardly dependent on the solvent polarity. Strained "trans" conformers, which implies twist angles greater than 90° must also be ruled out because, for such large twist angle, the ring strain energy (RCE) would be quite different for cycloheptene and cyclohexene derivatives leading thus to large differences in lifetime and absorption spectral range.

We thus conclude that this short live transient could be the orthogonal triplet, ^3D. For a 90° twist angle, the RCE, is very small for cycloheptenone as well as for cyclohexenone and acetyl 1-cyclohexene : thus the potential surfaces shown in fig. 5 are essentially the same for the three compounds. This is consistent with the fact that the transient absorption peaks at 280 nm in every case. The energy of such an orthogonal triplet should be very close to that of the orthogonal diradical ground state (^1D), and perhaps even lower, so that an orthogonal triplet cannot be quenched neither by piperylene nor by other low energy triplet acceptor, such as O_2.

Thus we still have no evidence for the metastable zwitterionic form of olefins. Can it be because we have not found the right condition to stabilise this elusive species or that we have not chosen the most appropriate molecule to look for it. Beside the compounds tested in this study we also have tried, without success, cinnamonitrile.

The question that cen be raised is whether this zwitterionic species can ever be seen since it requires that its lifetime be longer than the time necessary for the cage molecule to reorient themselves in order to stabilize it.

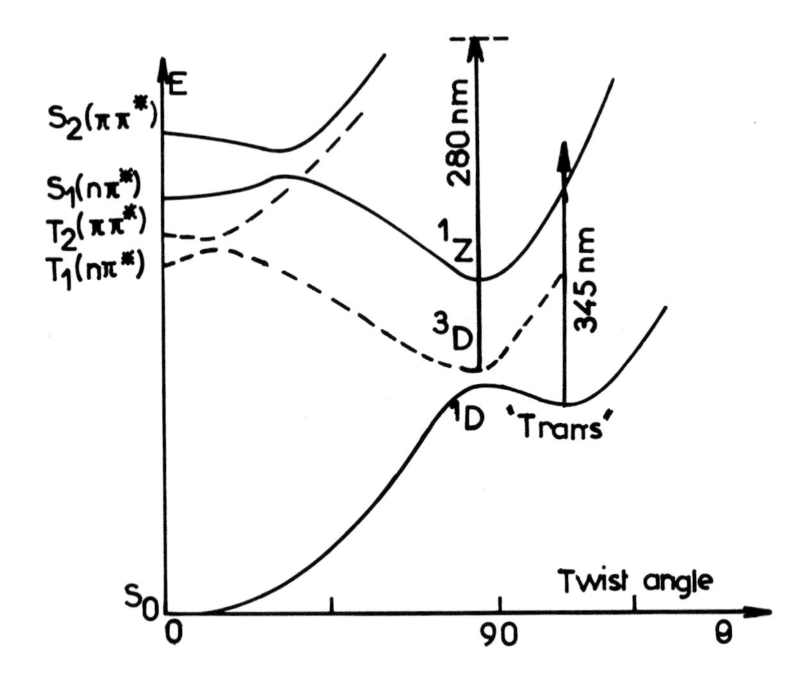

Figure 5 - Potential surfaces for acetyl-cyclohexene as a function of twist angle showing the "trans" absorption at 345 nm and the transient absorption tentatively assigned to the orthogonal triplet.

For comparison we have tabulated the spectral properties of medium sized cis and "trans" cycloalkenes.

	cis max (nm)	trans (nm)	Δ E (cm^{-1})
phenylcyclohexene	248	385	14.300
acetylcyclohexene	227	345	15.800
phenylcycloheptene	248	300	7.000
acetylcycloheptene	230	282	7.000
2cycloheptenone	222	265	7.200
cyclooctene	~185	196	~3.000

REFERENCES

1 - L. SALEM and W.D. STOHRER, J. Chem. Soc. Chem. Comm. p.140 (1975)

2 - R. BONNEAU, J. JOUSSOT-DUBIEN, L. SALEM and A.J. YARWOOD, J. Am. Chem. Soc. 98, 4329 (1976)

3 - R. BONNEAU, J. JOUSSOT-DUBIEN, A.J. YARWOOD and J. PEREYRE, Tetrahedron Letters. 77, 235 (1977)

4 - H.E. ZIMMERMAN, K.S. KAMM and D.P. WERTHEMANN, J. Am. Chem. Soc., 97, 3718 (1975)
M.C. BRUNI, F. MOMICCHIOLI, I. BARALDI and J. LANGLET Chem. Phys. Letters, 36, 484 (1975)

5 - P.J. KROPP, J. Am. Chem. Soc. 91, 5783 (1969)
P.J. KROPP, E.J. REARDON Jr., Z.L.F. GAIBEL, K.F. WILLARD and E.J. HATTAWAY ibid, 95, 7058 (1973)

6 - P.E. EATON and K. LIN, J. Am. Chem. Soc., 87, 2052 (1965)
P.E. EATON Acc. Chem. Res., 1, 50 (1968)

7 - E.J. COREY, M. TADA, R. LEMAHIEU and L. LIBBIT, J. Am. Soc. 87, 2051 (1965)

8 - T.D. GOLDFARB, Proceeding of the "First Chemical Congress of the North American Continent" Mexico Dec. 1975.

9 - R. BONNEAU, P. FORNIER de VIOLET and J. JOUSSOT-DUBIEN, Nouveau Journal de Chimie, 1, 31 (1977)

10 - H. NOZAKI, M. KURITA and R. NOYORI, Tetrahedron Letters, 16, 2025 (1968)

11 - R. BONNEAU and P. FORNIER de VIOLET, Compt. Rend. Acad. Sciences (Paris) accepted 1977.

12 - B.J. RAMEY and P.D. GARDNER, J. Am. Chem. Soc. 89, 3949 (1967) ; M.B. RUBIN Israël J. Chem. 7, 49 (1969) ; R. NOYORI and M. KATO, Bull. Chem. Soc. Japan 47, 1460 (1974)

ELECTRONIC STRUCTURE AND PHOTOPHYSICAL PROPERTIES OF PLANAR CONJUGATED HYDROCARBONS WITH A 4n-MEMBERED RING[1)]

Jakob Wirz

Physikalisch-chemisches Institut der Universität,
Klingelbergstrasse 80, CH-4056 Basel, Switzerland

Summary: The electronic absorption spectra of compounds 3 to 8 exhibit systematic features which differ radically from the well known pattern of benzenoid hydrocarbons. The observed regularities are discussed on the basis of a simple LCAO-scheme including first-order configuration interaction. The resulting assignments have been corroborated by polarization measurements for 5 to 7. Triplet state absorption spectra and energies of 6 to 8 were determined by flash photolysis and energy transfer experiments. The observed trend in fluorescence and triplet yields provides an illustrative example for the photophysical consequences of an avoided surface crossing.

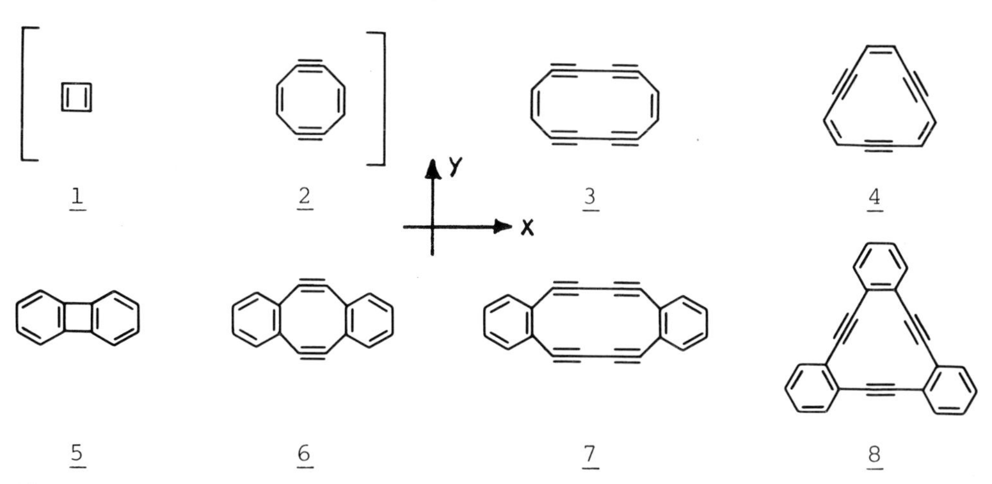

1) Part II, for part I see [1].

B. Pullman and N. Goldblum (eds.), Excited States in Organic Chemistry and Biochemistry, 283-294.

The benzenoid hydrocarbons outnumber by far the available representatives of another class of alternant hydrocarbons, those containing a planar 4, 8, or 12-membered ring (short: 4n-systems), which are the object of the present study. The latter suffer from the violation of Hückel's (4n+2)-rule for the π-system and, in most cases, from strain enforced on the σ-backbone by geometrical constraints. While photophysical data of the parent system, cyclobutadiene (1), are not at present available, its dibenzo derivative, biphenylene (5), has been studied in considerable detail. The analysis of the electronic spectrum of 5 due primarily to the work of Hochstrasser et al. [2] is in qualitative agreement with prediction from semiempirical PPP-calculations [3]. The assignment of the absorption bands below 35'000 cm^{-1} to two electronic transitions $^1B_{1g}$ (25'000 cm^{-1}) and $^1B_{3u}$ (28'000 cm^{-1}) is thus well secured. On the other hand, many authors have labelled the transitions of 5 in various ways with the symbols of Platt [4] (^1L_b, ^1L_a, ^1B_b, ^1B_a) or Clar [5] (α, p, β, β'). These labels are meaningless in 4n-systems since their absorption bands neither relate to the perimeter states defined by Platt for (4n+2)-systems nor obey the empirical rules found by Clar in benzenoid hydrocarbons. The electronic spectra of the 4n-compounds 3 (alkyl derivative) [6], 4 [7], 6 [8], 7 [9], and 8 [10] are given in the original papers reporting their synthesis but have not received further attention. We wish to point out the systematic pattern found in these spectra and to discuss these regularities in terms of a simple quantum mechanical model.

In principle, the free-electron perimeter model could be adapted to describe 4n-systems but its simplicity and heuristic value would then be largely lost because the quantum number of angular momentum, used by Platt in (4n+2)-systems to account for electron interaction (Hund's rules) and to derive selection rules, has to be abandoned in 4n-systems. We retain from the perimeter model only the notion of "shells", each consisting of a pair of orbitals which is well separated in energy from the other pairs, even after reduction of the symmetry by perturbations such as transannular bonds. We proceed along the lines of Dewar & Longuet-Higgins [11] who have developed an equally simple and essentially equivalent model for benzenoid systems in terms of LCAO-MO theory, assuming that electron repulsion has only the effect of mixing degenerate configurations (first-order CI). Fig. 1 shows the spectrum of orbital energies within the three relevant "shells" (highest filled, half-filled, and lowest vacant) of 4n-systems such as compounds 1 to 8 obtained by a Hückel or SCF calculation neglecting overlap. The orbital energies and coefficients obey the pairing theorem [12] if all AO basis energies are taken to be equal. For D_{2h} molecular symmetry the pairing relation always

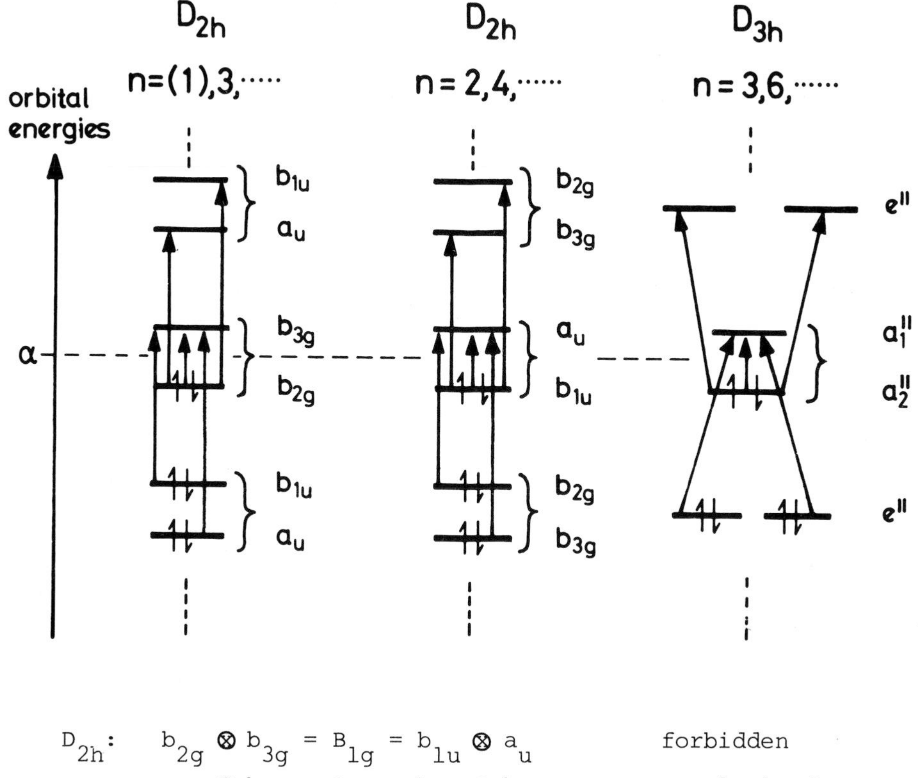

D_{2h}: $b_{2g} \otimes b_{3g} = B_{1g} = b_{1u} \otimes a_u$ forbidden

$a_u \otimes b_{3g} = B_{3u} = b_{2g} \otimes b_{1u}$ x-polarized

$b_{1u} \otimes b_{3g} = B_{2u} = b_{2g} \otimes a_u$ y-polarized

D_{3h}: $a_2'' \otimes a_1'' = A_2'$ forbidden

$e'' \otimes a_1'' = E' = a_2'' \otimes e''$ (x,y)-polarized

Fig. 1: Orbital diagram showing relevant one-electron excitations for 4n-systems.

connects orbitals of species b_{2g} with b_{3g} and a_u with b_{1u} but the relative energy within a given shell depends on the particular topology of each system.

The symmetry-forbidden one-electron excitation within the half-filled shell gives rise to the lowest excited configuration. By the symmetry-allowed one-electron excitations from the filled to the half-filled and from the half-filled to the vacant shell, we obtain four configurations of somewhat higher energy which are pairwise degenerate as a consequence of the pairing theorem. Though the lowest doubly excited configuration will have an energy of similar magnitude, it does not interact with the singly excited ones for reasons of symmetry, and does not give rise to a detectable transition in the absorption spectrum. Further singly excited configurations will be considerably higher in energy and are also ignored. Thus we arrive at five excited configurations relevant to a qualitative discussion of the absorption spectra. First-order configuration interaction will mix degenerate configurations with equal weight, the out-of-phase (minus) combination being stabilized when electron interaction is considered. The final result is depicted in Fig. 2: For D_{2h} molecular symmetry the model predicts five excited singlet states in the low-energy range, the lowest forbidden by symmetry, the next two symmetry-allowed but carrying no transition moment due to cancellation of the individual transition moments in the minus combination, and finally two allowed states of x and y polarization, respectively. These qualitative features are fully reproduced by extensive PPP SCF CI calculations [3,13] and are quite distinct from those predicted for (4n+2)-systems by the same model [11].

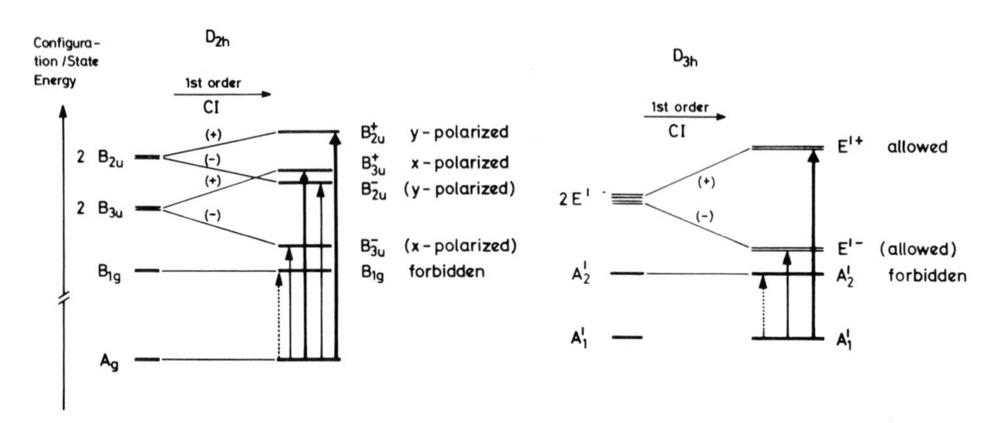

Fig. 2: Configuration diagram showing effects of first-order CI for 4n-systems. Plus/minus notation in symmetry labels following Pariser [14] indicates in-/out-of-phase combination of configurations.

For D_{3h} molecular symmetry the predicted spectrum is simplified because the B_{2u} and B_{3u} representations of D_{2h} merge to the degenerate E' representation (Fig. 2). This is indeed born out by the spectra of 4 and 8 (Fig. 3). The rather low intensity of the symmetry-allowed transition to the E'^- state indicates that the basic assumption of orbital pairing is a useful approximation in our model despite the fact that sp- and sp^2-hybridized carbon atoms may be expected to have somewhat different basis energies. In fact, the intensity of the $^1E'^-$ transition of 4 appears to arise largely by vibrational borrowing from the intense $^1E'^+$ transition, since the dominant 2000 cm^{-1} progression starts from a vibronic origin at 30'100 cm^{-1}, and the much weaker 0-0 band at 29'000 is distinctly revealed only in low-temperature spectra. We note that the band system belonging to the $^1A_2'$ transition is red-shifted upon cooling of the solvent, while that of the $^1E'^-$ transition is not appreciably affected. This will prove to be a useful empirical criterion to locate the origins of the B_{2u}^- and B_{3u}^- states in the more complicated spectra of 5 to 7. The shifts observed upon threefold benzo-anellation (4 → 8) are readily interpreted in terms of our model: The first transition ($^1A_2'$) arises from excitation within the half-filled shell. Hence, it is blue-shifted by increasing perturbation of the perimeter. The $^1E'$ transitions are due to excitations from one shell to another, and are thus red-shifted with increasing size of the perimeter. The difference between molecules of D_{3h} and D_{2h} symmetry is clearly revealed by a comparison of the spectra of 3 [6], a planar [12]annulene of D_{2h} symmetry (ignoring alkyl substituents), and 4. With the aid of a PPP-calculation, we tentatively locate the origins of the five predicted transitions as follows: $^1B_{1g}$ (19'500), $^1B_{3u}^-$ and $^1B_{2u}^-$ (25'000 to 28'500 ?), $^1B_{2u}^+$ (33'000), and $^1B_{3u}^+$ (40'000 cm^{-1}).

The variation of the transition energies in compounds 5 to 7 is shown in Fig. 4. The symmetry assignments are again in accord with the predictions of the simple model (Fig. 2) and are based on the following evidence: (i) Assignment of the first two singlet excited states in 5 by Hochstrasser [2], (ii) similarities in oscillator strengths, vibrational structure and energy, (iii) red-shift of the $^1B_{1g}$ relative to the $^1B_{3u}^-$ band system upon cooling, and (iv) band polarizations determined from linear dichroism of stretched polyethylene sheets doped with compounds 5 to 7. Our finding for the polarization of the two lowest transitions in 5 agrees with the unambiguous result of [2] if the usual assumption [15] of preferred long-axis orientation along the direction of stretching is made. As expected, the orientation preference increases from 5 to 7. The weak transitions assigned as $^1B_{1g}$ and $^1B_{2u}^-$ are both found to be x-axis polarized in all three compounds. The expected y-polarized origins of the $^1B_{2u}^-$ transitions could not be detected, the assignment of these transitions is therefore insecure. Details of the polarization measurements will be reported elsewhere [13].

It is established that both the intersystem crossing [16] and the fluorescence [17] quantum yields of 5 are very small. Since the lifetime of the lowest excited singlet state is extre-

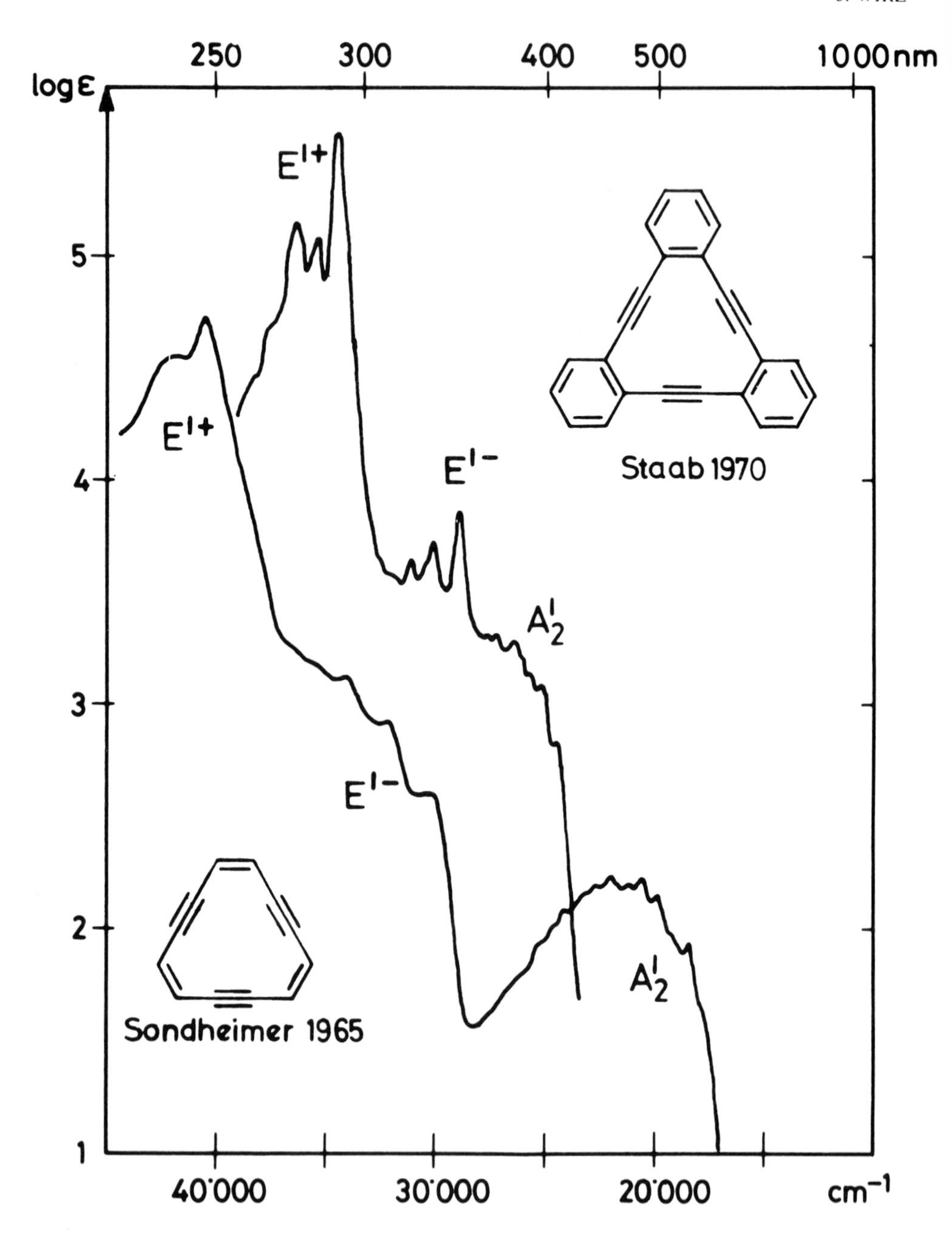

Fig. 3: Electronic absorption spectra of 4 and 8 redrawn from refs. [7] and [10].

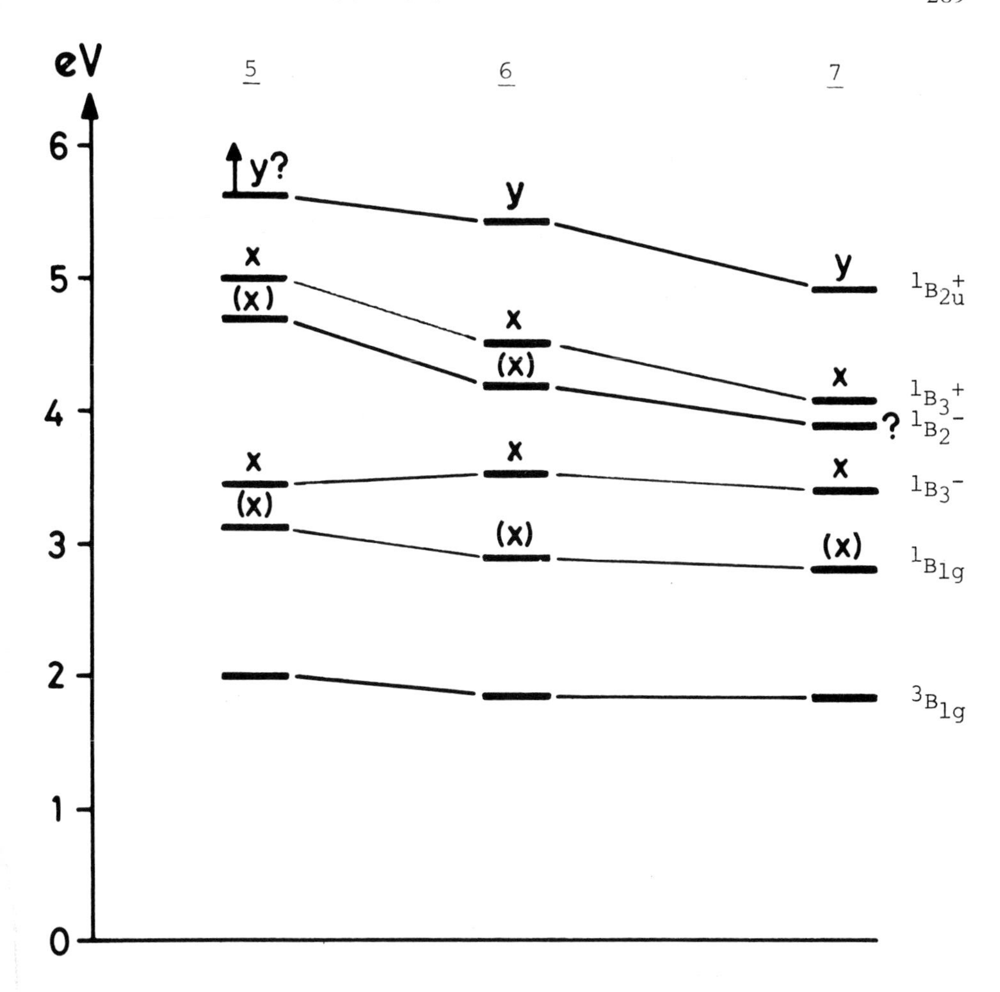

Fig. 4: Energy level diagram and symmetry assignments for electronically excited states of 5 to 7. The letters x and y indicate experimental polarizations of the corresponding absorption band systems. Triplet energy of 5 from [16], of 6 and 7 from energy transfer experiments (see text below).

mely short [18] and a photochemical conversion is not observed, it appears that radiationless conversion to the ground state is very rapid. Again this behaviour is in sharp contrast to that of the benzenoid hydrocarbons for which the predominant decay channels are fluorescence emission and intersystem crossing. We now report our results with the 4n-systems 4 and 6 to 8. We have not been able to detect any luminescent emission from solutions of 4. The fluorescence spectra of 6 to 8 are shown in Fig. 5, together with the spectrum of 5, reported recently by Shizuka et al. [17]. These authors assign the emission of 5 as fluorescence from the second excited singlet state. A number of arguments including "mirror-image" relation to the second electronic transition and long-axis fluorescence polarization were advanced to support this claim but in our opinion none of them is convincing. Though the lack of any mirror-image relation with the first electronic transition in 5 to 8 is striking, it is by no means surprising (see below). The similarity in the general appearance of the four spectra renders impurity emission unlikely. Two samples of 8 obtained by totally independent synthetic routes [10, 19] gave identical results. The emission of 8 has a lifetime of 41 ns in deaerated benzene solution and is quenched by one third in the presence of air. The fluorescence quantum yields differ by orders of magnitude. Preliminary estimates were obtained by comparing the intensity maxima with the intensity of a quinine sulfate standard at the same wavelength (Table 1).

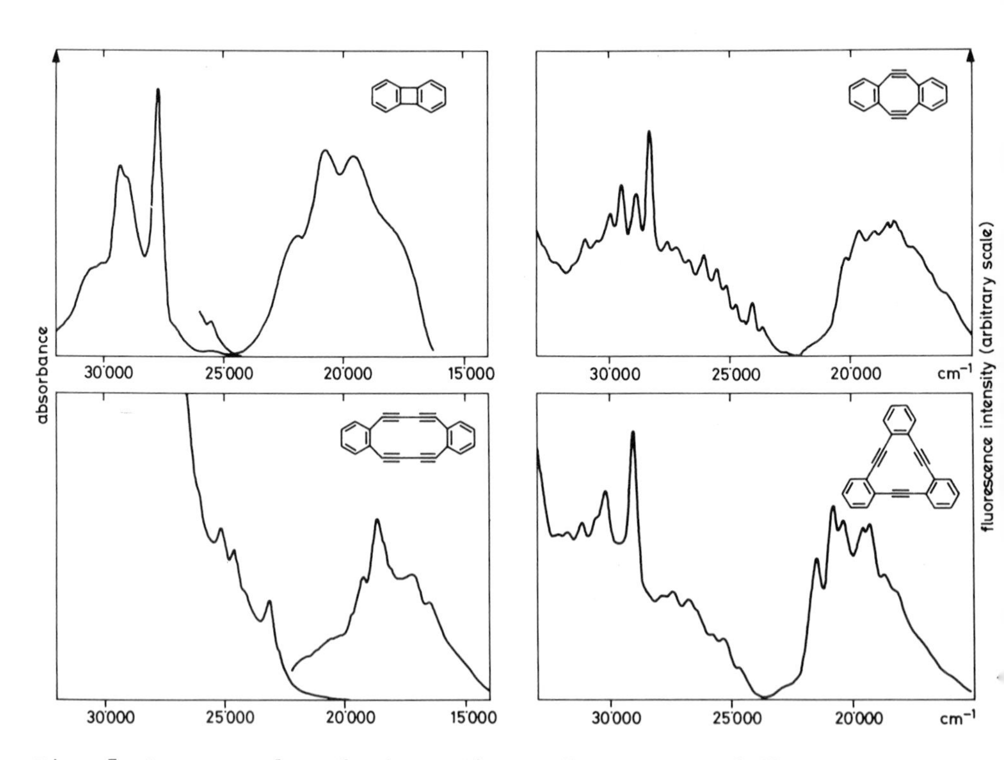

Fig. 5: Long-wavelength absorption and uncorrected fluorescence emission spectra of 5 to 8 in alkane solvents at room temperature. Spectrum of 5 redrawn from [17].

Table 1: Order-of-magnitude estimates for fluorescence and intersystem crossing quantum yields.

	4	5	6	7	8
ϕ_{Fl}	$< 10^{-4}$	$2 \cdot 10^{-4}$ [17]	10^{-3}	$5 \cdot 10^{-3}$	10^{-1}
ϕ_T	$< 10^{-2}$ (?)	$< 10^{-2}$ [16]	$< 10^{-2}$	$> 10^{-1}$	$> 10^{-1}$
$\phi_{Fl}+\phi_T$	$< 10^{-2}$ (?)	$< 10^{-2}$	$10^{-3} \div 10^{-2}$	$10^{-1} \div 1$	~ 1

None of the compounds 4 to 8 gave rise to a detectable long-lived phosphorescence emission on an Aminco-Bowman spectrometer which was equipped with a rotating can and an EMI 9558 QB photomultiplier tube. Therefore, we attempted to identify their lowest triplet state by conventional (μs) flash photolysis. Tetreau and coworkers [16] have observed the triplet-triplet absorption of 5 in the near UV (λ_{max} 340 nm; τ 100 μs in deaerated solution) by flash photolysis in the presence of naphthalene as a sensitizer, whereas direct excitation of 5 gave no transient absorption. With trisdehydro[12]annulene 4 we have not found any transient absorbtion in the range of 320 to 750 nm by either direct or sensitized excitation. The triplet absorptions of various sensitizers including tetracene (E_T 10250 cm^{-1}) were quenched by 4 with rates near 10^{10} M^{-1} s^{-1} while that of pentacene (E_T ca. 8000 cm^{-1}) was not affected. This allows to bracket the triplet energy of 4 in the range of 9000±1000 cm^{-1}. Energy transfer from triplet picene or fluorenone to pentacene is suppressed by the addition of 10^{-5} M 4 suggesting that the triplet lifetime of 4 in solution is less than ca. 10 μs. This will be verified by ns flash photolysis. The diffuse transient absorption spectra obtained with 6 (λ_{max} 555 nm; τ 60 μs; E_T 15000±500 cm^{-1}), 7 (λ_{max} 460 nm strong, λ_{min} 510 nm, λ_{max} 700 nm medium; τ 100 μs; E_T 15000±500 cm^{-1}), and 8 (λ_{max} 420 nm strong, λ_{min} 500 nm, λ_{max} 650 nm medium; τ ⩾ 500 μs; E_T 16800±300 cm^{-1}) were identified as triplet-triplet absorptions by direct observation of triplet energy transfer to and from these compounds in the presence of various benzenoid hydrocarbons with well-known [20] triplet spectra and energies. The triplet quantum yields given in Table 1 were estimated by comparing the intensities of the transient absorptions obtained after direct and sensitized excitation. Details will be reported elsewhere [13].

To summarize, the predominant pathway for the desactivation of 4 from its lowest excited singlet or triplet state appears to be radiationless conversion to the ground state, and this process gradually loses importance along the series 5 to 8. We interpret this behaviour as a consequence of the avoided surface crossing [21] which is predicted by MO theory to occur in [4n]annulenes at D_{4h} molecular symmetry. The case of the valence isomerization between two D_{2h} valence structures of singlet cyclobutadiene is well-known [22]. The "memory" for this avoided crossing fades away as the low-symmetry perturbations (transannular bonds, in-plane π-bonds) become more important and as the size of the system increases. The situation is qualitatively depicted for trisdehydro-[12]annulene 4 in Fig. 6. Note the change in shape and position of minimum energy of the ground state and lowest excited state hypersurface cross sections along a normal coordinate involving alternating compression and expansion of adjacent bonds. These factors are responsible for (i) the long vibrational progressions in the $S_1 \leftarrow S_0$ transitions (Figs. 3 and 5), (ii) the conspicuous absence of a mirror-image relationship between absorption and fluorescence (Fig. 5), and (iii) the large rate of radiationless desactivation in 4n-systems.

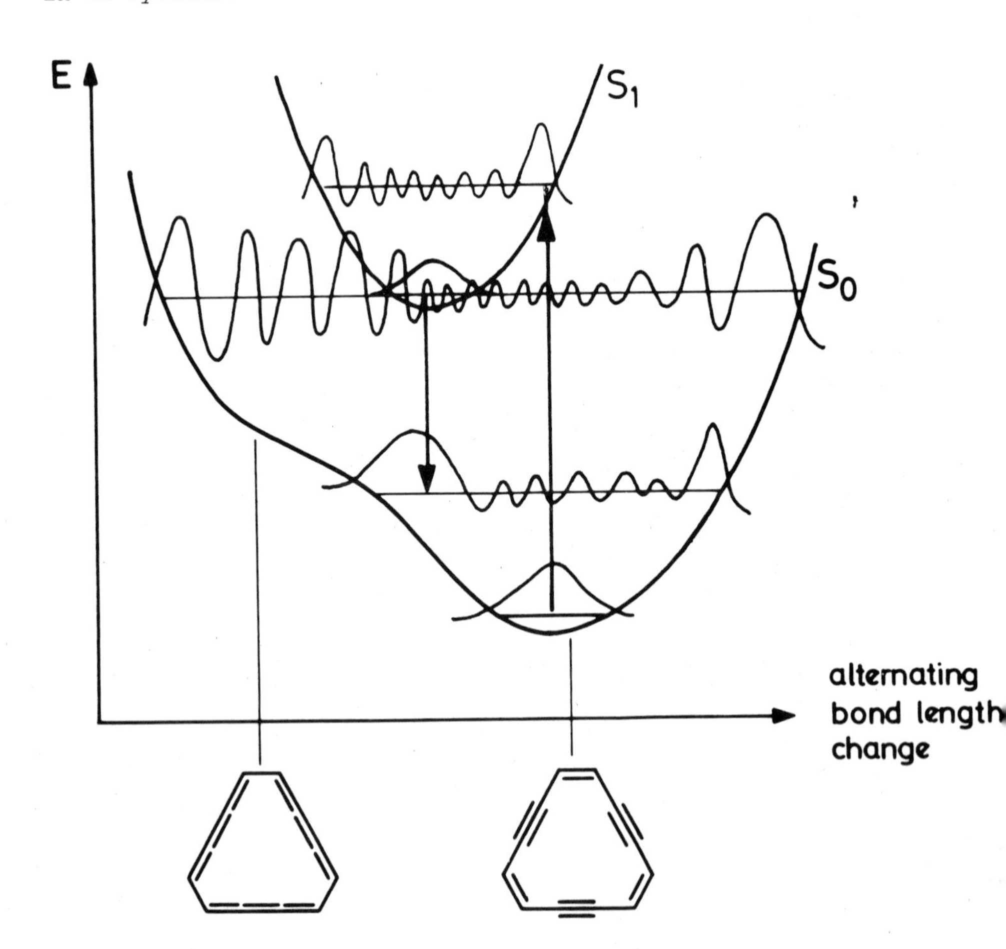

Fig. 6: Hypersurface cross section of a perturbed [4n]annulene.

This work is part of project No. 2.531.76 of the Swiss National Science Foundation. Financial support was obtained from Sandoz SA, Ciba-Geigy SA and Hoffmann-La Roche & Co. SA, Basel. The author is most grateful to Prof. F. Sondheimer, University College, London, and Prof. H.A. Staab, University of Heidelberg, for the generous gift of a sample of 6 and 8, respectively.

References:

[1] J. Wirz, Helv. Chim. Acta 59, 1647 (1976).
[2] R.M. Hochstrasser & R.D. McAlpine, J. Chem. Phys. 44, 3325 (1966) and references therein.
[3] N.S. Hush & J.R. Rowlands, Mol. Phys. 6, 317 (1963); F. Peradejordi, R. Domingo & J.I. Fernández-Alonso, Int. J. Quantum Chem. 3, 683 (1969); P. Francois, J. Chim. Phys. Fr. 67, 1063 (1970).
[4] J.R. Platt, J. Chem. Phys. 17, 484 (1949).
[5] E. Clar, Aromatische Kohlenwasserstoffe, Springer-Verlag (1952).
[6] G.M. Pilling & F. Sondheimer, J. Amer. Chem. Soc. 93, 1970 (1971).
[7] R. Wolovsky & F. Sondheimer, J. Amer. Chem. Soc. 87, 5720 (1965), "isomer B" corresponds to 4; see also UV-Atlas of Organic Compounds, Vol. V, Butterworths & Verlag Chemie (1971).
[8] H.N.C. Wong, P.J. Garratt & F. Sondheimer, J. Amer. Chem. Soc. 96, 5604 (1974).
[9] O.M. Behr, G. Eglinton, A.R. Galbraith & R.A. Raphael, J. Chem. Soc. 1960, 3614.
[10] H.A. Staab & F. Graf, Chem. Ber. 103, 1107 (1970).
[11] M.J.S. Dewar & H.C. Longuet-Higgins, Proc. Phys. Soc. (London) A67, 795 (1954).
[12] C.A. Coulson & G.S. Rushbrooke, Proc. Cambridge Phil. Soc. 36, 193 (1940); see e.g. L. Salem, Molecular Orbital Theory of Conjugated Systems, Benjamin (1966).
[13] E. Rommel & J. Wirz, Helv. Chim. Acta, in preparation.
[14] R. Pariser, J. Chem. Phys. 24, 250 (1956).
[15] J. Michl, E.W. Thulstrup & J.H. Eggers, Ber. Bunsenges. 78, 575 (1974).
[16] C. Tetreau, D. Lavalette, E.J. Land & F. Peradejordi, Chem. Phys. Letters 17, 245 (1972).
[17] H. Shizuka, T. Ogiwara, S. Cho & T. Morita, Chem. Phys. Letters 42, 311 (1976) and references therein.
[18] P.M. Rentzepis, Science 169, 239 (1970).
[19] J.D. Campbell, G. Eglinton, W. Henderson & R.A. Raphael, Chem. Commun. 1966, 87.

[20] G. Porter & M.W. Windsor, Proc. Roy. Soc. (London) A245, 238 (1958).

[21] L. Salem, C. Leforestier, G. Segal & R. Wetmore, J. Amer. Chem. Soc. 97, 479 (1975).

[22] R.J. Buenker & S.D. Peyerimhoff, J. Chem. Phys. 48, 354 (1968); G. Binsch, Jerusalem Symposia on Quantum Chemistry and Biochemistry, Vol. III, 25 (1971).

EXCITED STATES OF CHIRAL PYRAZINES

Günther Snatzke and György Hajós (1)

Lehrstuhl für Strukturchemie, Ruhruniversität
D 4630 Bochum 1 (FRG)

Absorption spectroscopy is one of the most used methods for investigation of the excited states of molecules. Although it is an extremely powerful tool for this purpose, it nevertheless has the disadvantage that weak bands - at least when recorded in solution - are in general not clearly discernible if they appear close to a strong band. A combination of this "isotropic" absorption measurements with that of the Circular Dichroism (CD) can often be of great help; especially the utilization of the so-called "g-factor" has been advocated by MASON (2) but this is still very seldom done. We would like to show here how this combination of UV- and CD-spectroscopy can be successfully used in assigning symmetry labels to four excited states of pyrazines. The preparative work, determination of the absolute configuration, and part of the CD-evidence has recently been published by us (3).

G-FACTOR

Two definitions for the g-factor have been given, $g = \Delta\varepsilon/\varepsilon$, which is a function of the wavelength, and $g = 4R/D$. In this latter equation R is the rotational strength, $R = F_0/4.\int\Delta\varepsilon/\lambda\ .d\lambda$, and D the dipole strength, $D = F_0\int\varepsilon/\lambda.d\lambda$, with the constant factor $F_0 = 91\cdot 8.10^{-40}$. These measurable quantities can be calculated (at least in principle) with the help of the equations

$$R = \mu.m.\cos(\vec{\mu},\vec{m}) \qquad \text{and} \qquad D = \mu^2$$

where μ and m are the electric and magnetic transition moment vectors, respectively, connected with the transition in question. From these definitions follows

$$g = (m/\mu).4.\cos(\vec{\mu},\vec{m})$$

and its order of magnitude can be estimated as follows. According

B. Pullman and N. Goldblum (eds.), Excited States in Organic Chemistry and Biochemistry, 295-302.

to MOSCOWITZ (4) chiral molecules belong to one of two classes,

i) this of molecules containing an inherently chiral chromophore, and

ii) that of molecules containing an inherently achiral chromophore which is, however, chirally perturbed by its invironment.

Only for the molecules of the first class $\vec{m}$ and $\vec{\mu}$ for a given transition (if present at all) can have components in the same direction. The term $4.\cos(\vec{\mu},\vec{m})$ is of the order of magnitude of 1, and the "natural" units for m and μ are the BOHR magneton (μ_B = $0{\cdot}927.10^{-20}$ emu) and the DEBYE ($\mu_D = 10^{-18}$ esu), respectively. Thus the g-factor for allowed transitions of inherently chiral chromophores will be of the order of magnitude of 10^{-2}.As an example may be cited 9(11)-dehydro boswellic acid (1), for which a g-factor of $0{\cdot}4.10^{-2}$ has been recorded (5), and which contains a nonplanar butadiene chromophore.

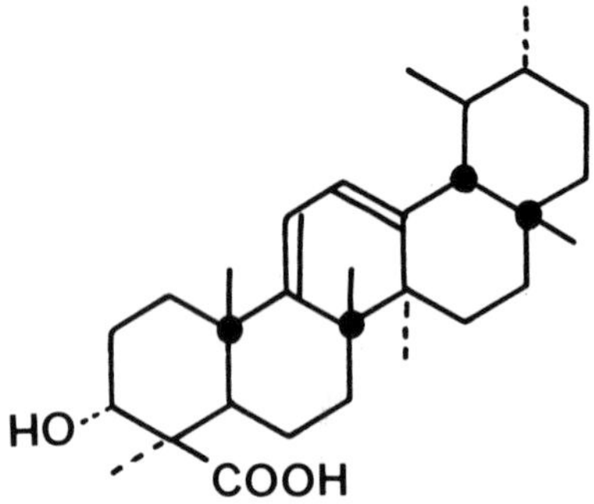

HO COOH

1

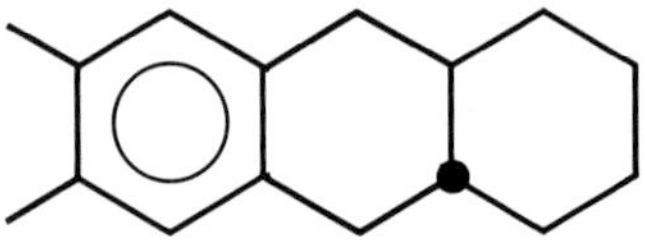

2

Transitions of molecules of the second class belong to one of 4 cases:

1) both moments vanish for symmetry reasons in first order; small values for μ and m are "stolen" from other transitions by vibronic coupling, and the magnitude of g can, therefore, not be predicted. Examples are the g-values for the B_{2u}- and B_{1u} -bands of benzene derivative 2 (D_{6h}-symmetry of the chromophore), which both were found to be $3{\cdot}4.10^{-4}$ (6).

2) Only the magnetic moment vanishes; in this case the numerator is very small, g is thus much smaller than 10^{-2}. As an example take the g-value for the B_{1u}-band of the pyrazine 3 (local D_{2h}-symmetry of the chromophore) which is $2{\cdot}5.10^{-4}$.

3) The transition is magnetically allowed but electrically forbidden; now the denominator is much smaller and the g-factor should be bigger than 10^{-2}. In agreement with these estimations $g = 2{\cdot}9.10^{-2}$ for "equatorial" methyl adamantanone 4 (7), and $g = 3{\cdot}0.10^{-1}$ for a 16-keto steroid (partial formula 5) (2)(local C_{2v}-symmetry for both).

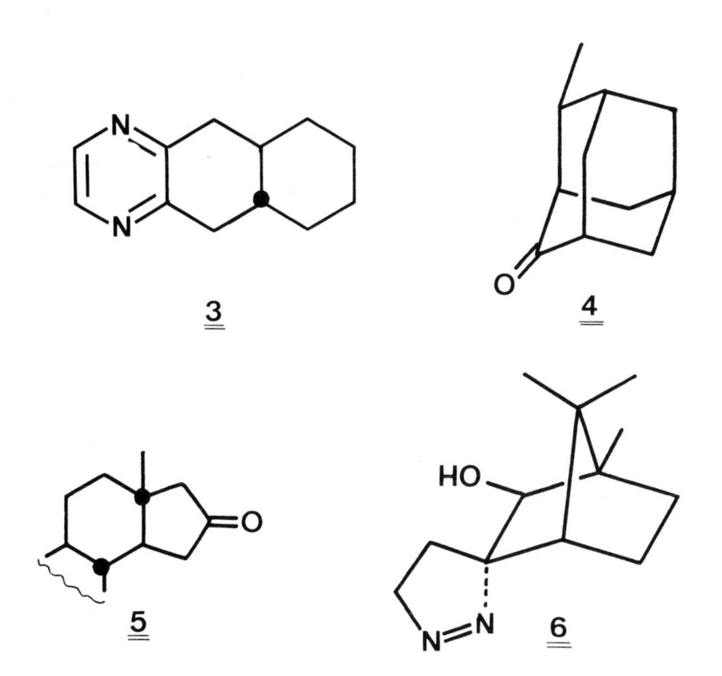

4) g is not well predictable in case that both m and μ are ≠ 0 (but then perpendicular one to the other), because the corresponding other moment in the direction of the allowed one is again obtained only by vibronic coupling. The $n \rightarrow \pi^*$ band around 330 nm of the pyrazoline derivative 6 of isoborneol may serve as an example (8): its g-factor has been found to be $2 \cdot 3.10^{-3}$ (local C_{2v}-symmetry). CD is thus especially valuable for the detection of bands corresponding to magnetically allowed but electrically forbidden transitions.

EXCITED STATES OF 1,4-PYRAZINES

1,4-Pyrazine comprises a strongly perturbed benzenoid chromophore which is thus treated better as belonging to point group D_{2h} than to D_{6h}. In addition to the 6π-system nonbonding electrons are present at the two nitrogen atoms which are coupled through bonds. Several calculations have been published for this system (cf. e.g. 9-11) and the energies of the occupied orbitals have been determined by photoelectron spectroscopy (cf.e.g. 12). The most reliable assignment (12) for the four uppermost MOs is

$$\pi_2\ (b_{2g}),\quad n^-\ (b_{1u}),\quad \pi_3\ (b_{1g})\quad \text{and}\quad n^+\ (a_g)$$

(rising orbital energy), the lowest virtual MOs are estimated to be

$$\pi_4^*\ (b_{3u})\quad \text{and}\quad \pi_5^*\ (a_u)$$

TABLE 1.

Calculated Transition Energies (11) for 1,4-pyrazine and Components of $\vec{\mu}$ and $\vec{m}$.

State	Transition	Energy (eV)	Components of $\vec{\mu}$	$\vec{m}$	UV-bands assigned (eV)	CD-bands assigned (eV)
$^1B_{3u}$	$n^+\rightarrow\pi_4^*$	3·91	x	-	3·83	3·79
^1A_u	$n^+\rightarrow\pi_5^*$	4·90	-	-		
$^1B_{2g}$	$n^-\rightarrow\pi_4^*$	5·12	-	y		5·46
$^1B_{2u}$	$\pi_3\rightarrow\pi_4^*$	5·26	y	-	4·69	4·47
$^1B_{1g}$	$n^-\rightarrow\pi_5^*$	6·41	-	z		
$^1B_{2g}$	$n^+\rightarrow\pi_6^*$	8·09	-	y		
^1A_g	$\pi_1\rightarrow\pi_4^*$	8·85	-	-		
$^1B_{3g}$	$\pi_1\rightarrow\pi_5^*$	8·90	-	x		
$^1B_{1u}$	$\pi_2\rightarrow\pi_4^*$	9·31	z	-	6·30	5·89
$^1B_{3u}$	$n^-\rightarrow\pi_6^*$	9·34	x	-		
$^1B_{2u}$	$\pi_2\rightarrow\pi_5^*$	10·36	y	-	7·6..7·7	

Transition energies have also been calculated by two groups (10,11) and the results of the more recent paper (11) are presented in TABLE 1 together with the components of $\vec{\mu}$ and $\vec{m}$ and the assigned UV-bands (13,14). For the $n\rightarrow\pi^*$ -transitions the agreement between calculated and found values is reasonably good, whereas $\pi\rightarrow\pi^*$ -transition energies inevitably are too high in these calculations. The absorption and emission spectra of 1,4-pyrazine have been investigated very thoroughly (cf. the references given in (11)) and all workers agree about the assignment of the electrically allowed transitions (cf. TABLE 1). Evidence for the $A_g\rightarrow B_{2g}$ ($n^-\rightarrow\pi_4^*$) transition on the other hand is controversial, and no unambiguous experimental identification has hitherto appeared. Our CD-spectra of 3 allow, however, localization of the corresponding band and assignment of the symmetry label from consideration of the g-factors.

CD-SPECTRUM OF 3

In isooctane solution 3 shows a negative band system "I" with pronounced fine structure around 320 nm (cf.FIG.1), another positive one, also with fine structure (II), a single band at 227 nm (III, $\Delta\varepsilon$=+2·58), and another positive CD at 211 nm (IV,$\Delta\varepsilon$=+2·19). The maxima for bands I,II, and IV coincide for the UV- and the CD-spectrum (both taken in isooctane), and by comparison with the UV-spectrum of 1,4-pyrazine can be assigned the singlet-singlet B_{3u}, B_{2u}, and B_{1u}-transitions, respectively. Similar positions of the CD- bands have also been found for the few other optically

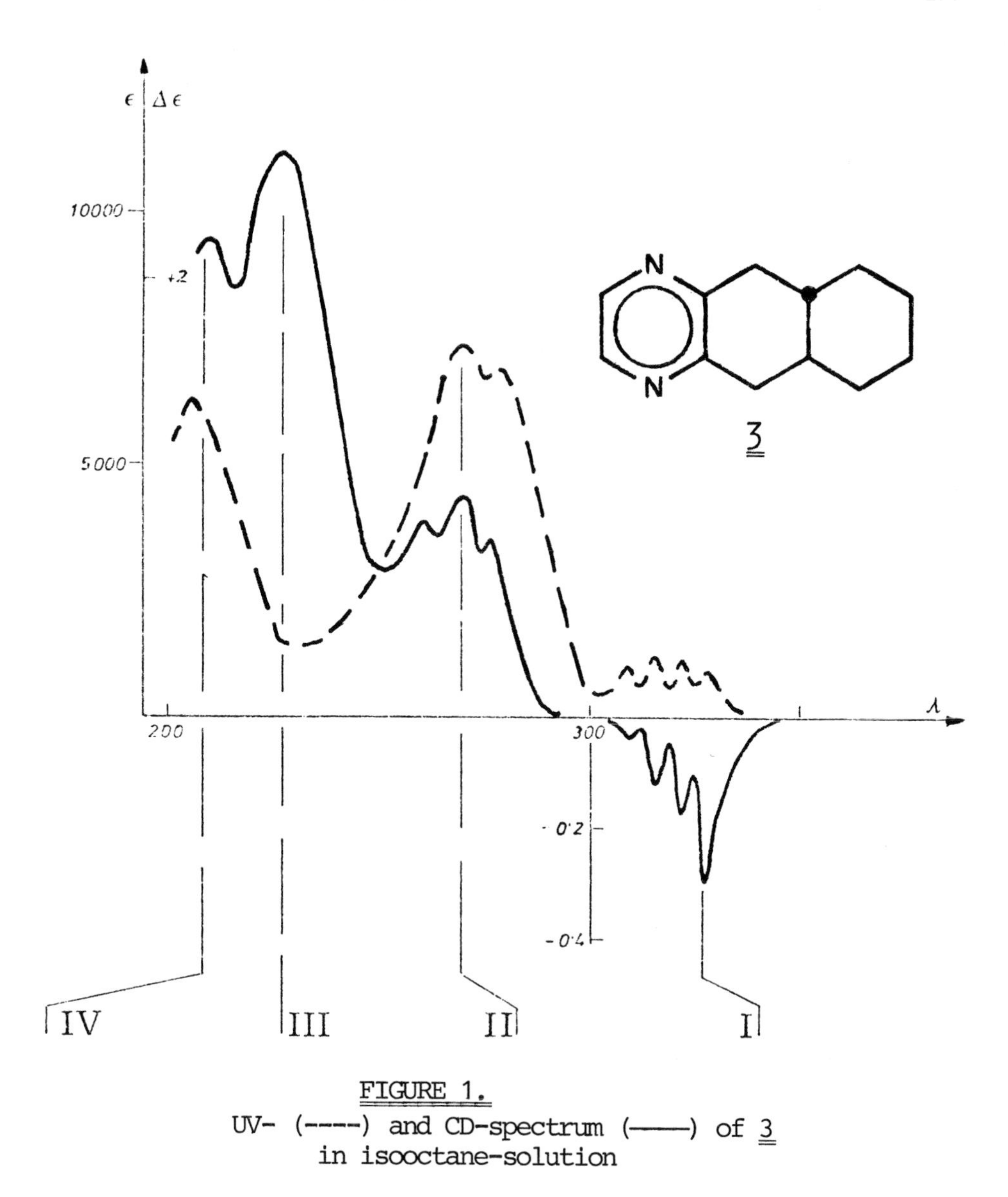

FIGURE 1.
UV- (----) and CD-spectrum (——) of 3 in isooctane-solution

active 1,4-pyrazine derivatives hitherto measured (15,16), and the changes of the CD-spectrum with change of solvent polarity or after mono- and diprotonation are analogous, too. The g-factors for these three bands are $3{\cdot}3.10^{-4}$ (I), $1{\cdot}4.10^{-4}$ (II), and $2{\cdot}2.10^{-4}$ (IV), and these are in accord with the assignments given (all three transitions electrically allowed but magnetically forbidden). The strongest band (III) in the CD-spectrum corresponds, however, not to a maximum but to a minimum in the UV-curve, its g-factor being at least one order of magnitude larger than the others ($1{\cdot}7.10^{-3}$; the actual value must even be larger as some of the UV-absorption around 227 nm is due to tailing of the strong bands II and IV). This then indicates an assignment to one of the transitions being magnetically allowed but electrically forbidden. By comparison with TABLE 1 this band around 227 nm must then be due to the $^1A_g \rightarrow {}^1B_{2g}$ ($n^- \rightarrow \pi_4{}^*$) excitation. The CD is too big to be connected with a singlet-triplet transision or a RYDBERG band. The usefulness of the combination of UV- with CD-spectroscopy for the determination of the parentage of excited states is thus again well demonstrated.

GENERALIZED SECTOR RULES

It is interesting to compare the obtained CD-spectra for 3 with those for the carbocyclic analogue 2 and for ketones. If we use the following coordinate system (only the "chiral" substitu-

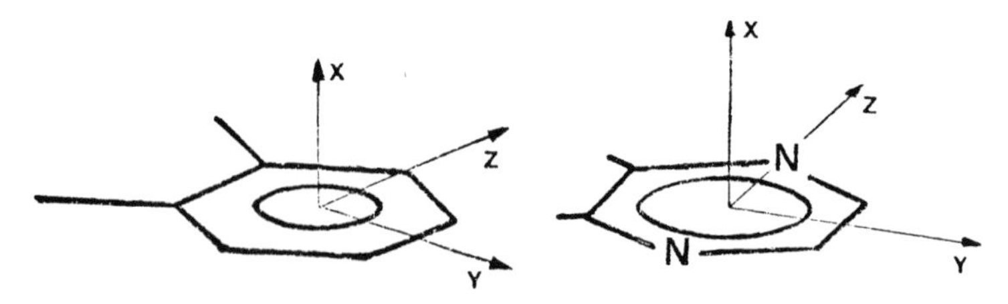

tion is shown) then for the first two $\pi \rightarrow \pi^*$ -bands the labels are the same under D_{6h}- and D_{2h}-symmetry:

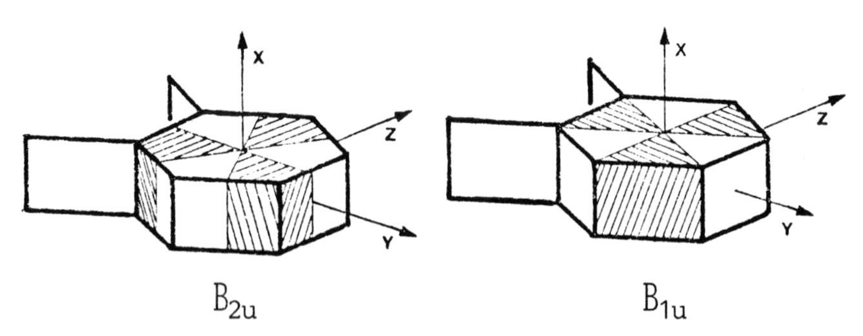

B_{2u} B_{1u}

It was found experimentally that for the same absolute configuration of both compounds 2 and 3 the B_{2u}- and B_{1u}-CD-bands have the same signs (6,17) which suggests the following generalization of sector rules:

<u>For two molecules of same (local) chromophore symmetry identical symmetry-determined sector rules hold, if the transitions in question have the same parentage.</u>

This generalization can also be applied to the $n^- \rightarrow \pi_4{}^*$ -CD-band of 3 . Under the local symmetry of a disubstituted 1,4-pyrazine (C_{2v}) the $n^- \rightarrow \pi_4{}^*$ -excitation is of $^1A_1 \rightarrow {}^1A_2$ ($b_1 \dot{\times} b_2$) type, the same as for the $n \rightarrow \pi^*$ - transition of a ketone. The signs for the respective bands of 3 and a cyclohexanone in the twist-form of same absolute conformation ¶ for the chiral rings are again identical (cf. 19). The same treatment further applies to the 240 nm band of chiral cyclic thio ethers, which also has local $^1A_1 \rightarrow ^1A_2$ ($b_1 \dot{x} b_2$) - symmetry (cf. 20).

ACKNOWLEDGEMENT

I would like to thank Mr. U. Wagner for careful measurements, the Deutsche Forschungsgemeinschaft and the Fonds der Chemischen Industrie for financial support.

REFERENCES and NOTES

(1) Fellow of the A.-v.-Humboldt-Stiftung 1974/5 at Bochum. Permanent address: Central Chemical Research Institutes of the Hungarian Academy of Sciences, Budapest.

(2) S.F.Mason, Quarterly Rev. **17**, 20 (1963).

(3) G.Snatzke and Gy. Hajós, Heterocycles **5**, 299 (1976).

(4) A. Moscowitz, Tetrahedron **13**, 48 (1961).

(5) G.Snatzke and L.Vértesy, Monatsh.f.Chem. **98**, 121 (1967).

(6) Badruddin, Thesis, Bochum (1975).

(7) G.Snatzke and G.Eckhardt, Tetrahedron **24**, 4543 (1968).

(8) G.Snatzke and J.Himmelreich, Tetrahedron **23**, 4337 (1967).

(9) J.D.Petke,J.L.Whitten, and J.A.Ryan, J.Chem.Phys. **48**, 953 (1968).

(10) M.Hackmeyer and J.L.Whitten, J.Chem.Phys. **54**, 3739 (1971).

(11) W.R.Wadt and W.A.Goddard III, J.Amer.Chem.Soc. **97**, 2034 (1975).

(12) R.Gleiter, E.Heilbronner, and V.Hornung, Helv.Chim.Acta **55**, 255 (1972).

¶ It should be kept in mind that actually the CD is determined by the absolute conformation of a molecule, i.e. only indirectly by its absolute configuration!

(13) K.K.Innes, J.P.Byrne, and I.G.Ross, J.Mol. Spect. **22**, 125 (1967).

(14) M.N.Pisanias, L.G.Christophorou , J.G.Carter, and D.L.Mc Corkle, J.Chem.Phys. **58**, 2110 (1973).

(15) H. Smith and A.A.Hicks, Chem.Commun. 1112 (1970).

(16) H.Rau, O.Schuster, and A.Bacher, J.Amer.Chem.Soc. **96**, 3955 (1974).

(17) G.Snatzke and P.C.Ho, Tetrahedron **27**, 3645 (1971).

(18) J.A.Schellman, J.Chem.Phys. **44**, 55 (1966).

(19) C.Djerassi and W.Klyne, Proc.Natl.Acad.Sci.U.S.A. **48**, 1093 (1962).

(20) G.Snatzke, Angew.Chemie, in preparation.

WAVELENGTH DEPENDENCE OF Q(F) AND $Q(e^-_{aq})$ OF SOME AROMATIC AMINES IN AQUEOUS SOLUTION

G. Köhler, Claudia Rosicky and N. Getoff

Institut für Theoretische Chemie und Strahlenchemie,
Universität Wien, A-1o9o Wien, Währingerstraße 38, Austria

1. INTRODUCTION

The photophysical and photochemical processes of highly excited molecules (e.g. in S_2 state) are relatively ill known in spite of their basic importance for the photodegradation of enzymes and other substances of biological interest.

For a number of benzene (1-9) and indole derivatives /1o-14/ it was recently established that the fluorescence quantum yield Q(F) is not independent of the excitation energy (λ_{exc}) (deviation from Wawilow's law /15/). In general, the Q(F) values for such systems are constant within the S_1 state, but decrease in the overlap region between S_1 and S_2. They remain constant in the S_2 state, but at higher energies a further decrease was observed for some systems /1,3,10/. This effect is very strong in hydrocarbons and less marked in polar, strongly associating solvents, like alcohols and ethers, with the exception of water /6,7,11/. For phenol and methylated phenols in apolar solvents a good correlation between $Q(H_2)$ and the decrease of Q(F) with λ_{exc} was established /7/. In alcoholic solutions, however, the formation of H_2 is suppressed and the decrease of Q(F) is very small /16/. In aqueous solutions of phenol excited in S_1 solvated electrons (e^-_{aq}) are formed /2,17/. Excitation in S_2 leads to an increase of $Q(e^-_{aq})$ and $Q(H_2)$, but the sum of both accounts only for half of the decrease of Q(F) /2,17/. For monophenylphosphate /19/ and indole /12/ in aqueous solutions enhanced electron formation from S_2 was as well held responsible for the decrease of Q(F).

Electron photoejection was found for a number of organic /8,12,18-29/ and inorganic substances /30-32/. For organic systems in aqueous solutions it was recently established that under steady state conditions(low light intensity) only monophotonic processes lead to electron formation /33/. Applying Cs^+ ions as heavy atom quenchers it was proved for aqueous phenol /17/ and monophenylphosphate /19/ that the electron ejection occurs from the singlet state. Similar observations were previously made for ß-naphtholate /27/.

B. Pullman and N. Goldblum (eds.), Excited States in Organic Chemistry and Biochemistry, 303-311.

It is well known that the formation of e^-_{aq} from aromatic amino acids is involved in the inactivation of enzymes by u.v.light /35/. It is also of fundamental importance to establish whether electron release from higher excited states can efficiently compete with the internal conversion to the lowest excited state. The aromatic amines seem to be most suitable for this kind of investigations, because their low ionization potential in the gas phase should favour photochemical electron formation. Aniline, benzylamine and phenylethylamine were investigated in this respect. Phenylethylamine is considered as a model for phenylalanine and also tyrosine, which are chromophors in enzymes. For comparison, toluene was also studied under equal conditions.

2. EXPERIMENTAL

2.1. Solutions

All chemicals were of the purest grade available (E.Merck, Darmstadt). Before use the amines were further purified by distillation under low pressure under argon and a purity control was performed by gas chromatography. Traces of O_2 in N_2O (Oxygena A.G., Vienna) were removed by bubbling through a pyrogallole-KOH solution. Four times destilled water was used for preparation of the solutions.

2.2. Fluorescence measurements

The setup for fluorescence measurements, the corrections, and the evaluation of the data have been previously described /6,7/. The excitation spectra were normalized to the fluorescence efficiency for the 0-0-transition, Qo(F), and were expressed as $\beta(\lambda exc)$:

$$\beta(\lambda exc) = Q_F(\lambda exc)/Qo(F) \qquad (2.1.)$$

The overall error was less than 5 %.

2.3. Actinometry

The amine solutions were irradiated at steady state conditions with monochromatic u.v.light at various wavelengths, namely: 276,7; 253,7; 228,8 and 213,9 nm. The light output from a thallium lamp (Philips) at 276,7 nm ((100 %) together with that at 291,7 nm (12,7 %) was determined by a differential actinometric method using the ferrioxalate actinometer /34/, taking $Q(Fe^{2+})$ = 1,245. A solution of 0,3 % V/V benzene in chloroform (d = 1 cm) served as a filter to cut off all light with λ= 265 nm. The light intensity at 253,7 nm (low pressure Hg lamp, Osram HNS 10 ofr, with incorporated filter for cutting off 184,9 nm line) was determined by means of the monochloracetic acid actinometer, $Q(Cl^-)$ = 0,36 /34,35/. The actinometry at 228,8 nm (cadmium lamp, Philips) and at 213,9 nm (zinc lamp, Philips) was carried out employing 1 M $ClCH_2COOH$ at 30°C with $Q(Cl^-)$ = 0,55 ± 0,05 for both lines. Appropriate filter solutions /34/ were used. Details concerning the calibration of the chloracetic acid at these two wavelengths will be published elsewhere.

2.4. Determination of $Q(e^-_{aq})$ and $Q(H_2)$

The aqueous amine solutions of 5.10^{-4} - 10^{-2} M amine (aniline at pH = 2 and 10; benzyl- and phenylethylamine at pH = 13) were saturated with N_2O as a specific scavenger for e^-_{aq} and irradiated with various u.v. doses at 30° C, where $k(e^-_{aq} + N_2O) = 5,6.10^9 M^{-1}s^{-1}$ /36/. The rate constants for the unprotonated amines, determined by pulse radiolysis, are: $k(e^-_{aq} + C_6H_5NH_2) = 3.10^7 M^{-1}s^{-1}$/37,38/, $k(e^-_{aq} + C_6H_5CH_2NH_2) = 1,8.10^8 M^{-1}s^{-1}$ /37/ and $k(e^-_{aq} + C_6H_5CH_2CH_2NH_2) = 3,3.10^8 M^{-1}s^{-1}$ /37/. The H atoms were scavenged by 1 M 2-propanol, $k(H + C_3H_7OH) = 5.10^7 M^{-1}s^{-1}$/39/. In acid solutions appropriate corrections were applied for the reaction of H with aniline, $k(H + C_6H_5NH_2) = 1,3.10^9 M^{-1}s^{-1}$ /38/. The resulting OH radicals are scavenged by 2-propanol and the amine. The nitrogen and hydrogen yields were measured by a standard vacuum line in combination with gas chromatography.

3. RESULTS

3.1. ß-values

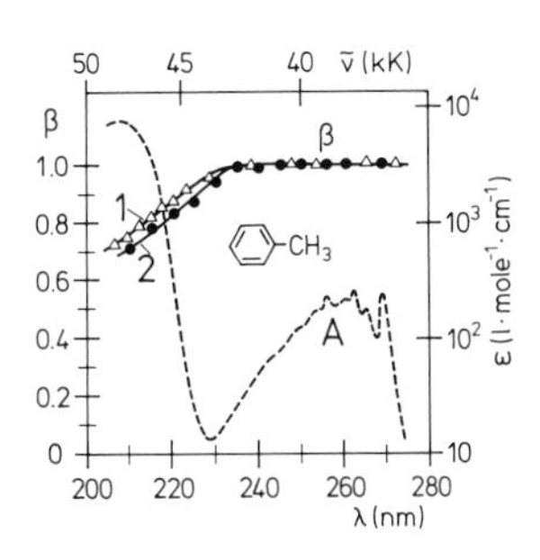

Fig. 1: ß-values of 0,2 M toluene as a function of λ_{exc} in methanol (1) and cyclohexane (2). (A) absorption spectrum of toluene in cyclohexane.

The wavelength dependence of the ß-values of toluene in methanol and cyclohexane is very similar (fig. 1). A strong decrease is observed at λ< 230 nm (at λ_{exc} = 213,9 nm ß = 0,82 in methanol and 0,78 in cyclohexane). A similar wavelength dependence of the ß-values was established for all three investigated amines. The unprotonated form of aniline shows a second decrease, at $\lambda \leq$ 205 nm (ß = 0,54). The absorption spectrum of aniline in acid solution is blue shifted ($\lambda \simeq$ 250 nm) compared to the basic solution and this is also reflected in the behaviour of the ß-values. At λ_{exc} = 213,9 nm the ß-values are: for protonated aniline 0,80, for the unprotonated form 0,70 and in cyclohexane 0,48. Aqueous benzylamine at pH = 11,8 (100 % unprotonated) shows likewise two steps of ß-decrease (ß = 0,84 at $\lambda \hat{=}$ 213,9 nm, fig. 3). The ß values of phenylethylamine in methanol, cyclohexane and the unprotonated form in aqueous solution are presented in fig. 4 as a function of wavelength. Here again the strongest ß-decrease is observed in cyclohexane. The ß-values for the investigated substrates at 213,9 nm are given in table 1.

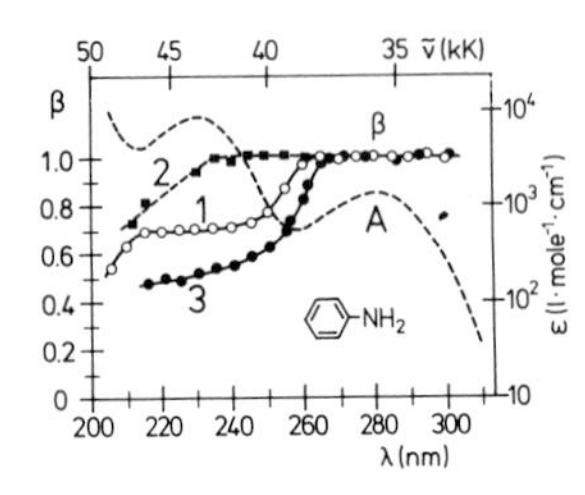

Fig. 2: ß-values of 10^{-5} M aniline as a function of λ_{exc} in water at pH = 7,5 ((1), at pH = 1,8 (2), and in cyclohexane (3).
(A) absorption spectrum in water (pH = 8).

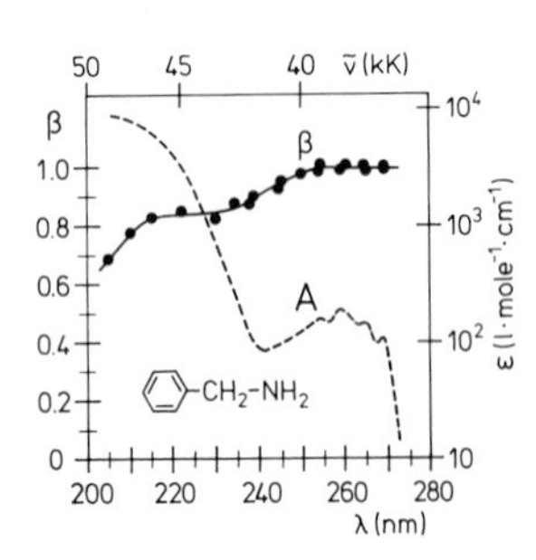

Fig. 3: ß-values of 3.10^{-5} M benzylamine in aqueous solution (pH = 11,8) as a function of λ_{exc}.
(A) absorption spectrum in water (pH = 12).

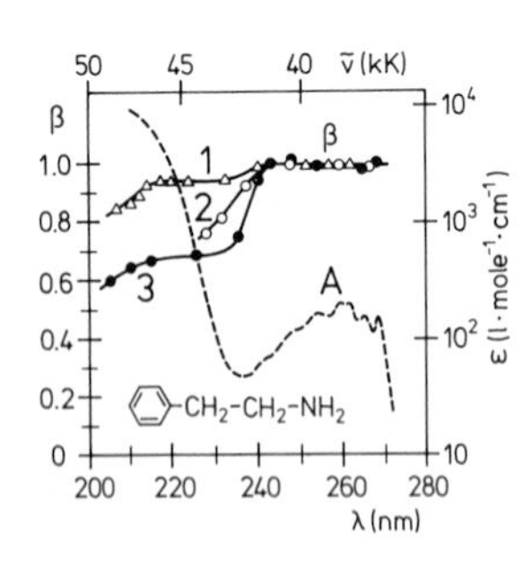

Fig. 4: ß-values of 10^{-5} M phenylethylamine as a function of λ_{exc} in methanol (1), water at pH = 12,5 (2), and cyclohexane (3).
(A) absorption spectrum in cyclohexane.

3.2. Quantum yields of e^-_{aq} and H

The $Q(e^-_{aq})$ values for the three amines at various wavelengths are summarized in table 1. They represent mean values for at least 5 measurements at each applied u.v.dose. $Q(H_2)$ of aniline at pH = 1 was also determined.

Solute	Q(e^-_{aq}) at λ (nm)				β at
	276,7	253,7	228,8	213,9	213,9
Aniline (prot.)	—	*) 0,14 /20/		*) 0,09	0,8
Aniline (unprot.)	0,06	0,27 /20/	0,28	0,32	0,7
Benzylamine	—	0,01	0,15	0,18	0,84
Phenylethylamine	—	0,02	0,22	0,27	**) 0,74

Table 1: Q(e^-_{aq}) for aniline, benzylamine and phenylethylamine in aqueous solution (text) at various wavelengths (total error ± 15 %)
*) Q(H_2), **) at 228,8 nm

4. DISCUSSION

The results given in figs. 2 to 4 show that the fluorescence quantum efficiency of the investigated aromatic amines aniline, benzylamine and phenylethylamine is not independent of the excitation energy in aqueous solutions. This is an exception to "Wawilow's law".

The excitation spectra are expressed in terms of β which is usually interpreted to be a measure for the overall efficency for the conversion of the primarily excited state, Sn, to the emitting levels of the S_1 state. The determined values are constant within the S_1 state but show a marked decrease in the region corresponding to the overlap between the first and second excited state. In the S_2 state β is again constant but decreases further below 215 nm for most of the considered systems. This behaviour is best demonstrated for aniline and benzylamine in basic medium (figs. 2 and 3).

The absorption spectrum of aniline in acidic solution (pH 1) is blue shifted compared to the basic solutions and interactions of the protonated amino group with the aromatic nucleus are very small /41/. The photophysical properties of protonated aniline can be compared with those of toluene to some extent and this is also reflected in the β-dependence. In cyclohexane solutions of aniline there is an even stronger decrease of β by excitation in the S_2 state than in aqueous solutions and this is in good agreement with the result for phenol /6/ and indole /11/.

The separation of the amino group from the ring by a -CH_2- group results in a smoothing of the β-curve (benzylamine fig. 3) but a longer side chain (phenylethylamine) again increases the energy effect on β. Because of the weak fluorescence of phenylethylamine in aqueous solutions, no measurements could be performed below $\lambda = 225$ nm, but for comparison the ß-values are also given for solutions in cyclohexane and methanol (fig. 3). It is interesting to point out that also in this compound a marked influence of the solvent occurs. This observation is somewhat unexpected in respect to the large separation of the amino group from the π-electron structure and the small solvent influence on the ß-values of toluene (fig. 1). The occurrence of a direct interaction of the amino group and the ring enabled by a favourable spatial configuration of the molecule is therefore suggested.

In order to explain the decrease of ß with λ_{exc} the wavelength dependence of electron ejection was studied for these aromatic amines in aqueous solutions. The values of $Q(N_2)$, which were taken to be equal to $Q(e^-_{aq})$, generally increase with excitation energy for the three amines in basic solution and correlate well with the decrease of Q(F). In this case no considerable hydrogen formation has been observed.

A reaction of N_2O with the excited molecules leading to formation of N_2 cannot be excluded, at least for excitation in S_1. It was shown by flash photolysis that this process is negligible for the S_2 state of aniline /20/. Since the light intensity was low ($I_o = 2.10^{-6}$ Einstein $l^{-1}s^{-1}$) a biphotonic electron ejection can be excluded /33/.

The onset of the decrease of ß for aniline at 265 nm is equivalent to an energy approximately 3 eV below the gas phase ionization potential (IP = 7,7 eV, /42/). The energy corresponding to this gap must be provided by the solvation energies of both charged entities formed in the electron ejection process. Hence, it must be concluded that the electron is formed directly in a solvated state. Therefore it is considered that high dipole moment changes, following excitation, polarize the molecular surroundings, and facilitate thereby the transfer of the electron to the solvent with a rate constant in the order of 10^{11} s^{-1}. This might also hold for the S_1 state, but with a much lower rate. In analogy to phenol /17/ it is assumed that electron ejection following excitation in S_1 of aniline occurs from the singlet manifold.

The photolysis of acidic solutions of aniline leads to formation of H_2 either by H cleavage from $-NH_3^+$ and subsequent reaction with 2-propanol or by electron ejection followed by fast reaction with H^+ to form H. But for protoanted aniline none of the two pathways can explain the λ-dependence of ß inasmuch as $Q(H_2)$ shows the same decrease than Q(F) (table 1). It should be mentioned that fast cage recombination under formation of e.g. ground state molecules could possibly lower the yields of the primary processes determined by scavenging methods. This might provide an explanation for the observed effect in the case of protonated aniline. Apart from this, it is more likely that here another process is dominant, namely the same which leads to the lowering of ß for toluene (e.g. formation of short lived isomers). It is to be noted that analysis of possibly formed stable isomers or of ammonia was not yet performed.

The good correlation between the wavelength dependence of electron ejection and fluorescence found for phenylethylamine in aqueous solution is likewise unexpected as the strong solvent influence on the ß-values (fig. 4). Intramolecular interaction between the aromatic ring and the amino group resulting from a favourable spatial configuration of the molecule were suggested above. The dipole moment associated with such an interaction might change by excitation and it appears that this could possibly facilitate the electron transfer by the polarization of the solvation shell.

Very similar interactions between the side chain and the ring were also suggested for the explanation of the increase of biphotonic electron ejection from aromatic carboxylic acids with the length of side chain /43/. These observations for phenylethylamine are of special interest for the explanation of the observed deviation of the ß-values from unity for the three aromatic amino acids, especially phenylalanine and tyrosine /10/. Hence it appears that for explanation of the photophysical properties of the aromatic amino acids not only the behaviour of the aromatic nucleus, e.g. toluene for phenylalanine, is to be held responsible, but it can also be influenced by interactions with the amino acid moiety of the molecule.

Summarizing the results, it can be concluded that electron ejection from suitable molecules like amines and alcohols in aqueous solutions is so rapid that it can compete favourably with internal conversion. In hydrocarbon solvents it was found for phenol that H formation might explain the ß-decrease and it is suggested that this applies also for the other substances. Both processes should be suppressed in solvents like alcohol and ether were no wavelength dependence of ß was observed. It appears that the solvent microstructure and the solute-solvent interactions determine the rate for the various processes possibly competing with the internal conversion to the fluorescing levels of the S_1 state. Very similar conclusions were also drawn to explain the solvent influence on the ß-values for various indoles from fluorescence measurements /44/.

The authors thank the Fonds zur Förderung der wissenschaftlichen Forschung in Austria for financial support.

REFERENCES

/1/ R.B.Cundall, S.McD.Ogilvie, p.33 in "Organic Molecular Photophysics", Vol.2, ed.J.B.Birks, J.Wiley & Sons, New York (1975)

/2/ J.Zechner, G.Köhler, G.Grabner and N.Getoff, Abstr. 1. Tagung d. Fachgruppe Photochemie, GDCh, Göttingen (1973)

/3/ C.Kaler and N.Getoff, Österr.Akad.Wiss., math-naturw.Kl., Sitzungsber. 183, 157 (1974)

/4/ G.Köhler and N.Getoff, Österr. Akad.Wiss., math-naturw.Kl., Sitzungsber. 183, 167 (1974)

/5/ G.Köhler and N.Getoff, Abstr. 2.Tagung d.Fachgruppe Photochemie, GDCh, Konstanz (1974)
/6/ G.Köhler and N.Getoff, Chem.Phys.Lett. 26, 525 (1974)
/7/ G.Köhler and N.Getoff, J.Chem.Soc., Faraday I, 72, 2101 (1976)
/8/ G.Köhler and N.Getoff, Abstr.Int.Conf.Photochem.Conv. and Storage of Solar Energy, London, Canada (1976)
/9/ Y.Ilan, M.Luria and G.Stein, J.Phys.Chem. 8o, 584 (1976)
/10/ I.Tatischeff and R.Klein, p.375 in "Excited States of Biological Molecules", ed. J.B.Birks (Int.Conf.Lisbon, April 1974), J.Wiley & Sons (1976)
/11/ I.Tatischeff and R.Klein, Photochem.Photobiol. 22, 221 (1975)
/12/ H.B.Steen, J.Chem.Phys. 61, 3997 (1974)
/13/ L.Kevan and H.B.Steen, Chem.Phys.Lett. 34, 184 (1975)
/14/ H.B.Steen, M.K.Bowman and L.Kevan, J.Phys.Chem. 80, 482 (1976)
/15/ S.J.Wawilow, Z.Phys. 22, 266 (1924); 42, 311 (1927)
/16/ G.Köhler, J.Zechner, G.Grabner and N.Getoff, Abstr. VII.Int.Congress on Photobiology, Rom (1976)
/17/ J.Zechner, G.Köhler, G.Grabner and N.Getoff, Chem.Phys.Lett. 37, 297 (1976)
/18/ N.Getoff and S.Solar, Monatsh.Chem. 105, 241 (1974)
/19/ G.Köhler and N.Getoff, Solar Energy (submitted)
/20/ J.Zechner, L.S.Prangova, G.Grabner and N.Getoff, Z.Phys.Chem. (N.F.) 102, 137 (1976)
/21/ T.R.Hopkins and R.Lumry, Photochem.Photobiol. 15, 555 (1972)
/22/ L.I.Grossweiner, Int.J.Radiat.Biol. 29, 1 (1976)
/23/ J.Jortner, M.Ottolenghi and G.Stein, J.Am.Chem.Soc. 85, 2712 (1963)
/24/ M.Ottolenghi, J.Am.Chem.Soc. 85, 3557 (1963)
/25/ G.Dobson and L.I.Grossweiner, Trans.Faraday Soc. 61, 808 (1965)
/26/ H.I.Joschek and L.I.Grossweiner, J.Am.Chem.Soc. 88, 3261 (1966)
/27/ C.R.Goldschmidt and G.Stein, Chem.Phys.Lett. 6, 299 (1970)
/28/ J.Feitelson and G.Stein, J.Chem.Phys. 57, 5378 (1972)
/29/ J.Zechner and N.Getoff, Int.J.Radiat.Phys.Chem. 6, 215 (1974)
/30/ N.Getoff, Z.Naturforschg. 17b, 87 (1962)
/31/ J.Jortner, M.Ottolenghi and G.Stein, J.Phys.Chem. 66, 2029; 2037; 2042 (1962)
/32/ M.S.Matheson, W.A.Mulac and I.Rabani, J.Phys.Chem. 67, 2613 (1963)
/33/ G.Grabner, G.Köhler, J.Zechner and N.Getoff, Photochem.Photobiol. (submitted)
/34/ J.G.Calvert and J.N.Pitts, Jr. "Photochemistry", J.Wiley & Sons, New York (1967)
/35/ L.I.Grossweiner, Curr.Topics Rad.Res.Quat. 11, 141 (1976)
/36/ M.Neumann-Spallart and N.Getoff, Monatsh.Chem. 106, 1359 (1975)
/37/ J.P.Keene, Rad.Research 22, 1 (1964)
/38/ N.Getoff, F.Schwörer and M.Pruchova, to be published
/39/ H.Christensen, Int.J.Rad.Phys.Chem. 4, 311 (1972)
/40/ J.Rabani, J.Am.Chem.Soc. 84, 868 (1962)
/41/ C.J.Seliskar, O.S.Khalil and S.P. McGlynn, p.231 in "Excited States" ed.E.C.Lim, Academic Press, New York (1974)
/42/ A.Streitwieser. Jr., p.1 in "Progress in Physical Organic Chemistry" ed.S.G.Cohen at al, Interscience, New York (1963)
/43/ L.J.Mittal, J.P.Mittal and E.Hayon, J.Phys.Chem. 77, 2267 (1973)
/44/ I.Tatischeff, R.Klein, T.Zemb and M.Duquesne, Nature (submitted)

DISCUSSION

DE SCHRYVER :

How does the absorption and emission spectra of phenylethylamine in cyclohexane compare with the respective spectra of toluene.

GETOFF :

There is no singificant difference between the absorption and emission spectra of toluene and phenylethylamine in cyclohexane.

ZACHARIASSE :

Is there an indication of intramolecular charge-transfer in the excited state of phenylethylamine, from the amine to the phenyl-ring, perhaps followed by proton transfer to the phenyl ring.

GETOFF :

For the explanation of the results concerning phenylethylamine we assume an interaction between the amino-group and the aromatic ring. At the moment we do not have any experimental hint for the nature of this interaction, but it is likely of charge transfer character.

INTRAMOLECULAR EXCITED STATE INTERACTIONS IN 1,3-DI(2-ANTHRYL) PROPANE

F.C. De Schryver*, J. Huybrechts, N. Boens, J.C. Dederen, M. Irie (1)
Dept.of Chemistry, University of Leuven, Celestijnenlaan 200 F, B-3030 Heverlee (Belgium)
A. Zachariasse
Max Planck Institut für Biophysical Chem.Spectroscopy Dept., Göttingen (Germany)

INTRODUCTION

The study of the photochemistry and photophysics of non conjugated bichromophoric systems (2) has contributed to different fields of interest ranging from organic synthesis, polymer chemistry and chain dynamics (3) to energy transfer and to complex formation in the excited state.
Previous studies have revealed the possibility to form intramolecular complexes in the excited state (2,4), even when the chain contains more than three links (5), they have given an insight into some of the factors determining complex formation and into the role of these complexes in intramolecular reactions (4f,5c,6). A model has been developed to predict the cyclization probability of chain ends (7) and its validity has been and is tested on the basis of data obtained in such intramolecular interactions. The study of bichromophoric systems can also contribute to a better understanding of the conformation of the excited state complex. Ferguson (6a) has studied recently the absorption and emission properties of the cleaved cyclomer of 1,2-di(9-anthryl)ethane in a host matrix of the cycloadduct and in a variety of rigid glasses. He has shown that under those experimental conditions a multitude of different conformations exist from which emission can occur. Mataga (6b) has observed two types of excimers in 1,2-di(9-anthryl)ethane and in 1,2-di(1-anthryl)ethane. In solution an excimer is observed with a peak at 460 nm and extending to 560 nm. Another excimer was observed in a rigid glass at 77 K with a

B. Pullman and N. Goldblum (eds.), Excited States in Organic Chemistry and Biochemistry, 313-322.

peak at 530 nm on cleavage of the photocyclomer. By comparison with the spectroscopic properties of a series of anthracenophanes, a conformation for each of these excited state complexes was proposed. In this paper we would like to point out some of the complexities involved in the study of these short chain bichromophoric systems using 1,3-di(2-anthryl)propane 1 as an example.

$(CH_2)_3$

RESULTS AND DISCUSSION

The absorption and emission spectra of 1 in methylcyclohexane (MCH) at 20°C are reported in fig.I in comparison with the spectra of 2-ethylanthracene under identical experimental conditions.

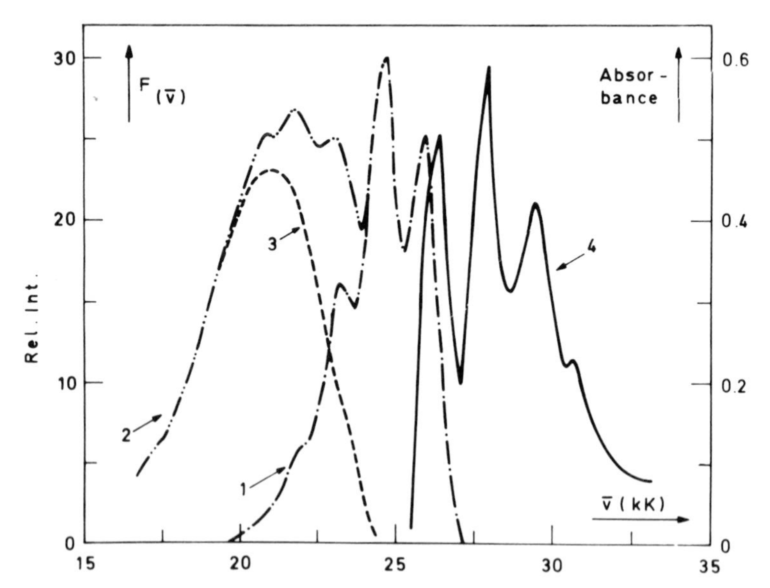

Fig.I : 1. Corrected emission spectrum of 2-ethylanthracene.
2. Corrected emission spectrum of 1 normalized at the 0,0 emission band.
3. Difference spectrum representing the complex emission.
4. Absorption spectrum of 1 in the 25-30 kK region in MCH at 20°C.

The maxima and relative peak heights of the absorption spectrum in the 25-30 kK region are identical for a solution of 1 and of 2-ethylanthracene containing the same chromophore concentrations. The emission spectrum of 1 shows an emission at longer wavelenghts with a maximum at 21 kK. The excitation spectrum of 1 is unchanged in shape over the whole range of the emission wavelengths. It is identical to the excitation spectrum of 2-ethylanthracene and in agreement with the absorption characteristics. The lifetimes of 2-ethylanthracene and the quantum yields of 1 and of 2-ethylanthracene in different solvents are reported in table 1.

Table 1 : Ouantum yields of fluorescence and lifetimes of 2-ethylanthracene and of 1 in different solvents at 20°C.

	1			2-ethylanthracene	
Solvent	ϕ^{F}_{total} (a)	% Mon.	% Comp.	ϕ_F	τ nsec (b)
MCH	0.21	48	52	0.21	4.5
dodecane	0.21	55	45	0.22	4.1
parafine	-	90	10	-	-
benzene	0.17	59	41	0.17	3.3 (c)

a. quantum yields measured relative to quinine sulfate in 0.1 N H_2SO_4. In view of the errors introduced in the reading of the optical density (below 0.04), the integration of the surface and the corrections for refractive index the values of ϕ_F are within 10%. The previous reported value (4f) is an overestimation due to the instrumental response.

b. measured by single photon time correlated spectroscopy. Single exponential decays were observed over more than three decades. The data were convoluted for the lamp curve and corrected for background noise.

c. Independent of the temperature between 20 and 70°C.

To ascertain that the emission at longer wavelengths was due to intramolecular complex formation in the excited state the time resolved emission spectra of 1 in MCH at 20°C were measured at different time windows.

The emission spectrum within the first time window is identical with that of 2-ethylanthracene as can be seen from fig.II. The broad complex emission grows in at further time intervals. Additional proof for complex formation in the excited state is found in the observation of the excitation and emission spectra of 1 in MCH at 77 K which are identical with the respective spectra of 2-anthracene (vide infra fig.IV). We previously reported (4f) that within the temperature range of 283 K and 223 K an isoemissive

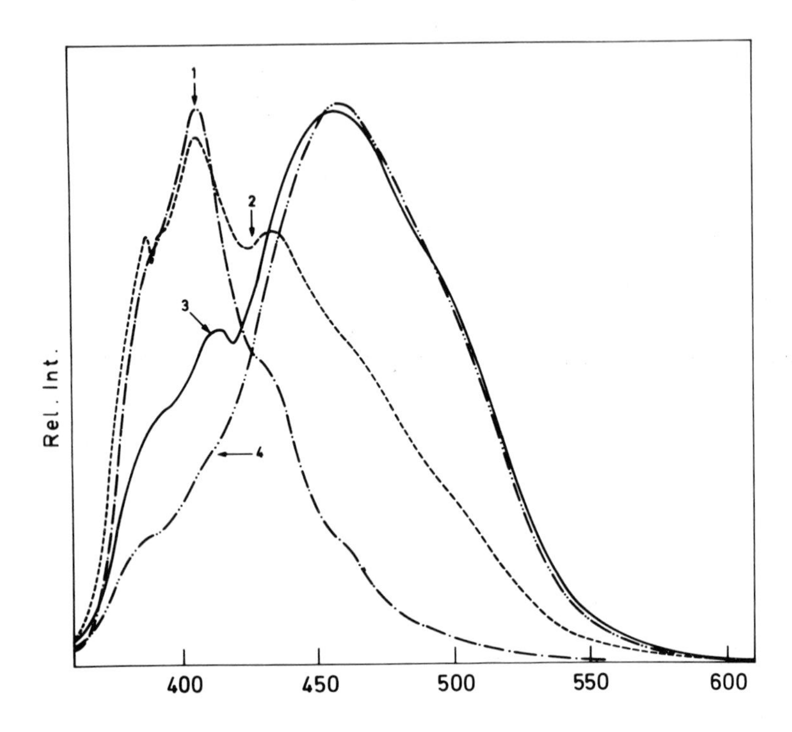

Fig.II : Time resolved uncorrected emission spectra of 1 in MCH at 20°C. Excitation wavelength 340 nm.
1. time window 0 - 4 nsec.
2. time window 12.7 - 16.9 nsec.
3. time window 19 - 27.1 nsec.
4. time window 30.8 - 43.5 nsec.

point could be observed in MCH. An isoemissive point is also observed at the same wavelength (440 nm) for 1 in parafine in the temperature range of 293 to 363 K.
In MCH it is now observed that in the high temperature region both monomer and excimer emission diminish, the latter one faster than the former one.
In the low temperature region the intensity of the long wavelength emission diminishes slower than expected. These results are presented in fig.III.

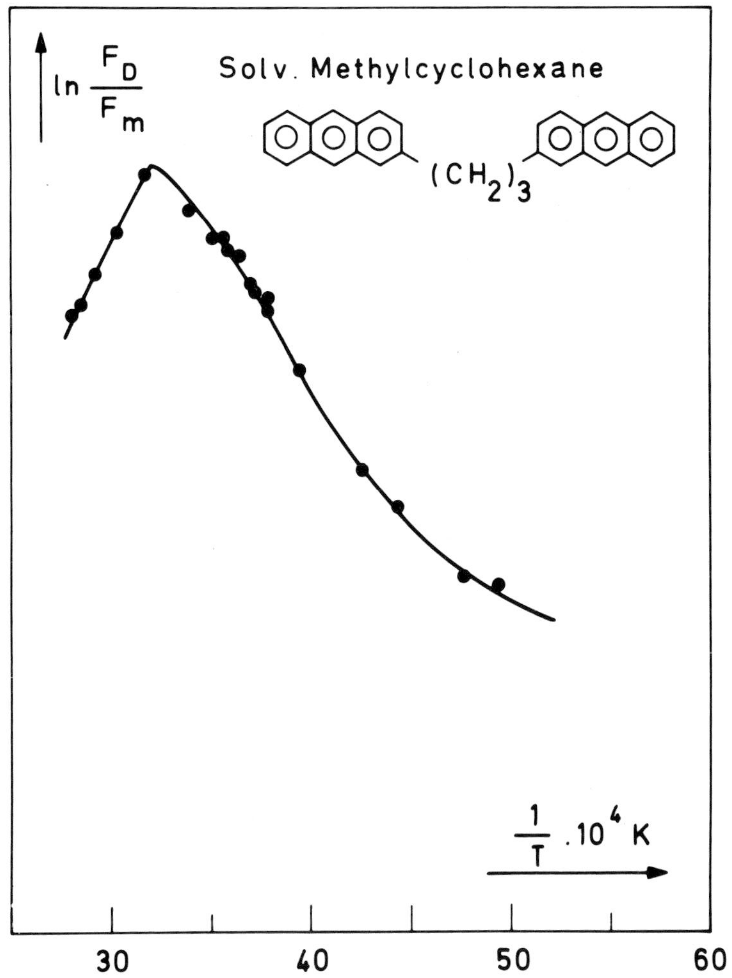

Fig.III : Plot of the logarithm of the ratio of the fluorescence quantum yield of the complex emission over the quantum yield of the monomer component as a function of one over the absolute temperature in MCH. The intensity of the monomer component was determined by comparison with the emission spectrum of 2-ethylanthracene at each temperature.

Those anomalies can be explained by careful analysis of the emission and excitation spectra of 1 in MCH between 77 and 353 K (fig. IV).

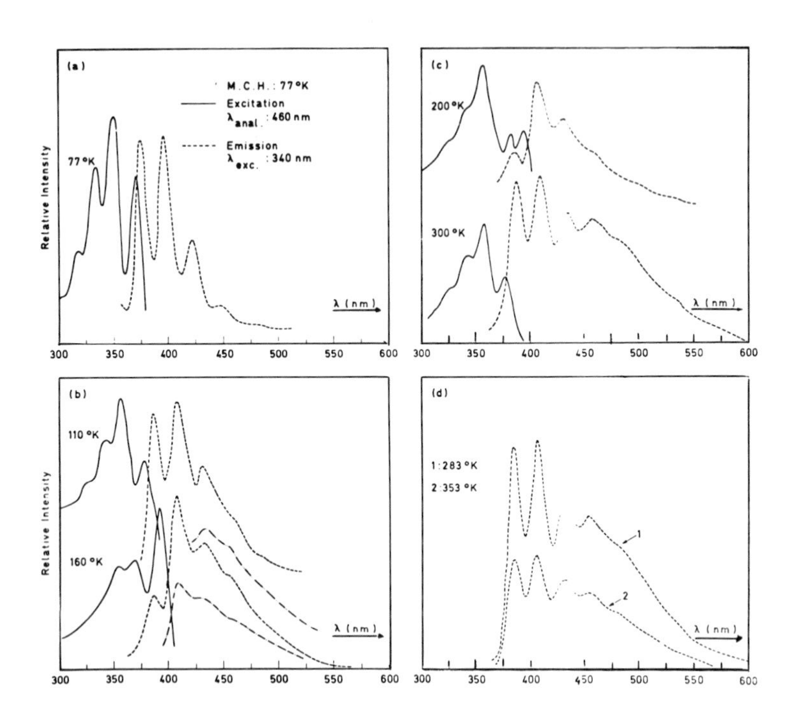

Fig.IV : Evolution of the corrected excitation and emission spectra of 1 in MCH from 77 K till 353 K.
Excitation analyzed at 460 nm; emission excited at 340 nm except in spectrum b at 160 K. The lower emission spectrum is excited at 370 nm, the upper emission spectrum is excited at 390 nm.

The spectra in MCH at 77 K are identical for 1 and 2-ethylanthracene as stated above. The 0,0 bands of the excitation and emission are respectively at 380 and 385 nm, no long wavelength component is observed.
Upon thawing of the glass, broadening of the excitation and emission spectrum occurs concurrent with the formation of a long wavelength emission. In the excitation spectrum at 160 K new bands (0,0 band at 394 nm) appear reminiscent of the "dimer" type excitation spectra (6b). A structured emission is observed upon excitation at 390 nm. Further increase of the temperature leads to a diminishing and, thereafter, disappearance of this "dimer" type component. Above 283 K the excitation spectrum of 1 is again identical to that of 2-ethylanthracene at the same temperature at all analyzing wavelengths. The emission spectrum shows however a broad complex component for which no ground state stabilized enti-

ty is observed and hence this emission can be ascribed to an excimer. The 0,0 band for excitation and emission are respectively at 376 and 385 nm. Further heating of the solution leads to a decrease of the total emission intensity as can be seen in fig.IVd. From the difference spectrum, obtained by substracting the emission of 2-ethylanthracene from the emission spectrum of 1 both at the same temperature and normalized at 385 nm maximum (fig.V), it can be seen that the maximum shifts to higher wavenumbers with increasing temperature in MCH and, to a lesser extent, in benzene.

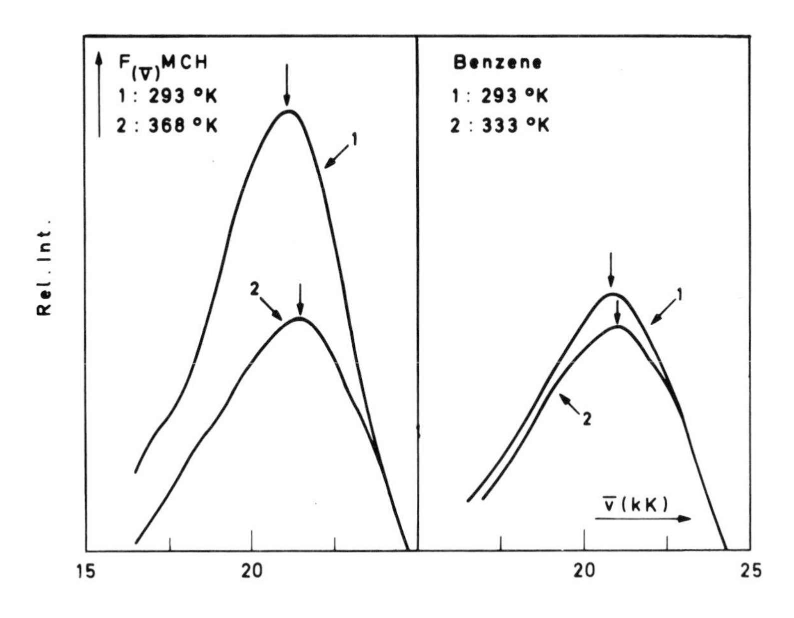

Fig.V : Corrected difference spectra of the excimer component of 1 in MCH at 293 and 268 K and in benzene at 293 and 333 K.

This shift cannot be related to the effect of the change of the refractive index of the solvent (table 2 a and b).
The variation of the maximum indicates a dependence of the complex on the viscosity and the polarity of the solvent as well as on the temperature. We suggest that the excimer emission is composed of emissions from more than one stable excited state conformation. The low energy end of the spectrum represents conformations which can react on increase of the temperature. This suggestion is supported by the strong temperature dependence of the cyclomerization of 1 in benzene as represented in table 3.

Table 2 : Position of the maximum of the excimer emission.

a. In MCH as a function of temperature

Temperature	$\bar{\nu}_{max}.K$
-50°C	21.4
20°C	21.0
95°C	21.5

b. In different solvents at 20°C

Solvent	n	$\eta_{e.p.}$	$\bar{\nu}_{max}.K$
MCH	1.43	0.69	21.0
dodecane	1.429	1.35	21.0
paraffin	1.48	108	21.4
benzene	1.52	0.6	20.8

Table 3 : Temperature dependence of the cyclomerization of 1 in benzene

T°C	$\phi_r^{(a)} \cdot 10^2$
20	0.4
40	0.65
60	1.35

(a) the reaction is concentration independent in the concentration range studied (10^{-4} to 5.10^{-5}M). Quantum yields were determined using 340 nm light for excitation and hexamethylenebismaleimide as an actinometer (5b).

Further support is found in the difference spectra obtained from time resolved emission spectroscopy in the 400 to 550 nm range. As can be seen from fig.VI the maximum of the excimer component shifts by changing the time window indicative of the structure of the emitting species within that time range. The maximum of curve 2 lies close to the one observed for the excimer component of 1 in paraffin at 20°C or in MCH at -50°C or at 95°C. It is also in the same range as one of the maximum observed by Mataga. This means that in these reaction conditions the conformation(s) responsible for emission at longer wavelengths are not so easily formed or that they do have a rate constant for radiationless decay to product or starting material which are substantially larger than that for emission.

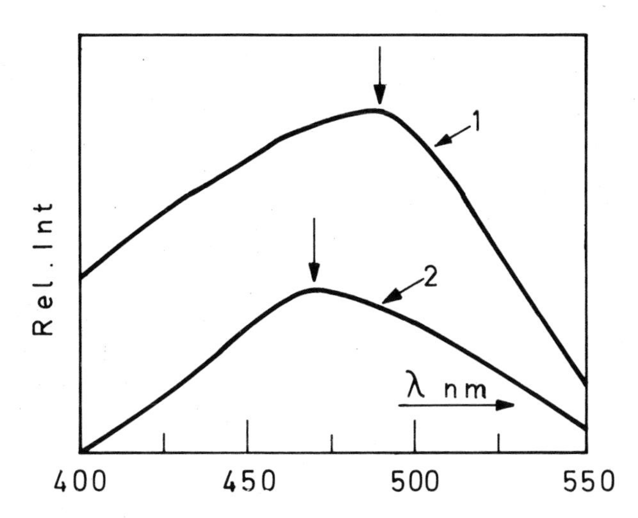

Fig.VI : Difference spectra obtained from the time resolved spectra normalized at 383 nm and corrected for the photomultiplier response. The sensitivity of the grating used does not vary substantially within the spectral range.
1. difference between curve 3 and 1 of fig.II
2. difference between curve 2 and 1 of fig.II

CONCLUSIONS

From the results presented a qualitative picture of the evolution in time of 1 upon excitation emerges. Emission from different excited state complexes, dimer or excimer, can be observed in this system indicating a not so stringent necessity of a single well defined conformation for such complexes. The possibility that different types of more or less stabilized excimers could exist was pointed out by Birks (8). The factors influencing these different species are solvent viscosity and polarity, temperature and reactivity of the system. A more detailed study of these factors will reveal the different stabilizing parameters involved.

ACKNOWLEDGEMENTS

The authors are indebted to the Ministerie voor Wetenschapsbeleid for financial support. The N.F.W.O. is thanked for support to the laboratory and for a post doctoral fellowship to N.B. The

Nato science programme has made collaboration of K.Z., F.D.S. and dr. Winnik possible. The technical assistance of Mr. R. De Boer and of I. Jans are gratefully acknowledged.

REFERENCES

(1) On leave of absence from Osaka University.
(2) For a review see F.C. De Schryver, N. Boens and J. Put, Advances in Photochemistry 10, 1977, in press.
(3) For a recent state of the art see M. Winnik, Acc.Chem.Res. 1977, in press.
(4) Recent contribution to the field are :
 (a) J.M. Borkent, J.W. Verhoeven and Th.J. De Boer, Chem. Phys.Lett., 42, 50 (1976)
 (b) Y.C. Wang and H. Morawetz, J.Am.Chem.Soc., 98, 364 (1976)
 (c) M. Inoh, Y. Kumano and T. Okamoto, Bull.Chem.Soc.Japan, 49, 42 (1976)
 (d) S. Tazuke, K. Sato and F. Banba, Chem.Lett., 12, 1321 (1976)
 (e) G.E. Johnson, J.Chem.Phys., 63, 4047 (1975)
 (f) F.C. De Schryver, N. Boens, J. Huybrechts, J. Daemen and M. De Brackeleire, IUPAC Conf. on Organic Photochemistry, Aix-en-Provence 1976, in press.
(5) (a) K. Zachariasse and W. Kühnle, Zeit.Phys.Chem. N.F., 104, 267 (1976)
 (b) J. Put and F.C. De Schryver, J.Am.Chem.Soc., 95, 137 (1973)
 (c) F.C. De Schryver, N. Boens, J. Huybrechts and M. De Brackeleire, Zeit.Phys.Chem. N.F., 104, 417 (1976)
(6) (a) J. Ferguson, M. Morita and M. Puza, Chem.Phys.Lett. 42, 288 (1976)
 (b) T. Hayashi, N. Mataga, Y. Sakata, S. Misumi, M. Morita and J. Tanaka, J.Am.Chem.Soc., 98, 5910 (1976)
 (c) T. Hayashi, T. Suzuki, N. Mataga, Y. Sakata and S. Misumi, Chem.Phys.Lett., 38, 599 (1976)
 (d) A. Castillan, R. Lapouyade and H. Bouas-Laurent, Bull. Chem.Soc. France, 210 (1976)
 (e) W. Gerhartz, R.D. Poshusta and J. Michl, J.Am.Chem.Soc. 98, 6427 (1976)
(7) M.A. Winnik, R.E. Tinamari, S. Jackowski, D.S. Saunders and S.G. Whittington, J.Am.Chem.Soc., 96, 4843 (1974)
(8) J.B. Birks, "The Exciplex", ed. M. Gordon, W.R. Ware, Academic Press, 1975, p.39

BEHAVIOUR OF EXCITED C-NITROSO COMPOUNDS IN THE PRESENCE AND ABSENCE OF OXYGEN

Th.J. de Boer, F.J.G. Broekhoven
and Th.A.B.M. Bolsman

Laboratory for Organic Chemistry, University of Amsterdam, Nieuwe Achtergracht 129, Amsterdam, The Netherlands

INTRODUCTION

Several years ago we showed that the primary photoprocess of most aliphatic C-nitroso compounds (RNO) upon irradiation with light in the 500-700 nm wavelength region is the formation of radicals R$^\bullet$ and NO$^\bullet$ by C-N bond homolysis (1).

Each of these radicals can in principle attack unchanged starting material, thus leading to a variety of products in yields strongly depending on the nature of the substrate and reaction conditions (2-7).

We have also found that the presence of oxygen can exert a drastic influence. In its triplet ground state molecular oxygen cannot react in the dark with nitroso compounds (either monomeric or dimeric), but it does so efficiently after irradiation of the nitroso compound.

MECHANISM OF PHOTO-OXIDATION

A suitable model for the study of the photo-oxidation of monomeric nitroso compounds is 2-methyl-2-nitroso-propane (t-BuNO), which is largely present in the monomeric form when dissolved in benzene. Its main photo-oxidation product is <u>tert</u>-butyl nitrate, formed in yields up to 73% when oxygen is present in excess. In the absence of oxygen the same product is formed but for no more than a few percents. Under such anaerobic

B. Pullman and N. Goldblum (eds.), Excited States in Organic Chemistry and Biochemistry, 323-329.

conditions, nitrate formation takes place *via* the long established mechanism of Donaruma and Carmody (3), involving two successive attacks of NO on the original starting material, followed by rearrangement to a diazonium nitrate.

$$RNO \xrightarrow{NO} RN(NO)O^{\bullet} \xrightarrow{NO} RN(NO)ONO$$
$$\longrightarrow RN{=}N{-}ONO_2 \longrightarrow RONO_2 + N_2$$

A characteristic feature of this process is the evolution of nitrogen.

Quantitative gas analysis (GLC) shows that during the photolysis in the presence of oxygen, there is hardly any nitrogen evolution (< 0.5%) i.e. it is even lower than in the absence of oxygen. Consequently the formation of large amounts of nitrate during the photo--oxidation must take place along an entirely different route than under anaerobic conditions.

It is known that the reaction of oxygen with NO (8) is relatively slow* while its reaction with alkyl radicals is very rapid and approaches diffusion control (9). This means that the t-butyl radicals formed by dissociation of the excited nitroso compound are rapidly attacked by oxygen thus generating the *tert*-butylperoxy radical, especially in solutions saturated with oxygen.

In recent studies Heicklen (10) and Pitts (11) have found that alkylperoxy radicals can react with NO -which is not destroyed too rapidly by oxygen- to yield nitrates *via* a somewhat controversial mechanism.

Heicklen (10) has suggested that methylperoxy radicals, as formed during photolysis (λ > 320 nm) of azomethane in the presence of oxygen, presumably react with NO initially to form a peroxy nitrite.

$$ROO^{\bullet} + NO \longrightarrow [ROONO]$$
$$\downarrow ? \qquad\qquad ? \downarrow (\pm\ O_2)$$
$$ROONO \longrightarrow RO^{\bullet}\ {}^{\bullet}ONO \longrightarrow RONO_2$$

---

*It is well established that the reaction between NO and O_2 is third order but there is still controversy as to whether the oxidation proceeds *via* $NO_3^{\bullet}$ (pernitrite or nitrate radicals) or *via* an NO dimer (8).

Because the isomerisation of peroxy nitrite to nitrate (with R=Me) was found to be a third order process, a molecule of oxygen was supposed to be involved, but Heicklen did not specify its precise function. Pitts found no evidence for an intermediate peroxynitrite under similar conditions and assumed that the peroxy radical ROO$^\cdot$ simply oxidises NO to NO_2, nitrate arising by combination of RO$^\cdot$ and NO_2. Whatever the precise mechanism, there seems little doubt that alkylperoxy radicals can be converted into nitrate by NO.

By reacting _tert_-butyl hydroperoxide with NOCl in the presence of $CaCO_3$, we obtained -presumably _via_ t-BuOONO- _tert_-butyl nitrate as the main product (52%). For this reason we are inclined to accept a mechanism involving homolysis of the weak O-O bond in the intermediate peroxynitrite. The resulting radical pair t-BuO$^\cdot$ $NO_2^\cdot$ recombines with preferential formation of the N-O bond (stronger than OO) thus leading to the nitrate.

Considering the key role of the t-BuOO$^\cdot$ radical, we attempted to trap it during photo-oxidation of t-BuNO. Although the nitroso compound itself can in principle act as a radical trap, spin adducts of ROO$^\cdot$

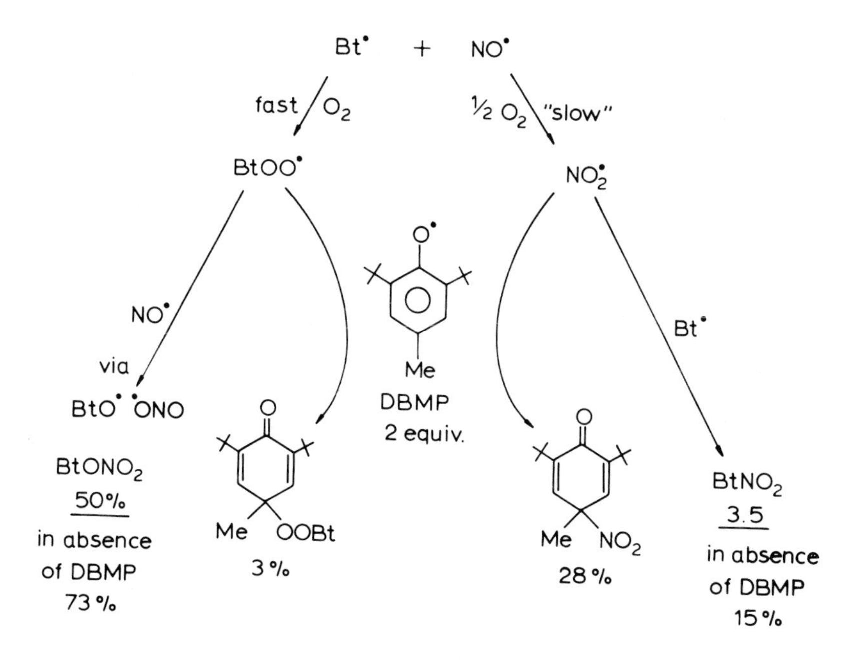

Scheme 1. Photo-oxidation of t-BuNO in the presence of DBMP (2,6-di-_tert_-butyl-4-methylphenol).

radicals have never been observed in ESR, presumably because corresponding nitroxides are highly unstable, if formed at all. So we followed a recommendation by Ingold (12) and employed 2,6-di-*tert*-butyl-4-methylphenol (DBMP) -or rather the phenoxy radical- as a scavenger of peroxy radicals in the presence of oxygen. Results are summarised in Scheme 1.

It is seen that the trapping of t-BuOO$^{\cdot}$ is not very efficient (cf. 3% of adduct). Under the reaction conditions the t-BuOO$^{\cdot}$ radicals are still captured mainly by NO, as judged from the relatively high yield of t-BuONO$_2$* (50% vs. 73% in the absence of DBMP). Moreover some of the t-BuOO$^{\cdot}$ radicals serve to abstract hydrogen from DBMP.

The main isolated adduct to the (substituted) phenoxy-radicals is the NO_2 derivative (28% yield). It should be noted that the trapping of NO is very slow under the experimental conditions, and therefore the NO_2 derivative is not formed by oxidation of an intermediate NO adduct. As shown in Scheme 1 there is a small yield of t-BuNO$_2$ in the presence of DBMP. In the absence of DBMP this byproduct is much more important (up to 15%).

Routes to t-BuNO$_2$.

Route a $\quad$ t-Bu$^{\cdot}$ + NO_2 $\longrightarrow$ t-BuNO$_2$

This route is well documented in the literature (13): the combination of (prim., sec. and tert.) allyl radicals with NO_2 leads almost entirely to the nitro compound, while nitrite formation is negligible.

Considering the high rate of reaction between *tert*-butyl radicals with oxygen it is difficult to explain the formation of t-BuNO$_2$ exclusively *via* route a. A logical alternative proceeds *via* unstable nitroxides.

Route b *via* nitroxides.

Alkoxy radicals can in principle be trapped by monomeric nitroso compounds at sufficiently low temperatures (-40°C), and the corresponding nitroxides are characterised by relatively high a_N values (27-30 Gauss) (14).

*Some of the nitrate may be formed by reaction of t-BuOOH + NO, cf. J.R. Shelton and R.F. Kopczewski, J. Org. Chem. 32, 2908 (1967).

$$\text{t-BuNO} + \text{t-BuO}^{\bullet} \longrightarrow \text{t-BuN(O}^{\bullet}\text{)-O-t-Bu}^{\bullet}$$

During our photo-oxidation no such species have been detected, presumably because of rapid deterioration of such nitroxides at room temperature.

This will be particularly true for the adducts of peroxy radicals with a very weak O-O bond in the nitroxide; this is probably the reason why they have never been observed in ESR experiments at the lowest temperatures. We are therefore inclined to assume that the transient peroxynitroxide plays an active role in the photo-oxidation mechanism as an intermediate of the overall oxidation of nitroso to nitro group.

$$\text{t-BuNO} + \text{t-BuOO}^{\bullet} \longrightarrow \text{t-BuN(O}^{\bullet}\text{)-OO-t-Bu} \longrightarrow \text{t-BuNO}_2 + \text{t-BuO}^{\bullet}$$

The t-butoxy radical thus generated may combine with NO to give (small amounts of) t-BuONO or it may add to the nitroso compound and thus provide an alternative route to the nitro compound.

$$\text{t-BuNO} + \text{t-BuO}^{\bullet} \longrightarrow \text{t-BuN(O}^{\bullet}\text{)-O-t-Bu} \longrightarrow \text{t-BuNO}_2 + \text{t-Bu}^{\bullet}$$

Formally the alkyloxy- and peroxy nitroxides can also be considered as t-Bu$^{\bullet}$ and t-BuO$^{\bullet}$ adducts to nitro compounds. Their easy decomposition reflects in a sense the poor radical scavenging properties of nitro compounds in general.

We have found no indications for yet an other route to the nitro compound _via_ a direct attack of oxygen on the excited nitroso compound with formation of a nitroxide which carries at the same time a peroxy radical function [RN(O$^{\bullet}$)OO$^{\bullet}$].

Earlier suggestions regarding the possible existence of such an intermediate (15) must now be considered dubious in the light of more recent data (16), which strongly support the view that a primary dissociative step triggers most if not all processes in the photo-oxidation of monomeric aliphatic nitroso compounds at long wavelengths.

Little is known about the precise nature of the

excited state when monomeric nitroso compounds are irradiated in the 600-700 nm region. Pulsed laser experiments have recently been carried out with 1-nitroso-adamantane (λ_{max} = 690 nm) in 3-methyl-pentane (c ~ 30 mM) at -180°C by de Busser and Rettschnick (17). Upon irradiation with pulses of 30 nsec at 693.3 nm an emission spectrum is observed with a maximum at 754 nm. The "energy gap" between the 0-0 absorption and emission band amounts to 900 cm^{-1} and the life time of the excited species is 70 nsec at -180°C and decreases to 20 nsec at -15°C. By flash photolysis it is found that the excited species has an absorption spectrum with a broad maximum at 540 nm.

Further studies are in progress to determine the nature of this species.

REFERENCES

1. J.A. Maassen, H. Hittenhausen and Th.J. de Boer, Tetrahedron Lett. 1971, 3213.
2. A. Mackor, Th.A.J.W. Wajer and Th.J. de Boer, Tetrahedron 24, 1623 (1968).
3. L.G. Donaruma and D.J. Carmody, J. Org. Chem. 22, 635 (1957).
4. A.H.M. Kayen, T.A.B.M. Bolsman and Th.J. de Boer, Recl. Trav. Chim. Pays-Bas 95, 14 (1976).
5. J. Dickson and B.G. Gowenlock, Justus Liebigs Ann. 745, 152 (1971).
6. A.H.M. Kayen, L.R. Subramanian and Th.J. de Boer, Recl. Trav. Chim. Pays-Bas 90, 866 (1971).
7. H.A. Morrison in "The Chemistry of the Nitro and Nitroso Group" (ed. H. Feuer), p. 165, Interscience, New York (1969).
8. J. Heicklen and A. Cohen in "Advances in Photochemistry", Vol. 5, p. 203, Interscience, New York (1968).
9. K.U. Ingold in "Free Radicals" (ed. J.K. Kochi), Vol. I, p. 37, John Wiley, New York (1973); and J.A. Howard, ibid. Vol. II, p. 15.
10. C.W. Spicer, A. Villa, H.A. Wiebe and J. Heicklen, J. Amer. Chem. Soc. 95, 13 (1973).
11. C.T. Pate, B.J. Finlayson and J.N. Pitts, Jr., J. Amer. Chem. Soc. 96, 6554 (1974).
12. K.U. Ingold in "Free Radicals" (ed. J.K. Kochi), Vol. I, p. 72, 80, John Wiley, New York (1973).
13. Reviewed by Y. Rees and G.H. Williams in "Advances in Free Radical Chemistry" (ed. G.H. Williams), Vol. III, p. 218, Logos Press, London (1969).

14. A. Mackor, Th.A.J.W. Wajer, Th.J. de Boer and J.D.W. van Voorst, Tetrahedron Lett. 1967, 385.
15. J.A. Maassen and Th.J. de Boer, Recl. Trav. Chim. Pays-Bas 91, 1329 (1972); ibid. 92, 185 (1973).
16. F.J.G. Broekhoven, Th.A.B.M. Bolsman and Th.J. de Boer, Recl. Trav. Chim. Pays-Bas 96, 12 (1977).
17. R. de Busser and R.P.H. Rettschnick, Dept. Phys. Chemistry, Univ. of Amsterdam, Unpublished results.

SOME POSSIBLE PRODUCTS OF THE REACTIONS OF $O(^1D)$ AND $O_2(^1\Delta)$ WITH UNSATURATED HYDROCARBONS

Peter Politzer and Kenneth C. Daiker

Chemistry Department, University of New Orleans, New Orleans, LA, 70122, U.S.A.

INTRODUCTION

There is a considerable and growing body of evidence which indicates that arene oxides, or epoxides, play a key role in the carcinogenic action of polynuclear aromatic hydrocarbons [1-5]. In at least one case, benzo[a]pyrene (I), the "ultimate carcinogen" (the active form of the molecule) may have been identified, as the 7,8-dihydroxy-9,10-epoxide [3,4]. These epoxides are formed metabolically, by a process about which relatively little is known. For example, there appears to be a general uncertainty regarding the nature of the oxygen-containing entity that interacts with the hydrocarbon in producing the epoxide; among the apparent possibilities are singlet oxygen atoms, singlet oxygen molecules, superoxide anions, and various peroxide species.

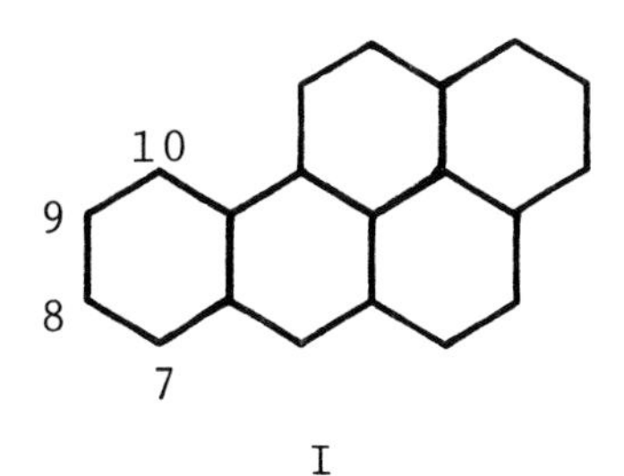

I

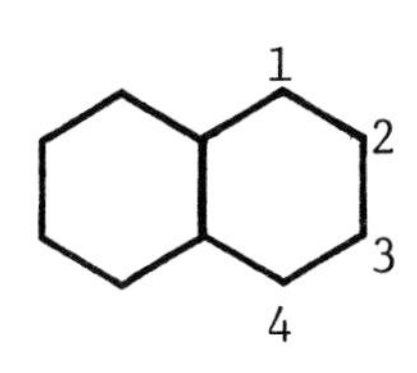

II

B. Pullman and N. Goldblum (eds.), Excited States in Organic Chemistry and Biochemistry, 331-344.

As part of our overall investigation of the carcinogenic actions of polynuclear aromatic hydrocarbons, we have studied some of the products of the interactions of singlet oxygen atoms and singlet oxygen molecules with several aromatic (and one non-aromatic) hydrocarbons. Our primary interest is of course in the epoxides that may form in these processes, whether as intermediate or as final products. The method of approach is to compute self-consistent-field wave functions for various configurations of the systems $O(^1D)$ + hydrocarbon and $O_2(^1\Delta)$ + hydrocarbon. These wave functions are of minimum-basis-set STO-5G quality [6]. From them can be obtained estimates of the energies involved in the interactions, the structures of intermediate and final products, and some of their interesting properties, such as charge distributions.

Because the program used in this work cannot handle a system bigger than O_2 + naphthalene, the aromatic hydrocarbons studied were benzene, toluene, and naphthalene (II). One small unsaturated but non-aromatic molecule, ethylene, was also included in this investigation, to serve as a model. The structures of the epoxide and the other products obtained from ethylene were determined fairly carefully, by optimizing most of the various bond lengths and bond angles. These results were then used as guides to the structures of the products in the cases of the larger molecules, thereby permitting satisfactory treatment of these with less extensive optimization than for ethylene. This use of ethylene as a model for epoxide formation seems quite reasonable, especially for the polynuclear aromatic hydrocarbons, because of the considerable degree of olefin-like character possessed by the more reactive bonds in the latter molecules. Thus, for example, the 7,8 and 9,10 bond lengths in benzo[a]-pyrene (I) are 1.374 A and 1.364 A, respectively [7]; the length of the 1,2 bond in naphthalene (II) is 1.361 A [8], and that of the 5,6 bond in 7,12-dimethylbenz[a]anthracene (III) is 1.335 A [9]. The C-C bond length in ethylene is 1.337 A [8].

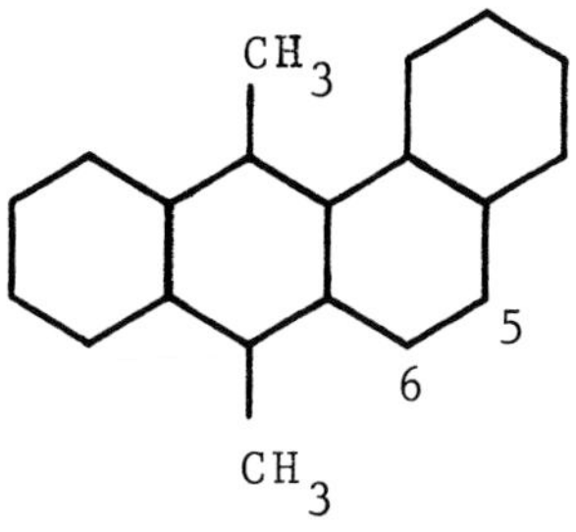

III

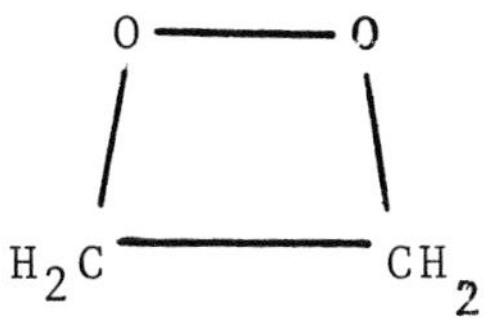

IV

RESULTS AND DISCUSSION

$O_2(^1\Delta)$, $O(^1D)$ + Ethylene:

Although a reaction between $O_2(^1\Delta)$ and ethylene has not yet been observed experimentally, there have been several theoretical studies of it, using semi-empirical techniques [10-13]. These indicate that the formation of the dioxetane (IV) is energetically favored, and that the preferred path of approach is that in which the system is in a peroxirane, or per-epoxide, type of configuration (Fig. 1a); one oxygen approaches the midpoint of the C-C bond, with the whole O_2 molecule being in the plane perpendicular to the original C_2H_4 plane and to the C-C bond at its midpoint. There has been some disagreement as to whether the system actually goes through a stable perepoxide intermediate in the course of forming the dioxetane [12,13]. Dewar and Thiel did find such an intermediate, although they did not fully report its structure [12].

It is this possible perepoxide intermediate that is of particular interest to us, since in addition to rearranging to the dioxetane, it can also form the epoxide. Two possible mechanisms for this latter step have been discussed by Dewar _et al_ [14]; they involve the interaction of the perepoxide with either another O_2 or another C_2H_4 molecule.

We have therefore made a fairly careful search for a stable perepoxide intermediate. We assumed the general perepoxide configuration, but we optimized the C-C, C-O, and O-O distances, as well as the angles of bending of the outer oxygen and of the hydrogens. We did find a definite perepoxide minimum on the energy

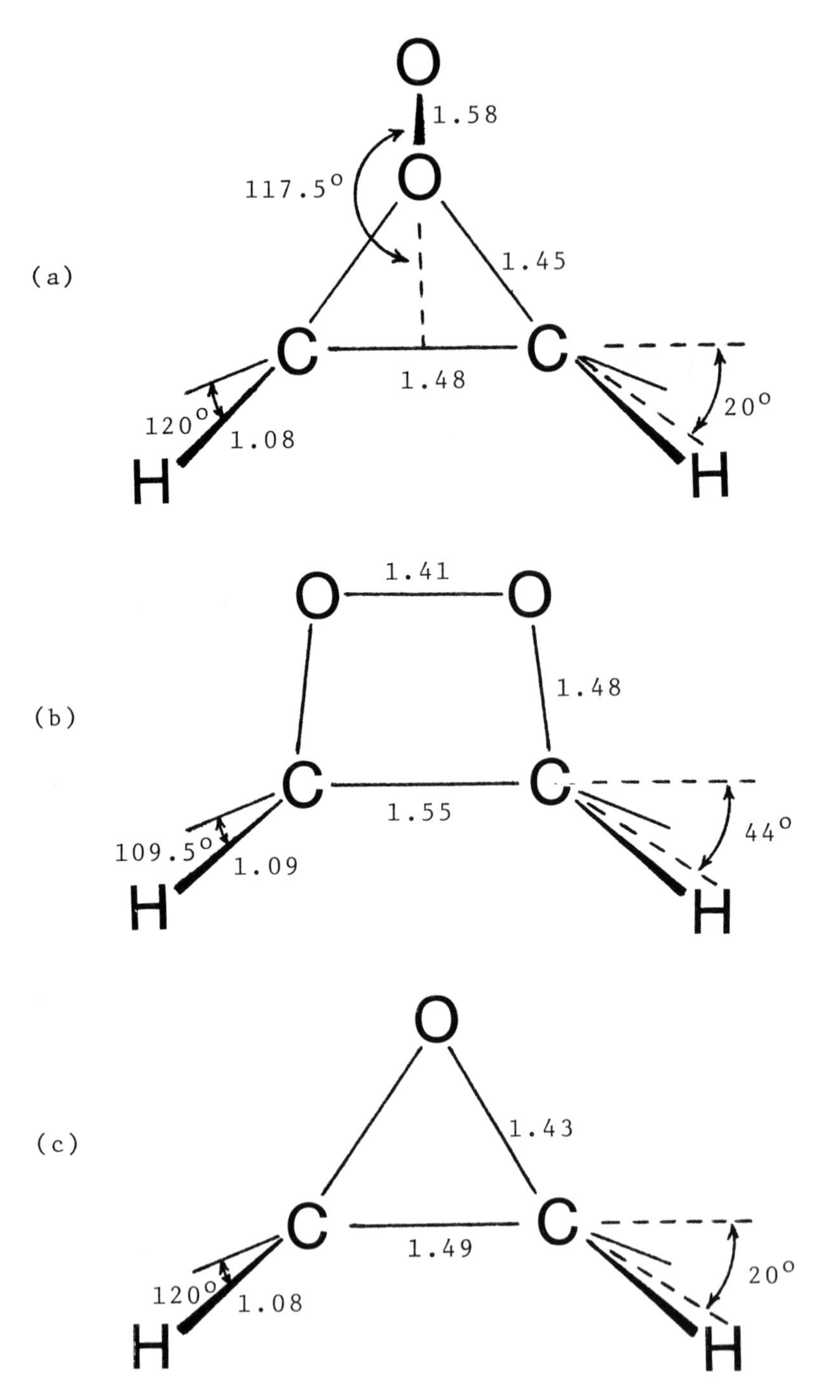

Figure 1. Calculated structures for ethylene perepoxide (a), dioxetane (b), and epoxide (c). All distances are in Angstroms.

surface, so that we conclude, in agreement with Dewar and Thiel, that there is a stable perepoxide intermediate. Our predicted geometry for it is shown in Fig. 1a.

An interesting feature of Fig. 1a is the increase in the C-C and O-O distances compared to the values in free C_2H_4, 1.337 A, and $O_2(^1\Delta)$, 1.2155 A. This is contrary to what has been assumed in some previous studies, in which these distances were kept at the free molecule values [11,13]. It seems reasonable, however, to expect these distances to increase, and indeed our calculated O-O bond length is in the neighborhood of its approximate value in various peroxides, 1.48 A, while our C-C distance is practically the same as in ethylene oxide, 1.47 A [8].

In forming the perepoxide, the pair of oxygens develops a degree of polarity; in the structure shown in Fig. 1a, the inner oxygen has a charge of -0.13, while that of the outer is -0.26. The overall dipole moment of the whole system is 3.88 D, which is similar to Dewar and Thiel's value of 4.26 D [12].

Although our results, and those of Dewar and Thiel, show the perepoxide to be a stable intermediate, it is much less stable than the dioxetane. Our calculated structure for the dioxetane is shown in Fig. 1b; this was obtained by optimizing the C-C, C-O, and O-O distances, and the angle of bending of the hydrogens out of the original C_2H_4 plane. The energy of this structure is 79 kcal/mole lower than that of the perepoxide, compared to Dewar and Thiel's 49.4 kcal/mole. The discrepancy between these two values may arise in part from a tendency of the MINDO/3 method (used by Dewar and Thiel) to exaggerate the stability of the perepoxide [12]. In any case, both results show that the rearrangement of the perepoxide to the dioxetane is energetically favored.

On the other hand, the process that is of interest to us, the conversion of the perepoxide to the epoxide, is not energetically favored. We obtained an energy for the latter by considering the $O(^1D)$ plus ethylene system. We optimized both the C-C and the C-O distances, and we verified that the oxygen is situated on the axis perpendicular to the original C_2H_4 plane at the midpoint of the C-C bond. The hydrogens were shifted 20° out of their original plane, following the example of the perepoxide. Our final structure is shown in Fig. 1c. It is in excellent and very gratifying agreement with the experimentally-determined structure of this molecule [8]. According to the latter, the C-C and C-O bond lengths are 1.470 A and 1.435 A, respectively, and the angle of bending of the hydrogens is 21.9°.

In investigating the energetics of the process perepoxide → epoxide, it is necessary to compare the energy of the perepoxide to that of the epoxide plus a single oxygen atom. For the optimized structures that have been described, we find that

$$E_{epoxide} + E_{O(^1D)} - E_{perepoxide} = 24 \text{ kcal/mole.}$$

Thus the process is <u>not</u> energetically a favored one. This suggests that the participation of some external agent is required for the conversion of the perepoxide to the epoxide. Such an external agent could conceivably be a second $O_2(^1\Delta)$ molecule, or a second C_2H_4 molecule. These two possibilities have been examined by Dewar <u>et al</u> [14], who concluded that the process involving a second C_2H_4 molecule is so highly favored that, in the specific case of ethylene, the main product of the interaction with $O_2(^1\Delta)$ should be the epoxide rather than the dioxetane. It should be noted that we find the epoxide to be very stable relative to free ethylene plus an $O(^1D)$ atom:

$$E_{epoxide} - E_{C_2H_4} - E_{O(^1D)} = -118 \text{ kcal/mole.}$$

$O(^1D)$, $O_2(^1\Delta)$ + Benzene:

In order to have a proper reference point, we first computed an energy for benzene itself, allowing the C-C distances to vary. Our optimum value is 1.39 A, virtually identical with the actual 1.397 A [8].

Using our ethylene epoxide as a guide, we proceeded to calculate the structure of the epoxide of benzene. To the best of our knowledge, this has not yet been determined experimentally. We varied the C-C and C-O distances, the angle between the benzene and the epoxide rings, and the angle by which the bridgehead hydrogens are shifted out of their original plane. The optimum structure is shown in Fig. 2a. An interesting feature is the apparent transformation from aromatic to diene-like character; four of the C-C bonds now have essentially single-bond lengths, while

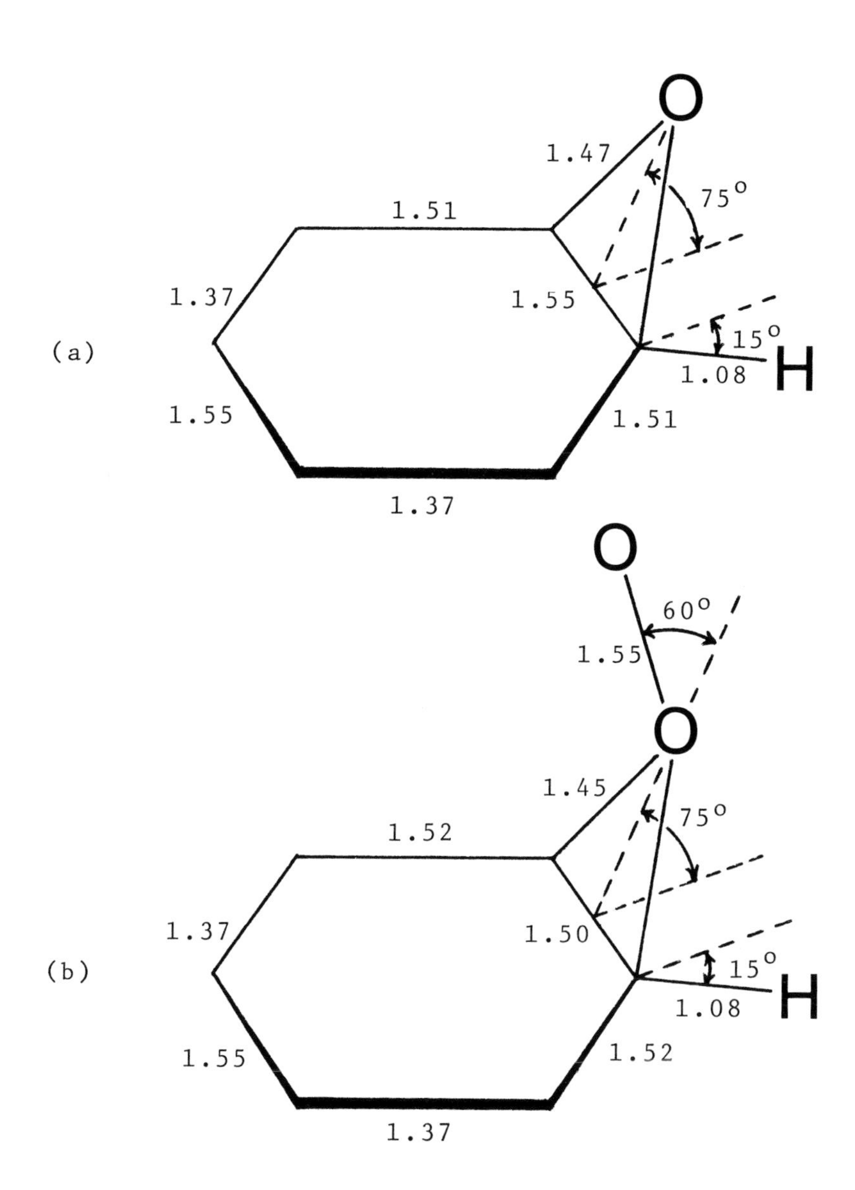

Figure 2. Calculated structures for benzene epoxide (a), and benzene perepoxide (b). All distances are in Angstroms.

the other two have moved in the direction of double bonds.

As in the case of ethylene epoxide, the formation of benzene epoxide from benzene and a singlet oxygen atom is energetically favored, by quite a substantial margin:

$$E_{epoxide} - E_{C_6H_6} - E_{O(^1D)} = -72 \text{ kcal/mole.}$$

An interesting property of many aromatic epoxides is their facile rearrangement to a phenol [1]. This is believed to occur in many instances via an intramolecular migration of a hydrogen atom, the so-called NIH shift. A keto intermediate would be produced, which would subsequently yield the phenol. For benzene, the sequence is [1]

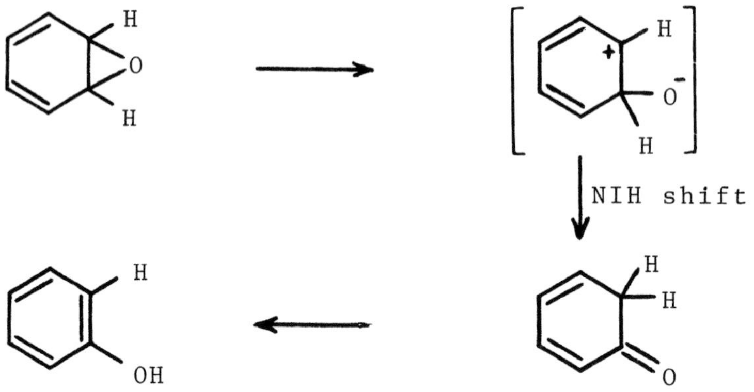

We tested this mechanism, on an elementary level, by computing energies for the keto intermediate and phenol, to compare with that of the epoxide, obtained earlier. The comparison must be regarded as a very approximate one, since we carried out almost no optimization of the structures of the intermediate and of phenol. For the latter, we used the C-C bond lengths of benzene plus essentially standard values for the C-O and O-H bonds of the hydroxyl portion; the O-H bond was taken to be perpendicular to the aromatic ring. In the keto intermediate, four of the C-C distances were as in benzene; the two adjacent to the keto group were taken from p-quinone (1.52 A) [8]. The C-O distance was 1.23 A.

We find the energy of this keto intermediate to be 14 kcal/mole lower than that of benzene epoxide, while phenol is 32 kcal/mole below the epoxide. Thus our results are fully consistent with the mechanism presented above.

With regard to the reaction of benzene with $O_2(^1\Delta)$, we were guided by the example of ethylene + $O_2(^1\Delta)$, and therefore investigated the structure of the corresponding perepoxide, as a possible intermediate in the formation of the epoxide. The previously-determined benzene epoxide structure was used to set the angle between the three-membered and the aromatic rings (105°), and for the angular displacement of the bridgehead hydrogens (15°). The benzene epoxide values were also taken for the lengths of all of the C-C bonds except that which forms part of the three-membered ring; this distance was varied. In addition, we optimized the C-O and O-O distances, and the angular displacement of the outer oxygen atom. The value obtained for the latter, 60° relative to the plane of the three-membered ring, is notable in that it is essentially the same as the analogous angle in the ethylene perepoxide. Our final structure is shown in Fig. 2b.

Further similarity to the ethylene perepoxide is shown by the oxygen charges and the overall molecular polarity. The inner and outer oxygens have charges of -0.14 and -0.27, respectively, while the total dipole moment is calculated to be 3.92 D.

Finally, just as was found in the ethylene case, the transition from the perepoxide to the epoxide is not energetically favored:

$$E_{epoxide} + E_{O(^1D)} - E_{perepoxide} = 24 \text{ kcal/mole.}$$

This suggests, as before, that some external agent must be involved in order for this process to occur. Perhaps that agent can again be a second $O_2(^1\Delta)$ or a second benzene molecule.

$O(^1D)$, $O_2(^1\Delta)$ + Toluene:

We were particularly interested in toluene (V) because of the possible effect of its methyl group. It is well-established that the presence and position of a methyl group can drastically affect the carcinogenic activity of a polynuclear aromatic hydrocarbon [2]. It is of great interest, therefore, to determine whether some effect upon the epoxide or perepoxide can be detected in the case of toluene. Somewhat surprisingly, we have found no significant effect!

For both the epoxide and the perepoxide, we considered two possible locations, the 2,3 bond and the 3,4 bond, and we carried out two sets of calculations. First, we modeled the epoxide or perepoxide portion of the molecule (including the entire three-membered ring) after the corresponding benzene product, obtained earlier, but we used the experimental benzene structure for the remainder of the six-membered ring and we introduced a standard methyl group. Second, we replaced this benzene-like portion of the six-membered ring by the diene-like structure that we determined for benzene epoxide. In all instances, this lowered the energy by roughly 12 kcal/mole.

We found no real preference for either the 2,3 or the 3,4 locations; the differences in energies were so small as to be essentially meaningless. This applies to both the epoxide and the perepoxide. It is of course conceivable that further optimization of our products' structures would reveal some significant preference for one of the two locations. It is also possible that any partiality that may be observed experimentally is due to other factors, such as solvent effects.

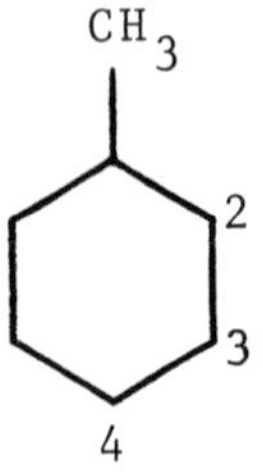

CH_3 2 3 4

V

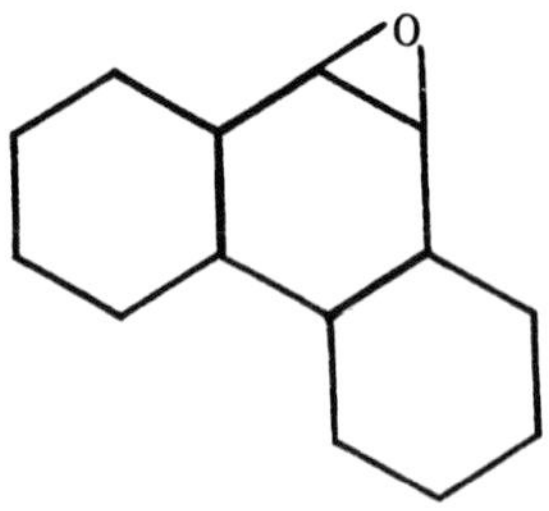

O

VI

From the close similarity in energies that was mentioned above, it follows of course that the energy associated with the process

$$\text{toluene perepoxide} \longrightarrow \text{epoxide} + O(^1D)$$

is virtually the same for both the 2,3 and the 3,4 isomers. Furthermore, it is also the same as for the analogous reaction involving the corresponding benzene derivatives, +24 kcal/mole. The parallel with benzene extends as well to the oxygen charges in the two toluene perepoxides, -0.14 for the inner oxygen and -0.27 for the outer one.

$O(^1D)$, $O_2(^1\Delta)$ + Naphthalene:

The 1,2 epoxide of naphthalene (II) is believed to be "the obligatory intermediate in the hepatic metabolism of naphthalene" [1]. We have therefore carried out calculations for both the 1,2 epoxide and also the 1,2 perepoxide. We modeled the structure after that of the 9,10-epoxide of phenanthrene (VI), which has been determined experimentally [15]; we simply removed one of the outer rings. All C-H bond lengths were given values of 1.08 A. The structure that we used is shown in Fig. 3. Note that the observed angle between the three-membered and the six-membered rings is 102.5°, essentially the same as our optimized value for benzene epoxide, 105°. There is similarly excellent agreement between the C-O distances in Fig. 3 and our optimized benzene epoxide value, 1.47 A. For the perepoxide, we simply added a second oxygen, at the same distance and angular displacement that we found to be optimum for benzene perepoxide.

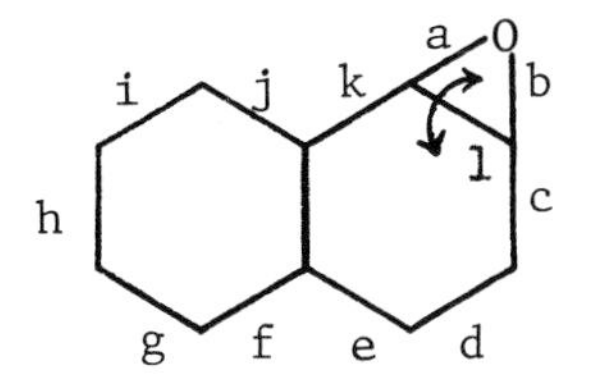

a=1.461 A	g=1.381 A
b=1.459 A	h=1.391 A
c=1.479 A	i=1.391 A
d=1.403 A	j=1.389 A
e=1.480 A	k=1.474 A
f=1.402 A	l=1.473 A

angle=102.5°

Figure 3. Structure used for naphthalene-1,2-epoxide, based upon experimentally-determined structure of phenanthrene-9,10-epoxide [15].

These naphthalene products present no important new features. Again the epoxide is very stable relative to the hydrocarbon plus a singlet oxygen atom,

$$E_{epoxide} - E_{naphthalene} - E_{O(^1D)} = -84 \text{ kcal/mole},$$

while the conversion of the perepoxide to the epoxide plus $O(^1D)$, as before, is not energetically favored:

$$E_{epoxide} + E_{O(^1D)} - E_{perepoxide} = +24 \text{ kcal/mole}.$$

The oxygen charges in the perepoxide follow the same pattern (-0.14 for the inner oxygen, -0.28 for the outer) as has been found throughout this study.

SUMMARY

Several general trends can be detected in the results that we have obtained. First, the stabilities of the epoxides relative to the hydrocarbons plus a singlet oxygen atom increase in the order:

benzene, toluene ($\Delta E = \sim -73$ kcal/mole) < naphthalene ($\Delta E = -84$ kcal/mole) < ethylene ($\Delta E = -118$ kcal/mole),

where

$$\Delta E = E_{epoxide} - E_{hydrocarbon} - E_{O(^1D)}$$

This order parallels an increase in the double-bond character of the C-C bond involved in the epoxide formation, as measured by its bond length, and is in accord with the fact that the epoxides of naphthalene and ethylene have been isolated, whereas those of benzene and toluene have not.

Second, the process

$$\text{perepoxide} \longrightarrow \text{epoxide} + O(^1D)$$

is not energetically favored in any of the four cases that have been investigated. Remarkably, $\Delta E = +24$ kcal/mole in each instance. It appears that some external agent must be involved in the perepoxide → epoxide conversion; two possibilities for such an agent are a second $O_2(^1\Delta)$ molecule (one $O_2(^1\Delta)$ having presumably participated in forming the perepoxide) or a second hydrocarbon molecule [14].

Finally, we find a consistent tendency for the two oxygen atoms in the perepoxide to have different charges, the one closer to the hydrocarbon being the more positive. This suggests that the $O_2(^1\Delta)$ molecule may develop a degree of polarity as it approaches the hydrocarbon, and that its electrophilic end leads the approach. Thus, if an $O_2(^1\Delta)$ - hydrocarbon interaction does play an important role in the carcinogenic actions of polynuclear aromatic hydrocarbons, then an effective tool for studying these processes should be the electrostatic potential that is produced in the neighborhood of a hydrocarbon molecule by its nuclei and electrons. This potential reveals the sites and regions on a molecule that an electrophile would tend to approach [16,17]. We are presently calculating the electrostatic potentials of various polynuclear aromatic hydrocarbons, with the purpose of gaining additional insight into their carcinogenic activities [18,19].

REFERENCES

[1] J. W. Daly, D. M. Jerina and B. Witkop (1972) Experientia 28 1129.
[2] J. C. Arcos and M. F. Argus (1974) Chemical Induction of Cancer, Vol. IIA, Academic Press, New York, pp. 135-236.
[3] P. G. Wislocki, A. W. Wood, R. L. Chang, W. Levin, H. Yagi, O. Hernandez, D. M. Jerina, and A. H. Conney (1976) Biochem. Biophys. Res. Comm. 68 1006.
[4] A. M. Jeffrey, K. W. Jennette, S. H. Blobstein, I. B. Weinstein, F. A. Beland, R. G. Harvey, H. Kasai, I. Miura and K. Nakanishi (1976) J. Amer. Chem. Soc. 98 5714.
[5] S. H. Blobstein, I. B. Weinstein, P. Dansette, H. Yagi and D. M. Jerina (1976) Cancer Res. 36 1293.
[6] GAUSSIAN 70, Quantum Chemistry Program Exchange, Indiana University, Bloomington, IN, 47401, program no. 236, developed by W. J. Hehre, W. A. Lathan, R. Ditchfield, M. D. Newton and J. A. Pople.
[7] J. Iball, S. N. Scrimgeour and D. W. Young (1976) Acta Cryst. B32 328.
[8] Tables of Interatomic Distances and Configuration in Molecules and Ions (1958, 1965) L. E. Sutton, ed., The Chemical Society, London, Spec. Pub. Nos. 11 and 18.

[9] J. Iball (1964) Nature 201 916.
[10] D. R. Kearns (1971) Chem. Rev. 71 395.
[11] K. Yamaguchi, T. Fueno and H. Fukutome (1973) Chem. Phys. Letters 22 466.
[12] M. J. S. Dewar and W. Thiel (1975) J. Amer. Chem. Soc. 97 3978.
[13] S. Inagaki and K. Fukui (1975) J. Amer. Chem. Soc. 97 7480.
[14] M. J. S. Dewar, A. C. Griffin, W. Thiel and I. J. Turchi (1975) J. Amer. Chem. Soc. 97 4439.
[15] J. P. Glusker, H. L. Carrell, D. E. Zacharias and R. G. Harvey (1974) Cancer Biochem. Biophys. 1 43.
[16] E. Scrocco and J. Tomasi (1973) in: Topics in Current Chemistry, New Concepts II, No. 42, Springer Verlag, Berlin, p. 95.
[17] P. Politzer and K. C. Daiker in: The Force Concept in Chemistry, B. M. Deb, ed., Macmillan of India, ch. 7, in press.
[18] P. Politzer, K. C. Daiker and R. A. Donnelly (1976) Cancer Letters 2 17.
[19] P. Politzer and K. C. Daiker in: Quantum Biology Symposium No. 4, P.-O. Lowdin, ed., John Wiley & Sons, New York, in press.

ACKNOWLEDGMENT

We are very grateful for the financial support of the Cancer Association of Greater New Orleans, and the University of New Orleans Computer Research Center.

CHEMICAL PRODUCTION OF EXCITED STATES: ADVENTITIOUS BIOLOGICAL CHEMILUMINESCENCE OF CARCINOGENIC POLYCYCLIC AROMATIC HYDROCARBONS

H. H. Seliger and J. P. Hamman

McCollum-Pratt Institute, The Johns Hopkins University, Baltimore, Maryland U.S.A.

INTRODUCTION

The chemical production of electronically excited states in biological systems falls into several categories. The first and most prominent is Bioluminescence, a late evolutionary selection for "luciferase"-catalyzed monooxygenations of "luciferin" substrates to produce light for signalling with efficiencies (photons emitted per substrate molecule reacted) close to 100% [1]. Second is the delayed luminescence of pre-illuminated chloroplasts at times (10^{-5} sec to 10^4 sec) too long for the process to be a primary fluorescence or slow fluorescence [2,3]. Delayed luminescence is thought to be the result of a back reaction (recombination) of photoreactants within the thylacoid that have undergone charge separation due to the primary illumination [3,4], and is, like the *in vivo* fluorescence of chlorophyll, non-functional. The third category which we have termed Adventitious Biological Chemiluminescence, ABC, is a non-functional, non-enzymatic light emission that occurs fortuitously and with low probability during aerobic metabolism. The literature is replete with observations, from the work of Gurwitsch beginning in the 1920's through the present, of ABC emitted from whole tissues, from isolated cells and cell homogenates, from microsomal extracts and from purified flavoprotein oxidases and peroxidases [5-12]. In those cases where the chemical precursors of the electronically excited states have been examined it appears that $O_2^{\dot{-}}$ and H_2O_2, released to varying degrees during the course of mixed function oxidase reactions, initiate radical reactions resulting in light emission, either a direct chemiluminescence of a number of susceptible molecules in the environs of the oxygen radicals or a sensitized chemiluminescence from a fluorescent acceptor molecule. The intermediacy of $O_2^{\dot{-}}$ and H_2O_2 or its

B. Pullman and N. Goldblum (eds.), Excited States in Organic Chemistry and Biochemistry, 345-359.

radical decomposition products has been demonstrated by the use of Luminol as a sensitive chemiluminescent probe [13-16] and by the inhibition of this ABC by superoxide dismutase and by catalase [17].

From an analysis of the steps in monooxygenase reactions of bioluminescent substrates, it was suggested that since metabolism of polycyclic aromatic hydrocarbons, PAH, also proceeded through monooxygenase reactions, certain metabolic intermediates, namely epoxides and their keto oxygen tautomers, might be subject to spontaneous oxygenation [1,18]. This low probability non-enzymatic oxidation could result in the release of sufficient free energy to leave the oxidized product molecules in excited states. This type of biological chemiluminescence, the spontaneous oxygenation of a <u>specific metabolic intermediate</u>, is therefore different from the nonspecific radical ABC initiated by the release of $O_2^{\dot{-}}$ and H_2O_2 during flavin oxidase and peroxidase reactions. The reactivity towards cellular nucleophiles of specific metabolites of these PAH's, which would be related to their mutagenicity or carcinogenicity, should also be reflected by their probability for reacting spontaneously with oxygen to produce chemiluminescence. Therefore, chemiluminescence accompanying aryl hydrocarbon hydroxylase, AHH, reactions of carcinogenic PAH's might be a measure of the mutagenicity or carcinogenicity of the metabolites. On the basis of these arguments, it was predicted that the PAH carcinogen benzo[a]pyrene, BP, should be chemiluminescent during microsomal metabolism [1]. We have verified the chemiluminescence of BP during its microsomal metabolism [19] and also that of 3-methylcholanthrene, and dibenz[a,h]anthracene, in liver microsomal preparations from methylcholanthrene-induced rats [20]. More recently we have shown that this is a specific ABC, not inhibited by superoxide dismutase or by catalase, and is produced from both diasteriomers of 7,8-dihydrodiol-9,10-epoxy-BP [21], the presumed ultimate carcinogenic metabolite of BP [22-24].

There is a large body of experimental evidence supporting the theory that enzymatic activation of parent carcinogenic polycyclic aromatic hydrocarbons (PAH) is required to produce ultimate carcinogenic metabolites [25,26]. In the case of BP three separate, consecutive, enzyme reactions must take place before the mutagenic diol epoxide is produced, *i.e.* before ABC can take place. None of these oxygenations or hydrations involves the K region of Schmidt [27] or of the Pullmans [28]. Parent carcinogenic PAH's are not intrinsically mutagenic; they must be metabolized either by a cell's AHH system or by microsomal extracts in order for mutagenicity to be expressed. Parent carcinogenic hydrocarbons do not bind covalently to protein or nucleic acid; they must be metabolized [29,30] or at least participate in photochemical reactions [31].

In view of all of this circumstantial evidence that the expression of carcinogenicity requires metabolism of the parent carcinogen, metabolism that takes place predominantly at some distance from the K region, it becomes difficult to understand why there should exist some partial correlation between the electron density or reactivity of the K region of a parent PAH and the carcinogenicity of one of its subsequent metabolites. In this paper we shall show that the electronic theories of carcinogenesis describe different parameters of the parent carcinogenic hydrocarbon that are involved in its metabolic conversion to a reactive bay region epoxide, the presumed ultimate carcinogenic metabolite. We shall describe nonspecific and specific chemical production of excited states in biological systems and propose that specific ABC is a measure of the chemical reactivity and rate of production of epoxide metabolites during drug metabolism of carcinogenic polycyclic aromatic hydrocarbons.

Electronic Theories of Carcinogenesis

The common thread binding all of the electronic theories that predict the carcinogenicity of polycyclic aromatic hydrocarbons is the implication of a specific region of high electron density in the molecule that acts as a "carcinogenophore." Schmidt's original hypothesis was that these electron-dense regions induced a keto-enol rearrangement, the precursor of a mutation leading to carcinogenesis. Svartholm [32] attempted to correlate carcinogenicity with reactivity of the mesophenanthrenic regions. Badger [33,34] showed a correlation between carcinogenic PAH's and the activity of mesophenanthrenic bonds to form addition compounds with OsO_4. The Daudels [35,36] and the Pullmans [28] extended the quantum mechanical calculations of the charge densities of these K (German *Krebs*, for cancer) regions and achieved success in correlating strong K regions (high charge density at mesophenanthrenic bonds) with carcinogenicity. However, high electron densities at mesoanthracenic regions (L regions) did not correlate with carcinogenicity. These L regions were therefore empirically accommodated in the numerical calculations as opposing the carcinogenicity of the parent molecule. Further circumstantial evidence for a carcinogenophore region of the parent molecule came from the correlations observed between photodynamic activity of PAH's and carcinogenesis [37-39; see also reference 40 for an early review], and the promotion by light of the carcinogenesis of PAH's contained in tar. This correlation between photodynamic action and carcinogenicity led Birks [41] to calculate resonance transfer overlap integrals between the photoexcited states of the parent hydrocarbons and the amino acid tryptophan, in the event that the electronically excited states of the parent carcinogens might be involved in the "carcinogenic reaction" with proteins. Again, a partial correlation was found between the energies of the 1st excited states and the carcinogenicities of the parent PAH's.

None of these theories takes into account the biochemistry of drug metabolism by enzyme systems. Therefore they cannot predict with any precision the carcinogenicity of subsequent metabolic products. Why then the tantalizing correlations between properties of parent compounds and carcinogenicity? A subtle distinction was recognized early by Anderson [42] who proposed that the K region was not the carcinogenophore but the *binding site* of the carcinogen to its metabolizing enzyme, and that hydroxylation took place at a different site from the K region, anticipating Boyland's independent confirmation of this metabolic region. The implications of Anderson's model are major. The reactivities of K regions as measured chemically, in addition reactions with OsO_4, or biologically, in photodynamic action, or the relative shapes and angularities of parent carcinogens as evidenced by energy levels of excited electronic states, are *initial* screening criteria that relate to the *binding* of the xenobiotic compounds to the body's drug metabolizing enzymes, but not necessarily to the carcinogenicities of the subsequent metabolites.

Seliger and Hamman [20] extended this binding site-metabolic site model to the remaining bonds of the molecule. The K region, by virtue of its highest electron density, would still be the most probable binding site. However, from examination of the spectrum of the products reported [43] for the metabolism of BA and BP they proposed a minimum of three linear geometries that could account for the relative yields of all of the known metabolites. These were, in relation to the benzo[a]pyrene molecule:

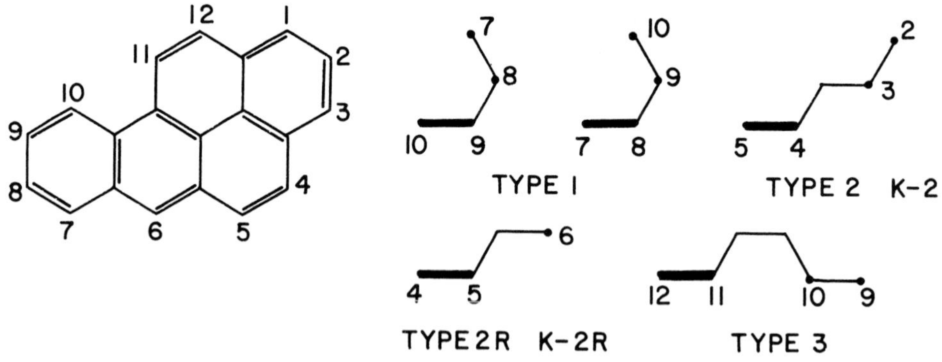

where the thick solid line represents the binding site and the solid dots represent the sites for oxygenation (epoxidation). Where binding coincides with a K region the geometry is called K-2. The restricted (R) Type 2 geometry represents all of the L region reactions.

The common geometry mechanism is an empirical theory that was constructed to fit the observed distributions of metabolites of PAH's with only two initial criteria: (a) metabolic oxygenation of the parent carcinogen must occur and therefore the binding site of the substrate to the enzyme must be different from the site for oxygenation. (b) the large number of PAH's in the environment are not produced biochemically. They are the result of high temperature pyrolysis, the burning of carbonaceous material-volcanic action, forest fires, the industrial revolution, cigarette smoking. They therefore arose late in evolution subsequent to the emergence of green plants from the seas onto the land. Therefore the first step in detoxification, the oxygen step, should be culled from those chemical reactions already evolved for biological aromatic hydrocarbons, *i.e.* sterols. The optimum efficiency in handling the myriad of PAH's would be achieved if, due to the structures of the PAH's, they exhibited common geometrical relationships between the binding sites and the sites of oxygenation. This would permit a degree of nonspecificity for the detoxifing enzymes that would be consistent with the observations that induction of AHH systems by any specific carcinogen such as BA or MC results in the ability of the AHH system to metabolize a wide range of carcinogenic PAH's. This nonspecificity would be reflected in an economy of binding-active site configurations for the detoxifying enzymes, *i.e.* a selection for a minimum number of common geometries.

A corollary of this second criterion is that there are some enzymatic oxygenation reactions that, for certain PAH's, result in strongly electrophilic metabolites. These PAH's would be the carcinogenic PAH's. Since these same detoxification reactions would not be expected to have been selected during evolution for producing carcinogenic metabolites from their natural biochemical substrates, there must be something "different" about the structure of carcinogenic PAH's that permits the fortuitous production of carcinogenic metabolites during the "normal", predominant production of soluble, noncarcinogenic detoxification products. It is this difference, the structure of the parent carcinogenic PAH, that provides the key to the partial correlations between electronic theory parameters and carcinogenicity.

The cynosure of all of these empirical models was provided by the work of Jerina *et al.* [44] who proposed that epoxides in the "bay region" formed by angular benzo rings are much more likely to form reactive carbonium ions and thus are much stronger electrophiles than epoxides at other positions. With this description of the reactive site of the proposed ultimate carcinogenic metabolite the entire pattern of electronic theories becomes clear. Consider the case of BP. Drug metabolism of a potential carcinogen involves a "recognition" by one of the hydroxylase enzymes, most probably the non-carcinogenic K-2 pathway:

However, there is also some small probability for the following competitive, consecutive reactions

resulting in the mutagenic 7,8-diol-9,10-oxide-BP, a bay region epoxide. In this case the saturated 7,8 bond inhibits epoxide hydratase attack on the 9,10 epoxide and thus effectively increases the probability that the epoxide will react with a nucleophile [45].

The original correlation between strong K regions and carcinogenicity is therefore nothing more than the confirmation of the presence of a mesophenanthrenic region, an angular benzo ring, a bay region and therefore a possible production of a bay region epoxide. A PAH with a strong K region would be more likely to react photochemically to produce photodynamic damage and would therefore also be likely to be mutagenic.

The L region is involved in competing reactions leading apparently to noncarcinogenic metabolites. It was recognized quite early that steric suppression of L region competing interactions increased the carcinogenicity of the substituted molecule. For example, 7-methyl BA, benzo[a]pyrene and cholanthrene can be considered as BA derivatives where the L region has been suppressed. The structures of 7-methyl-BA, BP and cholanthrene respectively are shown below.

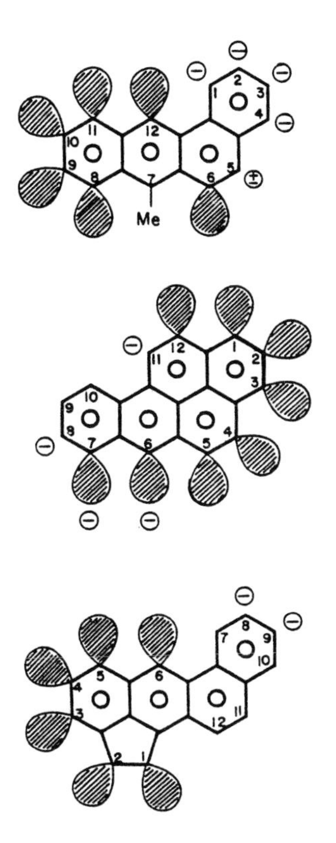

The carbon positions at which mono-methylation retains the carcinogenicity of the parent molecule [46] are shown as the diagonally hatched orbitals. The minus signs (-) represent the carbon positions at which methylation or hydroxylation eliminates the carcinogenicities of the molecules. The plus-minus signs (±) represent conflicting experimental observations. In BA therefore the major reactive epoxide is 1,2-epoxide. In BP it appears that interference with the common geometry steps leading to 7,8-dihydrodiol-9,10-epoxy-BP inhibits the carcinogenicity of the molecule. The similarity between cholanthrene and 7-methyl-BA is evident from the structures above. Again the "1,2-epoxide," in the cholanthrene case the 7,8-epoxide, is implicated as the carcinogenic epoxide.

In the common geometry mechanism the correlation of Type 3 geometry with carcinogenicity is equivalent to the identification of the production of a bay region epoxide. However the K-2 geometry, which uses the Pullman's K region as a binding site, along with the K-2R geometry, appear to be the major noncarcinogenic detoxification pathways for PAH's. Only Type 1 and Type 3 geometries lead to bay region epoxides.

If angular benzo rings are subject to "bay region" epoxidation, the molecule is a potential carcinogen. If the competing and more probable K-2 and K-2R reactions leading to soluble noncarcinogenic metabolites occur, the xenobiotic molecule can be excreted rapidly and is not normally a "carcinogen." If it is not excreted rapidly it may be further metabolized to a bay region epoxide. This may be one mechanism whereby cytotoxic damage affecting the excretion of metabolites may promote carcinogenesis.

The poor correlation between excited state energy levels and carcinogenicity is evident in the following example: Although there are essentially no differences in the energies of the 1st excited states of the noncarcinogenic 1,2,3 and 4 monomethyl derivatives of BA compared with the carcinogenic 6,7,8 or 12 monomethyl derivatives [47], the differences in the carcinogenicities are tremendous. The implication of the proposed correlations would invoke charge transfer complexes or a photochemical binding of the PAH to cellular macromolecules. These mechanisms would be more likely to correlate with photodynamic action than with carcinogenicity.

A possible correlation between carcinogenicity and photodynamic action has recently been re-investigated [48]. It follows that any cytotoxicity due to photoexcitation would be related to the electron densities of the K and L regions and therefore to the probability for photochemical reactions at these bonds. The arguments against the validity of attempts to correlate photodynamic action with carcinogenicity are essentially the same as those that point out the limitations of the K region calculations; further enzymatic reactions must occur before the carcinogenicity is expressed. The K region appears to be the site at which a minimum of metabolic activity occurs (a binding site rather than a reactive site). In reference 48, BA, a borderline carcinogen, exhibited a photodynamic activity 18 times that of BP whose photodynamic activity was exceeded even by naphthacene. The reactivity of K and L regions of parent compounds leading to photodynamic activity is only incidently related to subsequent metabolism to bay region epoxides.

In Table 1 are shown the proposed common geometry pathways for producing bay region epoxides from all of the known unsubstituted carcinogenic PAH's. The reactive bay region epoxides are produced fortuitously amidst a spectrum of much more probable noncarcinogenic metabolites, all resulting from a minimum of three steric configurations for epoxidation reactions.

Adventitious Biological Chemiluminescence

The influence of light on biological systems is most obvious. Sunlight provides the free energy for all life on earth and regulatory systems in almost all organisms depend in some manner on light. Life and even the organic substrates produced during primitive

TABLE 1

Parent Hydrocarbon	Geometry[a]	Reactive Epoxide
benzo[a]phenanthrene	Type 1	1,2-E
benzo[b]phenanthrene		
benz[a]anthracene	Type 1	1,2-E
benzo[c]phenanthrene	Type 1	3,4-E
chrysene	K-3	3,4-E
dibenzo[a,c]phenanthrene	(Type 1)(EH)[b] (K-3)	1,2-D[c]-3,4-E
dibenzo[a,i]phenanthrene	Type 1	3,4-E
	Type 3	3,4-E
dibenzo[c,i]phenanthrene	Type 1	1,2-E
picene	Type 3	1,2-E
dibenz[a,h]anthracene	Type 1	1,2-E
dibenz[a,j]anthracene	Type 1	1,2-E
benzo[a]pyene	(Type 1) (EH) (Type 3)	7,8-D-9,10-E
benzo[e]pyrene	Type 1	4,5-E
dibenzo[a,e]pyrene	(Type 1) (EH) (K-3)	3,4-D-1,2-E
dibenzo[a,h]pyrene	(Type 1) (EH) (K-3)	3,4-D-1,2-E
dibenzo[a,i]pyrene	(Type 1) (EH) (K-3)	3,4-D-1,2-E
dibenzo[a,l]pyrene	Type 1	1,2-E
	Type 1	11,12-E
dibenzo[c:d,e]pyrene	Type 3	8,9-E
tribenzo[a,e,i]pyrene	Type 1	1,2-E
naphtho[a]pyrene	Type 1	1,2-E
phenanthra[a]pyrene	Type 1	1,2-E
cholanthrene	Type 1	7,8-E

[a] For symmetric molecules only one reaction is shown.
[b] Epoxide hydratase
[c] D = dihydrodiol

photochemical evolution have centered about the potential for electronically excited states of molecules to do chemical work. The advantage of photochemistry is in its selectivity. Specific pigment molecules in a uniform ambient temperature bath can be raised to energy levels corresponding to 20,000-50,000°K, permitting selective reactions to occur that are not accessible otherwise. Since nature is by far the most efficient chemical machine it would be of interest to ask whether and under what conditions it might be of some advantage to an organism to produce excited states chemically for some type of biochemical regulation. The energetics of this reaction would be quite different from the ordinary commerce of biochemistry that proceeds in units of ATP hydrolysis. This would require free energy releases of 70-90 kcal $mole^{-1}$ in order for the "pigment" molecule involved in the control process to be produced in its 1st excited state. There appear to be only two possible functions for chemically-produced excited states in biological systems. The first and the only known function for chemically-produced excited states in biological systems is for bioluminescence, the efficient emission of light for signalling. In this case there are specific enzymes and substrates and, except for bacteria and yeasts where the control is nutritional, there have evolved concurrently, complex regulatory and triggering systems for the bioluminescent reactions. The second would have to be related to a control process, either essential to the replication or division of a cell and which by natural selection remains at so high an activation energy that only chemically-produced excited states may participate, or for DNA repair and excision, analogous to photoreactivation, in which case the energy requirements again may be high.

In neither of these two hypothetical cases for chemically-produced excited states would there be a selection for light emission. If the low *in vivo* fluorescence quantum yield of system I chlorophyll can be used as an example of efficient utilization of the energy of electronically excited states to do chemical work, a negligible fraction of these very specific chemical regulation reactions would be detectable by light emission. It is doubtful therefore that significant biological chemiluminescence would be emitted from organisms as the result of control processes without the partial disruption and/or insertion of acceptor fluorescent probes to detect the presence of excited states.

There is, however, a significant amount of light emitted from many chemical oxidations. The statistical term "significant" refers to whether a signal can be detected above the background noise [49]. In some cases with low noise phototubes and good light collection the background can be equivalent to the emission of only 200 photons s^{-1}. Therefore it is quite possible to detect chemiluminescent reactions where the quantum yields are 10^{-12} or below. In chemiluminescent reactions of organic molecules the free energy for light emission is usually the result of oxidations involving oxygen directly. As the result of chemical reactions of oxidases and peroxidases,

$O_2^{\dot{-}}$ or H_2O_2 may be released. These in turn can result in $OH^{\cdot}$ or $CO_3^{\dot{-}}$ radicals or in some cases O_2^*. During metabolism therefore it is quite possible for minute releases of these reactive oxygen species to produce an adventitious radical chemiluminescence among a variety of molecules in the immediate vicinity of the release. These fortuitous chemiluminescent reactions represent a negligible pathway for degradation ($<10^{-6}$) and as such are non-functional. Nevertheless they can be observed because of the extreme sensitivity of present day photon detection. It should not be surprising therefore that rapidly metabolizing tissues, cells, organelles, homogenates, and even purified enzyme systems emit a detectable nonspecific Adventitious Biological Chemiluminescence, ABC [1,21]. The microsomal chemiluminescence of lipids [10] and the chemiluminescence of Luminol in xanthine oxidase, horseradish peroxidase and myeloperoxidase reactions [12-16] are examples of such nonspecific ABC. These are strongly inhibited by superoxide dismutase and by catalase.

Specific ABC and Reactivity

One of the predictions of the common geometry mechanism was that the metabolism of carcinogenic PAH's should produce reactive intermediates (epoxides ⇌ carbonium ions) that would be subject to spontaneous oxygenation to produce ring-opened carbonyl product molecules in excited electronic states, an Adventitious Biological Chemiluminescence. This specific ABC was proposed to be different from the radical-pathway chemiluminescences initiated by $O_2^{\dot{-}}$ and the radical dissociation products of H_2O_2; it should not be inhibited by superoxide dismutase or by catalase. We have verified that the microsomal metabolism of BP is accompanied by chemiluminescence [19]. This was shown to be a specific ABC produced from 7,8-dihydrodiol-9,10-epoxy-BP, the latter the product of the microsomal metabolism of 7,8-dihydrodiol-BP [21]. The specific ABC accompanying the microsomal metabolism of 7,8-dihydrodiol-BP is not inhibited by superoxide dismutase or by catalase, and shows saturation kinetics as a function of substrate concentration.

The epoxide can spontaneously rearrange to the phenol. In some cases rearrangement to the phenol is inhibited and the oxirane ring may be induced to open to form the carbonium ion. At this point the carbonium ion can react with water to produce the diol, with amino-nitrogens or sulfhydryls to produce covalent conjugates or, with low probability, with the $^{\ominus}O\text{-}O^{\oplus}$ form of oxygen to produce chemiluminescence. The oxygen reaction is a negligible pathway. However if it results in chemiluminescence it should be readily detectable. We have measured the chemiluminescent quantum yield for the microsomal metabolism of 7,8-dihydrodiol-BP to be *ca.* $10^{-8}/\phi_{f\ell}$, where $\phi_{f\ell}$ is the as yet unknown fluorescence quantum yield of the product molecule. The oxygenation step proposed for the ABC of the carbonium ion of the diol epoxide of BP implies that the probability of the carbonium ion reaction towards amino nitrogens, *i.e.* the 2-amino group of guanine, should also be proportional to the probability of the reaction with molecular oxygen. If the mutagenicity (carcinogenicity) of the diol epoxide of BP is related to its covalent binding to protein or to nucleic acid (reactivity of the carbonium ion), the mutagenicity should also be related to the rate of reaction of the carbonium ion with oxygen to produce chemiluminescence. Therefore chemiluminescence can be an assay for the enzyme pathway leading to the production of the diol epoxide and as a tracer for the concentrations and total amounts of reactive epoxides.

We have already reported that the total light emitted per mole of BP metabolized is the same as that for 3-OH-BP [21], the latter isolated by high pressure liquid chromatography. The temporal kinetics of ABC from the metabolism of BP are the same as from the metabolism of 3-OH-BP, showing a peak in chemiluminescence at around 10 minutes (Fig. 1 of ref. 19). The kinetics of ABC from an equimolar concentration of 7,8-dihydrodiol-BP show a rapid rise of chemiluminescence, peaking at 45 seconds and the total light emitted per mole is more than 3 X that for BP. From these data it can be inferred that the metabolic production of 3-OH-BP is not accompanied by chemiluminescence and that the production of 3-OH-BP is not necessary to the production of 7,8-dihydrodiol-9,10-epoxy-BP.

For BP, reacted with microsomes and NADPH in the presence of nucleic acid or poly-guanine, the covalent binding should be proportional to the integral of the ABC. In the *Salmonella typhimurium* bacterial system [26] the mutagenicity should also be proportional to the total chemiluminescence emitted. Glutathione and cysteine inhibit the microsomal ABC of BP and 7,8-dihydrodiol. However they have no effect on the rate of microsomal production of 3-OH-BP from BP. If we assume that the reaction of the sulfhydryl is with the 9,10 arene oxide of 7,8-diol-9,10-epoxy-BP, the fractional inhibition of microsomal ABC should be correlated with the inhibition of the microsome-activated mutagenicity of BP and 7,8-diol-BP in the bacterial system and with the inhibition of the microsome-activated binding of BP and 7,8-diol-BP to nucleic acid and protein. If these correlations hold it should be possible to correlate the microsomal ABC of a polycyclic aromatic hydrocarbon to its potential mutagenicity by a chemiluminescent assay parallel to the scoring of bacterial revertants. The microsomal ABC of 9,10-dihydrodiol-BP is less than 4 percent of the 7,8-diol. A corollary of the production of microsomal ABC and mutagenicity is the prediction that the microsomal ABC of 7-methyl-BA should be much greater than that of BA alone. Since the 1,2,3 and 4 monomethyl derivatives of BA are non-carcinogenic their microsomal ABC should also be negligible. Reactions of oxygen with carbonium ions in general are exergonic but may result in non-fluorescent products. It is conceivable therefore that non-fluorescent excited states might be detected by means of sensitized fluorescence of suitable acceptor molecules.

The mechanism proposed for specific ABC in the case of BP is a reaction

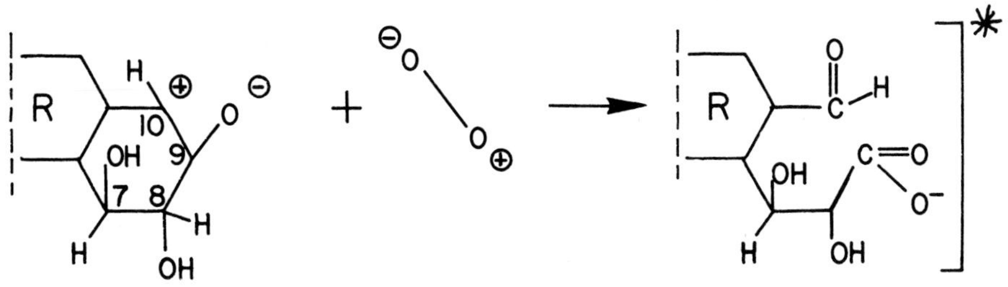

to produce a carbonyl excited state. In the case of the PAH's the product aromatic aldehydes have a finite probability for fluorescence. The reaction of the carbonium ion with the amino nitrogen or sulfhydryl nucleophiles will not necessarily be chemiluminescent. Apart from the ground state carbonium S_N1 reactions with macromolecular nucleophiles, it is possible that the carbonyl excited state product molecules of the oxygen reaction are even stronger electrophiles than ground state carbonium ions. They might then participate in a class of nucleophilic substitution reactions having high activation energies, not normally accessible even to carbonium ions. If these are specific "carcinogenic" reactions, differentiated from the background of cytotoxic and mutagenic ground state nucleophilic substitution reactions, the measurement of specific ABC may take on even further significance. It would be important therefore to compare the distributions of intrinsic binding of the diol epoxides of BP to nucleic acid and to chromatin in the presence and absence of oxygen; 7,8-dihydrodiol-9,10-epoxy-BP has been shown to require oxygen for chemiluminescence [21].

ACKNOWLEDGEMENTS

Supported by U. S. Energy Research and Development Administration under contract E4-76-S-02-3277. Contribution No. 896 of the McCollum-Pratt Institute and Department of Biology of The Johns Hopkins University.

REFERENCES

1. Seliger, H. H. (1975) Photochem. Photobiol. 21:355.
2. Strehler, B. and Arnold, W. (1951) J. Gen. Physiol. 34:809.
3. Arnold, W. and Azzi, J. (1971) Photochem. Photobiol. 14:233.
4. Kraan, G. P. B., Amesz, J., Velthuys, B. R. and Steemers, R. G. (1970) Biochem. Biophys. Acta 223:129.
5. DeMent, J. (1945) Fluorochemistry. Chemical Pub. Co., New York.

6. Barenboim, G. M., Domanskii, A. N. and Turoverov, K. K. (1969) Luminescence of Biopolymers and Cells. Plenum Press, New York.
7. Quickenden, T. I. and Que Hee, S. S. (1974) Biochem. Biophys. Res. Commun. 60:764.
8. Stauff, J. and Wolf, H. (1964) Z. Naturforsch. 19B:87.
9. Stauff, J. and Ostrowski, J. (1967) Z. Naturforsch. 226:734.
10. Howes, R. M. and Steele, R. H.(1971) Res. Commun. Chem. Pathol. Pharmacol. 2:619; *ibid* (1972) 3:349.
11. Arneson, R. M. (1970) Arch. Biochem. Biophys. 136:352.
12. Hodgson, E. K. and Fridovich, I. (1976) Arch. Biochem. Biophys. 172:202.
13. Totter, J. R., Medina, V. J. and Scoseria, J. L. (1969) J. Biol. Chem. 235:238.
14. Totter, J. R., diGros, E. D. and Riveiro, C. (1969) J. Biol. Chem. 235:1839.
15. Greenlee, L., Fridovich, I. and Handler, P. (1962) Biochem. 1: 779.
16. Fridovich, I. and Handler, P. (1962) J. Biol. Chem. 237:916.
17. Kellogg, E. W. III and Fridovich, I. (1975) J. Biol. Chem. 250: 8812.
18. Seliger, H. H. (1975) Fed. Proc. 34:623.
19. Hamman, J. P. and Seliger, H. H. (1976) Biochem. Biophys. Res. Commun. 70:675.
20. Seliger, H. H. and Hamman, J. P. (1976) J. Phys. Chem. 80:2296.
21. Hamman, J. P. and Seliger, H. H. (1977) Biochem. Biophys. Res. Commun. (1977) in press.
22. Newbold, R. F. and Brookes, P. (1976) Nature 261:52.
23. Wislocki, P. G., Wood, A. W., Chang, R. L., Levin, W., Yagi, H., Hernandez, O., Jerina, D. M. and Conney, A. H. (1976) Biochem. Biophys. Res. Commun. 68:1006.
24. Huberman, E., Sachs, L., Yang, S. K. and Gelboin, H. V. (1976) Proc. Nat. Acad. Sci. U.S.A. 73:607.
25. Miller, J. A. and Miller, E. C. (1971) J. Nat. Cancer Inst. 47: V-XIV.
26. Ames, B. N., Durston, W. E., Yamasaki, E. and Lee, F. D. (1973) Proc. Nat. Acad. Sci. U.S.A. 70:2281.
27. Schmidt, O. (1939) Z. physikal. Chemie (B) 42:83; 44:194.
28. Pullman, A. and Pullman, B. (1955) Adv. Cancer Res. 3:117; (1969) in The Jerusalem Symposia on Quantum Chemistry and Biochemistry Vol. 1. eds. E. D. Bergmann and B. Pullman. Academic Press, N. Y. p. 9.
29. Grover, P. L. and Sims, P. (1968) Biochem. J. 110:159.
30. Gelboin, H. V. (1969) Cancer Res. 29:1272.
31. Ts'o, P. O. P., Caspary, W. J., Cohen, B. I., Leavitt, J. C., Lesko, S. A., Lorentzen, R. J. and Schechtman, L. M. (1974) in "Chemical Carcinogenesis, Part A" eds. Ts'o, P. O. P. and Di Paolo, J. A. Marcel Dekker, New York. p. 113.
32. Svartholm, N. (1941) Arkiv. Kemi. Min. Geol. A15 No. 13.
33. Badger, G. M. (1949) J. Chem. Soc. 456, 1909.
34. Badger, G. M. (1950) J. Chem. Soc. 1726, 1809.

35. Buu-Hoi, N. P., Daudel, P., Daudel, R., Lacassagne, A., Lecocq, J., Martin, M. and Rudali, G. (1947) C. R. Acad. Sci. Paris 225:238.
36. Daudel, P. and Daudel, R. (1950) Biol. Med. 39:201.
37. Epstein, S. S., Small, M., Falk, H. L. and Mantel, N. (1964) Cancer Res. 24:855.
38. Small, M., Mantel, N. and Epstein, S. S. (1967) Exp. Cell Res. 45:206.
39. Malling, H. V. and Chu, E. H. Y. (1970) Cancer Res. 30:1236.
40. Santamaria, L. and Prino, G. (1964) in Res. Prog. in Org. Biological and Med. Chem. eds. Gallo, V. and Santamaria, L. Societa Editoriale Farmaceutica, Milano. p. 259.
41. Birks, J. B. (1959) in "General Discussion" Discussions Faraday Soc. 27:232.
42. Anderson, W. (1947) Nature (London) 160:892.
43. Selkirk, J. K., Croy, R. G. and Gelboin, H. V. (1974) Science 184:169; (1975) Arch. Biochem. Biophys. 168:322.
44. Jerina, D. M. Lehr, P. E., Yagi, H., Hernandez, O., Dansette, P., Wislocki, P. G., Wood, A. W., Chang, R. L., Levin, W. and Conney, A. H. (1976) in "*In vitro* Activation in Mutagenesis Testing" eds. De Seres, F. J., Bond, J. R. and Philpot, R. M. Elsevier, Amsterdam.
45. Sims, P., Grover, P. L., Swaisland, A., Pal, K. and Hewer, A. (1974) Nature (London) 252:326.
46. Dipple, A. (1976) "Polynuclear Aromatic Carcinogens" in Chemical Carcinogens, ed. Searle, C. E., ACS Monograph 173, ACS, Washington, D. C. p. 245.
47. Morgan, D. D. Warshawsky, D. and Atkinson, T. (1977) Photochem. Photobiol. 25:31.
48. Morgan, D. I . and Warshawsky, D. (1977) Photochem. Photobiol. 25:39.
49. Seliger, H. H. (1973) in "Chemiluminescence and Bioluminescence" eds. Cormier, M. J., Hercules, D. M., and Lee, J. Plenum Press, N. Y. p. 461.

HIGHER EXCITED STATES AND VIBRATIONALLY HOT EXCITED STATES: HOW IMPORTANT ARE THEY IN ORGANIC PHOTOCHEMISTRY IN DENSE MEDIA?

Josef Michl, Alain Castellan, Mark A. Souto, and Jaroslav Kolc

Department of Chemistry, University of Utah, Salt Lake City, Utah 84112, U.S.A.

INTRODUCTION

In photochemical reactions of organic molecules in fluid or rigid solutions, and even in the gas phase at moderate to high pressures, it is usually assumed without question that internal conversion to the lowest excited electronic state and thermal equilibration of vibrational motion occur much more rapidly than any chemical transformations, so that direct irradiation first produces cool S_1, or possibly directly cool T_1 if special structural features greatly enhance the rate of intersystem crossing relative to thermalization (e.g., heavy atoms), and sensitization produces cool T_1, before any chemical reactions occur. The reacting species is therefore assumed to be thermalized S_1 or T_1, except perhaps if S_2 (or T_2) is so close in energy as to be thermally accessible, and kinetic schemes are set up accordingly. These assumptions represent an analogy to Kasha's rule which states that thermalized S_1 and T_1 states are the only ones from which significant light emission is observed in dense media.

Similarly as Kasha's rule, the above general statement has well-established exceptions. In some molecules, typically those with large S_2-S_1 and/or T_2-T_1 gaps, internal conversion $S_2 \rightarrow S_1$ and/or $T_2 \rightarrow T_1$ is unusually slow, so that photochemical reactions from S_2 and/or T_2 can be observed (e.g., azulene derivatives[1] and thioketones[2]; some molecules with two widely separated chromophores in which internal conversion actually corresponds to energy transfer can also be placed in this category). Although these are very interesting cases, we shall not be concerned with them here. Rather, we shall consider the more representative case of organic molecules for which there is no reason to expect exceptionally slow internal conversion.

B. Pullman and N. Goldblum (eds.), Excited States in Organic Chemistry and Biochemistry, 361-371.

Kasha's rule is quite compatible with other sources of information on the rates of the potentially competitive processes. Radiative rate constants from excited states of organic molecules correspond to lifetimes of nanoseconds or longer, while the rates of vibrational relaxation and of internal conversion usually correspond roughly to lifetimes of picoseconds. The analogous organic photochemical rule has a much less obvious physical background: in the limiting case of a repulsive excited surface, dissociation will occur in a single vibrational period and has a good chance of being competitive with any other molecular process. *A priori*, one would expect a whole spectrum of behavior types up to the other limiting case, in which the photochemical reaction rate is limited by some quite slow step such as translational diffusion of reaction partners. In inorganic photochemistry, wavelength-dependence of quantum yields and even nature of products is common and is usually attributed to reactions from higher excited states. The questions then are, why is organic photochemistry different, and is it really completely different?

The literature already contains numerous reports which implicate higher and/or hot excited states of organic molecules as the reacting entities, but few of these are unambiguous. One of the first challenges to the conventional view came from Becker *et al.*[3] who noted that the fluorescence quantum yields of several molecules undergoing unimolecular photochemical transformations in rigid media fall off with increasing energy of the initial excitation, even within a single electronic transition, and proposed that this is due to a competition between thermal relaxation and chemical rearrangement ("vibronic photochemistry"). While this appears to be a likely explanation, it is also possible that the wavelength-dependent process is intersystem crossing to the triplet manifold from which the products are then formed. Only very recently has an example been found in which the latter alternative can be safely excluded[4]: 1,4-dewarnaphthalene shows fluorescence both from the starting material and the product, naphthalene, and the ratio of their intensities changes with wavelength of excitation in the way expected for a hot excited state reaction.

Another early challenge was posed in a review article by Ullman[5], who collected several examples in which reactions from higher excited states and from hot ground states appeared plausible. Other such proposals appeared and some were later retracted. Clearly, it is not an easy matter to establish these mechanisms beyond reasonable doubt, and this perhaps accounts for the general skeptical attitude towards new claims of this type. Wavelength dependence of the nature and/or quantum yield of photoproducts in itself does not provide sufficient proof, since it can be due to other causes: in addition to the possibility that intersystem crossing competes with vibrational relaxation and/or internal conversion, mentioned above, factors such as presence of several conformers in the starting material can be responsible[6]. Still, it appears possible that at least in some of the cases, e.g. the rigid enones investigated in detail by Schaffner *et al.*[7], higher and/or hot excited states undergo chemical transformations.

Some of the cleanest examples of chemical reactions from higher and/or hot triplet states in dense media came from studies of reactions which proceed by successive absorption of two photons, $S_0 \overset{h\nu}{\rightarrow} S_n \rightarrow S_1 \rightarrow T_1 \overset{h\nu}{\rightarrow} T_n \rightarrow$ product. In these, S_0, S_n, S_1, and T_1 are unreactive and only excitation of T_n yields products. This mechanism was first established by Joussot-Dubien and Lesclaux[8] in the case of photoionization of aromatic hydrocarbons. A study of photodissociation of the benzylic C-H bond in toluene by Schwarz and Albrecht[9] showed that the two-photon mechanism operates for long-wavelength initial irradiation but that it changes into a one photon mechanism if somewhat more energetic initial photons are used, suggesting that not only a higher and/or hot excited triplet, but also a hot excited singlet, are reactive. Similar results were communicated for the more complex pericyclic reactions of electrocyclic ring-opening and retrocycloaddition of condensed cyclobutenes and cyclobutanes[10-15].

We feel that such two-photon processes offer a unique opportunity to characterize the reactivity of higher and/or hot excited states. We are particularly interested in the pericyclic reactions of cyclobutenes whose complexity is more representative of the bulk of elementary reaction steps in organic photochemical reactions than a simple photoionization or a bond dissociation are. Presently, we summarize the results of our efforts in this area, first, to establish the nature of the two-photon mechanism beyond any doubt, and second, to obtain a semiquantitative idea of quantum yields from the higher and/or hot excited states on systems for which there is no indication that the rate of internal conversion and vibrational relaxation is anything but normal.

RESULTS

The two reactions investigated were $\underline{1}\rightarrow\underline{2}$ and $\underline{3}\rightarrow\underline{4}$, reported briefly previously[10-13]. Their course was followed by u.v. spectroscopy (3-methylpentane glass, 77°K). Standard light sources, filters, monochromators and procedures of ferrioxalate and uranyl oxalate actinometry were used.

Spectroscopy

The absorption, fluorescence, and phosphorescence spectra of $\underline{1}$ and $\underline{3}$ are quite unexceptional and are closely related to the spectra of the parent chromophores, naphthalene, benzene and

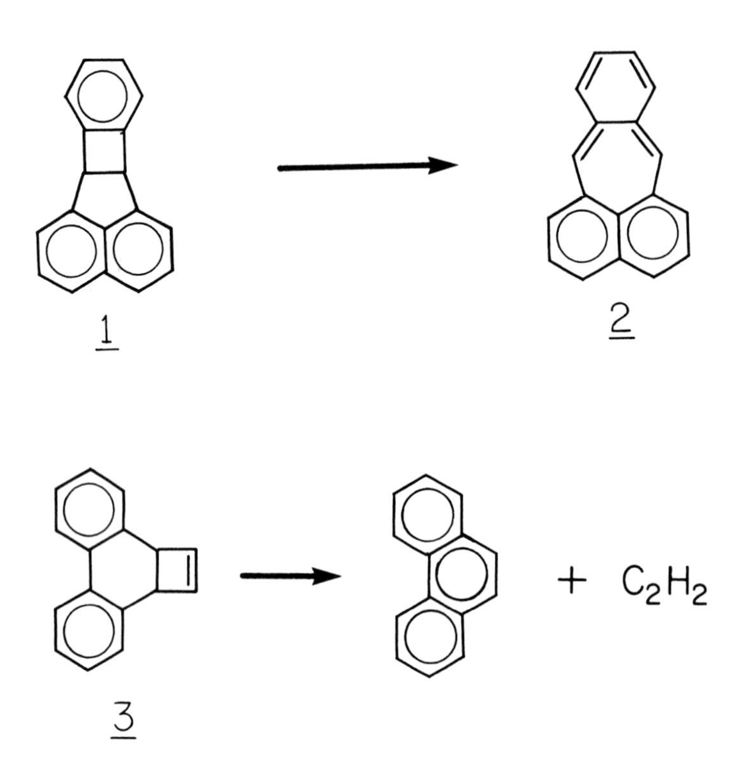

biphenyl (fluorene). The only emissions observed are $S_1 \rightarrow S_0$ fluorescence and $T_1 \rightarrow S_0$ phosphorescence. The spectrum of 2 has been analyzed elsewhere.[16]

Corrected excitation spectra of both fluorescence and phosphorescence (naphthalene and fluorene standards) are identical with the absorption spectrum within the experimental error of a few per cent, the ratio of fluorescence to phosphorescence intensity is independent of excitation wavelength down to 214 nm. The data for the two compounds are summarized in Table 1.

Table 1

Spectroscopy of 1 and 3 (3-methylpentane, 77^oK)

	ν(0-0)(FL,cm^{-1})	ϕ_{FL}	ν(0-0)(PH,cm^{-1})	ϕ_{PH}	τ_{PH}(sec)
1	31 200	0.41±0.03	20 800	0.041±0.005	3.0±0.1
3	32 000	0.51±0.03	22 800	0.067±0.005	3.45±0.1

Conclusion. Photophysical properties of 1 and 3 are comparable to those of other simple naphthalene and biphenyl derivatives. Excitation to S_n produces S_1 with approximately 100% efficiency. Significant intersystem crossing to the triplet manifold occurs, but only after internal conversion and thermalization in S_1.

Single-Beam Photochemistry

Broad-band irradiation of a rigid glass containing 1 or 3 with 1kW xenon arc filtered to remove wavelengths below 300 nm rapidly produces 2 and 4, respectively. Initial rate of product formation is proportional to the second power of light intensity I^2.

When monochromatic light of 303 nm is used for irradiation of 1, the initial rate of formation of 2 is again proportional to I^2, but at 229 nm and 214 nm, it is proportional to I within experimental error. With 254 nm light, the reaction proceeds too slowly for measurement. The quantum yields for the one-photon processes are $\phi_{214} = 0.009$, $\phi_{229} = 0.006$, $\phi_{254} < 0.0003$ (i.e., zero within experimental uncertainty).

Similarly for 3, the initial rate is proportional to I^2 at 310 nm, to I at 214 nm ($\phi_{214} = 0.012$), and to I^N, $1 < N < 2$, at 229 and 254 nm.

Conclusion. Photochemical products from 1 and 3 are 2 and 4, respectively. Absorption of a single photon of wavelength 254 nm or longer by 1, i.e. excitation into the lower vibrational levels of the L_b and L_a states of its naphthalene part and the L_b state of its benzene part, and absorption of a 310 nm photon by 3, corresponding to the $S_0 \rightarrow S_1$ transition, produce S_1 and T_1, as shown by the observed emissions, but produce no observable photoproduct until a second photon is absorbed. With the light intensities used, the second photon absorption cannot be simultaneous with the first one and it also cannot be due to $S_1 \xrightarrow{h\nu} S_n$. It could be due to $T_1 \xrightarrow{h\nu} T_n$ or to absorption by some other intermediate produced from either S_1 or T_1, which reverts to the starting material if it does not absorb a second photon.

If the initial photon has energies corresponding to 229 nm or shorter wavelengths in 1 and 254 nm or shorter wavelengths in 3, it causes production of the final products, 2 and 4, respectively, with quantum yields of the order of 0.5-1 per cent. This should be reflected in decreased quantum yield of fluorescence and phosphorescence at these shorter wavelengths, but a 1 per cent change is well within our experimental error.

Double-Beam Photochemistry

If 1 is excited by a beam of 303 nm light, which is sufficiently weak to cause essentially no reaction by itself, and simultaneously with a second beam of intense 330-410 nm light, which is not absorbed by ground-state 1, formation of 2 is observed, with initial rate proportional to the intensity of the 330-410 nm light beam. If the 303 nm beam is turned off, there is no reaction.

Initial rate of product formation was measured using the 303 nm light beam and a second beam of quasi-monochromatic light (circular wedge interference filter), as a function of the wavelength of the second beam, correcting for its absolute intensity, in order to obtain the action spectrum of the second photon. This is zero within experimental error at long wavelengths and has a broad peak between 345 and 435 nm, coinciding with the region of intense triplet-triplet absorption of naphthalene[17].

Similarly for 3, using a weak beam at 310 nm and a strong one at 330-450 nm, we find that the initial reaction rate is proportional to the intensity of each of the two light sources. The action spectrum resembles the triplet-triplet absorption spectra of biphenyl and fluorene[18].

In another series of experiments, a microsecond flash was used to populate the triplet state of 1, and after a delay of Δt, an intense beam of 330-410 nm light was turned on. After all phosphorescence disappeared (~20 sec), the procedure was repeated. A log plot of the initial rate of formation of 2 in this mode of operation against Δt had a negative slope of 2.4 ± 0.5 sec, and provided an estimate of the lifetime of the intermediate responsible for the absorption of the second photon in the product-producing process. Within experimental error, this agrees with the observed phosphorescence lifetime of 1.

Similarly for 3, the intermediate lifetime determined by this procedure was 3.0 ± 0.5 sec in 3-methylpentane and 0.9 ± 0.3 sec in a 2-methyltetrahydrofuran-1,2-dibromoethane mixture (74:26), while the phosphorescence lifetimes of 3 in these solvents were 3.45 ± 0.1 sec and 1.31 ± 0.1 sec, respectively, again only slightly longer. In the presence of the visible light beam, these phosphorescence lifetimes are actually shortened by about 20% and are in excellent agreement with the observed lifetimes of the intermediates.

Conclusion. Triplet T_1 is the intermediate species responsible for the absorption of the second photon in the photochemical transformation. Thus, thermalized S_1 and T_1 are unreactive, and it appears reasonable to assume that the rate of $S_1 \rightarrow S_0$ internal conversion is similar as in other aromatic hydrocarbons, i.e. negligible compared with the rates of fluorescence and intersystem crossing to the triplet manifold. This yields $\phi_{isc} \simeq 0.59$ for 1 and $\phi_{isc} \simeq 0.49$ for 3.

Quantum Yield of the $T_1 \overset{h\nu}{\rightarrow} T_n \rightarrow$ product Step

The steady double-beam arrangement described above was used along with standard actinometry to estimate the quantum yield for the process $T_n \rightarrow$ product as 0.005 for 1→2 and ~0.01 for 3→4. These numbers are only approximate, the main source of uncertainty being the possible error in the estimate of the number of photons per second absorbed to T_1. We have used the expression of Keller and Hadley[19], using the above estimates of ϕ_{isc}, and literature values for triplet-triplet extinction coefficients for related chromophores[17,18].

Conclusion. The efficiency of the process $T_n \rightarrow$ product is of the order of 0.5 per cent, comparable with the efficiency of the $S_n \rightarrow$ product process for similar total energies of S_n and T_n. Almost all of the molecules apparently undergo the competing process $T_n \rightarrow T_1$ instead.

The experimental data are summarized in Fig. 1 for 1; the numbers for 3 are quite similar.

DISCUSSION

The results described above leave no doubt that the production of 2 and 4 from 1 and 3, respectively, does not proceed from thermalized S_1 nor T_1 states, but only from molecules which contain a considerable amount of additional energy. Initially, a large part of this extra energy is in the form of electronic energy, and it is quite possible that the reactions observed proceed in higher excited states, S_n and T_n. It is also possible that the actual chemical transformation is preceded by internal conversion from the initially reached S_n or T_n states to one of the lower states, possibly even S_1 or T_1, and that the vibrationally hot molecule then undergoes a chemical transformation. The distinction between these two possibilities is fuzzy once the Born-Oppenheimer approximation is removed, and since our experiments do not permit us to differentiate between the two possibilities anyway, we prefer to state that the observed reactions proceed in higher and/or hot excited states.

Although formally 1 is bichromophoric, the reacting bond is in benzylic position with respect to both chromophores; as

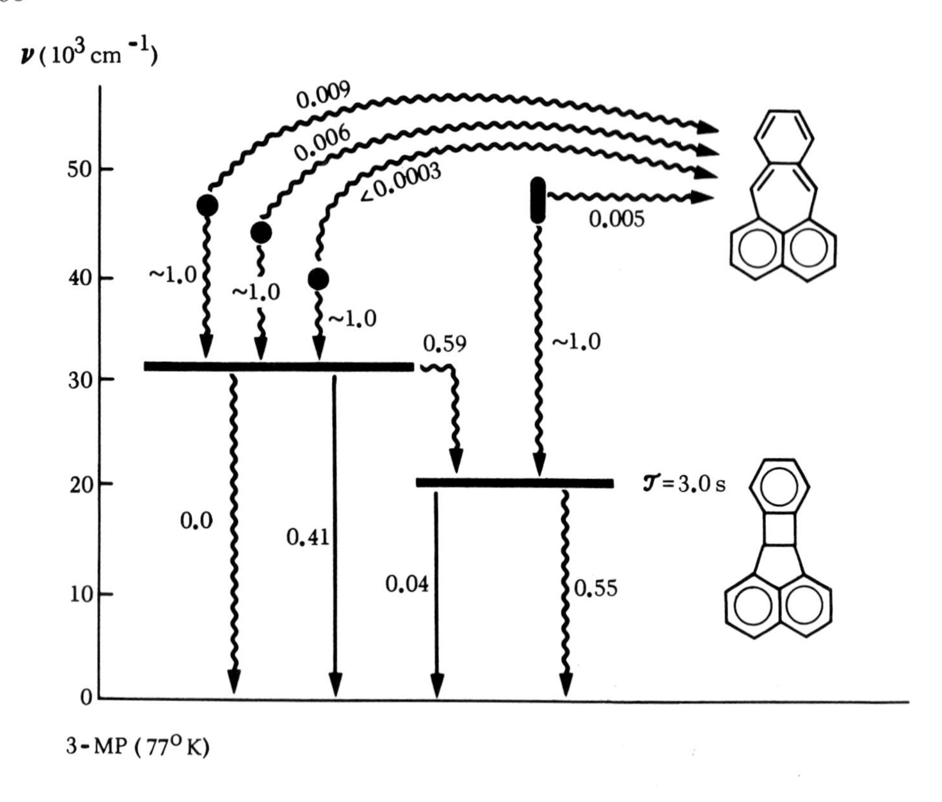

Fig. 1. Summary of photophysics and photochemistry of 1.

soon as motion along the reaction coordinate starts, the two chromophores are in conjugation. Indeed, the reactivity is not given by the location of the initial excitation: exciting S_1 or S_2 of the naphthalene chromophore or S_1 of the benzene chromophore are equally without effect, and it appears that it is the total amount of the extra energy available which is important.

It is quite likely that the rate of internal conversion in 1 and 2 is unexceptional, and the order of magnitude of 1 per cent may be a good estimate for the amount of reaction to be expected from a higher and/or hot excited state of other photoreactive organic molecules, too. How important a "hot contribution" of this magnitude will be in practice will clearly depend on the efficiency with which the product is formed from thermalized S_1 or T_1 states. If this is zero, as happens in 1 and 2, the "hot contribution" is very important indeed, being responsible for all of the reaction. If S_1 or T_1 reacts with good efficiency, the "hot contribution" may be quite difficult to detect.

Although we presently propose that a quantum yield of one per cent may well be a typical order of magnitude for the "hot contribution", large variation with molecular structure is not out of the question. At any rate, it seems to us that the possible existence of a significant "hot contribution" should always be considered in kinetic studies of reactions which proceed with relatively small quantum yields, unless the light wavelength used in the study corresponds to the lowest vibrational level of the lowest excited state.

At this point, we wish to mention the existence of another question raised by the present experiments, namely, why is "extra energy" required in the reactions $\underline{1}\rightarrow\underline{2}$ and $\underline{3}\rightarrow\underline{4}$? An answer based on simple consideration of molecular orbital and state correlation diagrams has been proposed elsewhere[20] and will not be discussed here.

Finally, we wish to suggest that the discrepancy between what is considered as normal behavior in organic as opposed to inorganic photochemistry may be only apparent and due to the circumstance that the photochemist typically works with wavelengths longer than 200 nm. At shorter wavelengths, many excited states of organic compounds will be of $\sigma\sigma^*$ type. These may frequently be dissociative[21], and thus formally analogous to the more easily reached charge-transfer states of metal complexes. In this sense, the $n\pi^*$ and $\pi\pi^*$ states of organic molecules bear certain resemblance to the d-d states of complexes.

CONCLUSIONS

The experiments described here suggest that the answer to the question posed in the title is, "Sometimes, higher and/or hot excited states are very important even in organic photoreactions in dense media, and their participation must not be dismissed automatically."

ACKNOWLEDGEMENT

This work was supported by the donors of the Petroleum Research Fund, administered by the American Chemical Society. One of us (A.C.) acknowledges a NATO post-doctoral fellowship.

REFERENCES

1. C. M. Lok, J. Lugtenburg, J. Cornelisse, and E. Havinga, Tet. Lett., 4701 (1970).

2. A. Couture, K. Ho, M. Hoshino, P. de Mayo, R. Suau, and W. R. Ware, J. Am. Chem. Soc., 98, 6218 (1976); P. de Mayo, Acc. Chem. Res., 9, 52 (1976); and references therein.

3. R. S. Becker and J. Michl, J. Am. Chem. Soc., 88, 5931 (1966); R. S. Becker, E. Dolan and D. E. Balke, J. Chem. Phys., 50, 239 (1969).

4. R. V. Carr, B. Kim, J. K. McVey, N. C. Yang, W. Gerhartz, and J. Michl, Chem. Phys. Lett., 39, 57 (1976).

5. E. F. Ullman, Acc. Chem. Res., 1, 353 (1968).

6. For recent examples, see T. Wismonski-Knittel, G. Fischer, and E. Fischer, J. Chem. Soc., Perkin Trans. II, 1930 (1974); J. W. J. Gielen, H. J. C. Jacobs, and E. Havinga, Tet. Lett., 3751 (1976).

7. J. Gloor and K. Schaffner, Helv. Chim. Acta, 57, 1815 (1974); J. Am. Chem. Soc., 97, 4776 (1975); and references therein; F. Nobs, U. Burger and K. Schaffner, Helv. Chim. Acta, 60, in the press (1977).

8. For a review, see R. Lesclaux and J. Joussot-Dubien, in "Organic Molecular Photophysics", Vol. 1, J. B. Birks, Ed., Wiley, New York, N. Y., 1973.

9. F. P. Schwarz and A. C. Albrecht, J. Phys. Chem., 77, 2411, 2808 (1973).

10. J. Michl and J. Kolc, J. Am. Chem. Soc., 92, 4148 (1970).

11. J. Kolc and J. Michl, Abstracts, IVth IUPAC Symposium on Photochemistry, Baden-Baden, Germany, July 16-22, 1972, p. 167.

12. J. Kolc and J. Michl, J. Am. Chem. Soc., 95, 7391 (1973).

13. J. M. Labrum, J. Kolc, and J. Michl, J. Am. Chem. Soc., 96, 2636 (1974).

14. K. Honda, A. Yabe, and H. Tanaka, Bull. Chem. Soc. Japan, 49, 2384 (1976).

15. J. Meinwald, G. E. Samuelson, and M. Ikeda, J. Am. Chem. Soc., 92, 7604 (1970); N. J. Turro, V. Ramamurthy, R. M. Pagni, and J. A. Butcher, Jr., J. Org. Chem., 42, 92 (1977).

16. J. Kolc, J. W. Downing, A. P. Manzara, and J. Michl, J. Am. Chem. Soc., 98, 930 (1976).

17. Y. H. Meyer, R. Astier, and J. M. Leclercq, J. Chem. Phys., 56, 801 (1972).

18. J. B. Birks, "Photophysics of Aromatic Molecules", Wiley, New York, N. Y., 1970.

19. R. A. Keller and S. G. Hadley, J. Chem. Phys., 42, 2382 (1965).

20. J. Michl, J. Am. Chem. Soc., 93, 523 (1971); Mol. Photochem., 4, 287 (1972); in "Chemical Reactivity and Reaction Paths", G. Klopman, Ed., Wiley, New York, N. Y., 1974, p. 301.

21. Not all such states are purely dissociative, as indicated by recent observations of weak Stokes-shifted fluorescence. See, e.g., P. Ausloos, Mol. Photochem., 4, 39 (1972); A. M. Halpern, Mol. Photochem., 5, 517 (1973).

Solvent Effects on Excited States

G.G. Hall and C.J. Miller

Mathematics Department, University of Nottingham
Nottingham NG7 2RD, England

1. Introduction

In the biochemical context it is the molecule in solution which is important rather than the free molecule. The solvent interactions may change its most stable configuration, will shift its spectra and may modify its reactions.

The theoretical treatment of solvent effects has been dominated until recently by 'continuum' models in which the solvent is assumed to be spatially homogeneous and characterised by its macroscopic properties. These models become less satisfactory for solutes which are far from spherical and for solvents such as water which have specific short-range interactions with solutes. For these solvents a 'supermolecule' approach is being developed in which at least one solvent molecule is associated with the solute so that specific interactions can be studied. A recent survey of the field has been given by Pullman (1976).

In this contribution the problems of treating aqueous solutions will be discussed, with particular reference to the excited states of the solute.

2. Aqueous solution

Supermolecule calculations of NH_4^+ - H_2O (Pullman & Armbruster, 1974) and of $HCONH_2$ - H_2O (Alagona et al., 1973) show very clearly the energetic importance of maintaining the linear hydrogen bond between water and solute where this can be formed. This becomes a first principle in determining the geometrical structure of the first hydration shell. More generally, these

B. Pullman and N. Goldblum (eds.), Excited States in Organic Chemistry and Biochemistry, 373-379.

calculations show that the electrostatic contribution to the total interaction energy is the dominant one in angular variations of the energy. This suggests very strongly that a suitable point charge model for the solvent molecule may provide interaction energies for systems containing too many solvent molecules to be calculated as supermolecules. A calculation on the $HCONH_2$ - H_2O system by both methods (McCreery, Christoffersen & Hall, 1976b) leads to almost identical optimum configurations, though the point charge model usually exaggerates their attractiveness. A similar conclusion is reached by Noell and Morokuma (1976) from their calculations on hydrates of Li^+.

Many different point charge models of molecules have been proposed from time to time. In this context it is possible to judge between them using the criterion that they should enable the electrostatic potential around the molecule to be approximated closely. Since the asymptotic expansion of the potential depends solely on the spherical moments of the electron density, it becomes a condition that the spherical moments should be reproduced by the point charges as far as possible. This discriminates against models which do not reproduce the total dipole moment. For wavefunctions expressed using FSGO basis functions a point charge model of H_2O can be derived (Tait & Hall, 1973) which preserves all the spherical moments but involves an unreasonably large number of charges. A simpler model by Shipman (1975) preserves only the total charge and dipole but has many fewer charges. An extension of this discussion to more general wavefunctions (Hall & Martin, 1977) shows that other point distributions such as point dipoles and point quadrupoles are then needed.

The use of point charges and possibly point dipoles to represent solvent molecules has one danger when the solute is treated quantum mechanically, viz. it is possible for such a point charge to remove electrons from the solute molecule. This danger can be prevented by including a constraint on the molecular orbitals to make them orthogonal to appropriate basis functions on the solvent molecules. This procedure rapidly becomes difficult to apply. An alternative procedure which is sufficient to prevent any collapse of the solute molecule is to introduce a penalty function

$$P_\alpha = \sum_i \left(\frac{a_\alpha}{r_{i\alpha}} \right)^n$$

for each nucleus α in the solvent. Here $r_{i\alpha}$ is the distance between the nucleus α and the *i*th electron in the solute, a_α is a radius round the nucleus α and is approximately its van der Waals radius, while *n* is approximately 12. These penalty functions are added to the solute Hamiltonian before calculating the wavefunction and subsequently their average value is subtracted from the total energy (see also McCreery, Christoffersen & Hall, 1976a).

One feature which characterises the liquid state as distinct from the solid state is the freedom for molecular groupings to rotate. To achieve this it is natural to surround the solute molecule with hydration shells until it is approximately spherical, since the sphere is the only structure that can rotate freely. The first hydration shell may be sufficient for this purpose on some occasions, but for larger molecules more may be required. The resulting sphere can then be treated as the 'cavity' for the purpose of applying macroscopic theories to estimate the effect of the remaining part of the solvent.

3. Spectral shifts

In a microscopic model of a solvated molecule, such as that proposed above, wavefunctions can be calculated for both ground and excited states of the molecule, and the spectral shift is found by comparing the results with those for the free molecule. Such a calculation automatically includes a variety of effects. Whether or not all the significant effects are present can be judged by comparing with the perturbation theory analysis. A review of this has been given by Amos and Burrows (1973).

The perturbation theory of the solute molecule in the presence of solvent molecules produces three leading terms in the spectral shift. One of these is present for all solvents and represents the change in dispersion energy between solute and solvent when the former is excited. It is always a red shift. The other two terms arise only for polar solvents. One is due to the change in polarizability of the solute and is a red shift if the excited state is more polarizable. The third term requires the solute molecule to change its dipole moment from ground to excited state and is geometry-dependent.

A point charge model assumes a rigid solvent and consequently cannot include the dispersive term. The other terms arise from the flexibility of the solute and these will be properly calculated in a point charge model, especially if any long-range effects are treated by applying a macroscopic theory to the hydrated sphere. In principle, the dispersive term could be calculated using the Casimir-Polder integral and the dynamic polarizabilities (see Mahanty and Ninham, 1976), but this is likely to be difficult. Thus, to obtain agreement with experiment some allowance for the dispersive terms will be necessary.

4. Hydrated benzene

These general principles can be illustrated using the benzene molecule as a solute. Hydrogen bonding is not present and benzene is non-polar, so the largest effects will be due to the polarization of the molecule by the polar water molecule. A large interaction can be achieved by placing a water molecule above the benzene plane with its oxygen over the centre of the ring and the hydrogens closer to the ring. To allow the hydrogens, which carry a positive point charge, to interact with the carbon electrons they will be in a vertical plane through two opposite carbon nuclei. A second water molecule can then be placed below the benzene in a similar configuration except that, to reduce the inter-hydrogen repulsions, this molecule should be in a plane with a different pair of opposite carbon nuclei. The distance between the oxygens and the centre of the ring is 6.6 Bohrs.

The spectrum of benzene in the presence of a perturbing point charge has been considered by Bishop and Craig (1963) in their study of the anilinium ion. They show that, although each state shows an appreciable drift of π electrons towards the added charge, the effect on the spectrum is very small indeed since both states are equally stabilized.

5. Applications

Ab initio calculations to test these models have been initiated. Preliminary results have been obtained for the benzene model. One advantage of the quantum mechanical model compared with macroscopic models is that more details of the electronic structure of the solute can be given. Table I shows the π orbital energies for the free molecule compared with the molecule associated with $2H_2O$. It is clear that solution stabilizes all the orbitals. The σ electrons are also stabilized, but to a lesser extent. The effect on the excitation energies is shown to be small in Table II. The contrast between these tables is

Table I π Orbital Energies for Benzene

		ε_1	ε_2	ε_3
occupied	free	-0.35112	-0.18222	-0.18222
	+ $2H_2O$	-0.37712	-0.20842	-0.20728
		ε_4	ε_5	ε_6
virtual	free	0.34317	0.34317	0.59653
	+ $2H_2O$	0.31789	0.31863	0.57219

worthy of note. The stabilisation of the orbital energies is approximately 0.68eV for all π orbitals whereas the shift of the excitation energies is approximately 0.019eV to the blue for all excitations.

Table II Excitation Energies for Benzene

Singlets				
free	0.27552	0.30088	0.40477	0.40477
$+2H_2O$	0.27624	0.30160	0.40541	0.40557
Triplets				
free	0.19835	0.22835	0.22835	0.25835
$+H_2O$	0.19907	0.22903	0.22908	0.25911

Preliminary calculations, without the macroscopic terms, have also been made on a model of the solvated amphetamine ion. In this model two water molecules are added above and below the benzene ring and three more are hydrogen-bonded to the ammonium head. Table III shows the π orbital energies for this model with two, three and five water molecules added to the ion. This shows that the effects of the two groups of water molecules are additive.

Table III π Orbital Energies for Amphetamine Ion

	ε_2	ε_3	ε_4	ε_5
free	-0.28308	-0.27973	0.24057	0.24230
$+2H_2O$	-0.30907	-0.30450	0.21612	0.21715
$+3H_2O$	-0.27079	-0.26551	0.25412	0.25472
$+5H_2O$	-0.29683	-0.29013	0.22916	0.23012

The two waters have an effect closely similar to those of hydrated benzene. The three on the ammonium head have an opposite effect and raise the orbital energies. The excitation energies for the singlet states are shown in Table IV. Again the shifts are very much smaller than the orbital energy shifts. The blue shift produced by the waters over the benzene ring dominates the results and, though the three additional waters give a red shift, its magnitude is very small. The macroscopic potential terms for this ion are expected to be large and to produce a substantial red shift.

6. Conclusions

The combination of a point charge model for the nearest solvent molecules inside a solvation sphere and the extra potential terms given by the macroscopic electrostatic model gives a model of the solvated molecule economical enough for *ab initio* calculations. The full effect of the solvent on the electronic spectra cannot be obtained unless the dispersion contributions are added. In some circumstances, when the solute and solvent have comparable excitation energies, it will be necessary to calculate the excited states using a configuration interaction of single replacements from both molecules.

Table IV Singlet Excitation Energies for Amphetamine Ion

free	0.27471	0.29923	0.40152	0.40284
$+2H_2O$	0.27540	0.29989	0.40215	0.40338
$+3H_2O$	0.27469	0.29921	0.40124	0.40258
$+5H_2O$	0.27538	0.29986	0.40183	0.40311

Acknowledgment

The authors wish to thank the North Atlantic Treaty Organisation for financial support.

References

Alagona,G., A. Pullman, E. Scrocco and J. Tomasi (1973), Int.J. Pept.Res. 5, 251.

Amos,A.T. and B.L. Burrows (1973), Adv. Quantum Chem. 7, 289.

Bishop,D.M. and D.P. Craig (1963), Mol.Phys. 6, 139.

Hall, G.G. and D. Martin (1977), in preparation.

Mahanty, J. and B.W. Ninham (1976), "Dispersion Forces", Academic Press, London.

McCreery,J.H., R.E. Christoffersen and G.G. Hall (1976a), J.Am. Chem.Soc. 98, 7191.

McCreery,J.H., R.E. Christoffersen and G.G. Hall (1976b), J.Am. Chem.Soc. 98, 7198.

Noell,J.O. and K. Morokuma (1977), J.Phys.Chem. in press.

Pullman,A. (1976), "The New World of Quantum Chemistry", Reidel Publishing Co., Dordrecht p. 149.

Pullman,A. and A.M. Armbruster (1974), Int.J. of Quant.Chem. 58, 169.

Shipman,L.L. (1975), Chem.Phys.Lett., 31, 361.

Tait,A.D. and G.G. Hall (1973), Theor.Chim.Acta, 31, 311.

DISCUSSION

DUNNE :

Do you consider that it would be moderately feasible to extend your interesting model to calculate chemical shift data in N.M.R.?

HALL :

I hope that it will be possible to do this since I anticipate an appreciable reduction of the ring current due to the polarization of the π orbitals.

PHOTOCHEMISTRY OF VICINAL POLYKETONES

Mordecai B. Rubin

Technion-Israel Institute of Technology, Haifa

This discussion of photochemistry of vic-polyketones will be concerned mainly with the occurrence of α-cleavage as a primary process of their excited states. This reaction, sometimes referred to as the "Norrish Type I Reaction", involves homolytic fission of one of the bonds to the -CO group in an excited state of a ketone and produces two free radical centers. It was recognized many years ago as a primary process in vapor phase photolysis of simple ketones

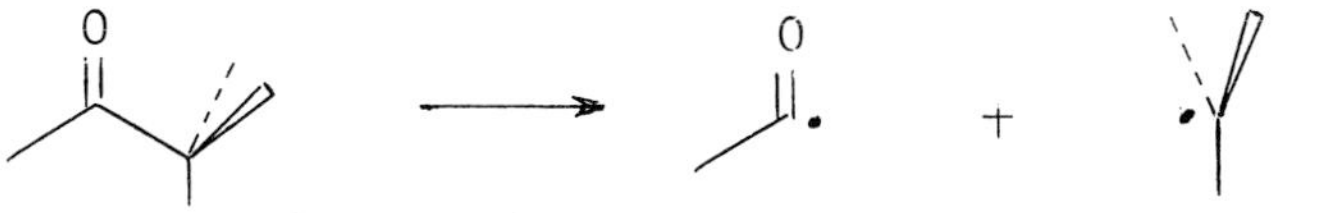

and was later shown to be an important process in solution photochemistry of saturated and unsaturated ketones (involving mainly triplet, but also singlet, states). In general, unsymmetrical ketones cleave regioselectively so as to afford the more stabilized of the two possible pairs of radicals.

This regioselectivity and the subsequent fate of the resultant biradical are illustrated below for 2-methylcyclohexanone. The intermediate biradical may revert to starting ketone thereby reducing the chemical quantum yield, hydrogen atom transfer can lead to a

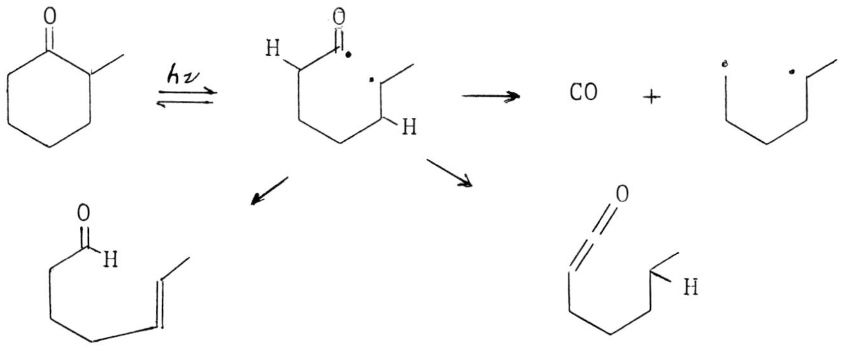

$h\nu$ CO +

B. Pullman and N. Goldblum (eds.), Excited States in Organic Chemistry and Biochemistry, 381-385.

ketene (often trapped by reaction with the medium) or unsaturated aldehyde, and loss of carbon monoxide affords a new biradical which may cyclize or fragment to stable products.

Intuitively, it would appear highly probable that α-cleavage could be an important process of excited states of vic-polycarbonyl compounds:

$$\mathrm{RCOCOR} \rightarrow \mathrm{R\cdot} + \mathrm{RCO\dot{C}O} \quad \text{or} \quad 2\ \mathrm{R\dot{C}O}$$

$$\mathrm{RCOCOCOR} \rightarrow \mathrm{R\dot{C}O} + \mathrm{RCO\dot{C}O}$$

$$\mathrm{RCOCOCOCOR} \rightarrow \mathrm{R\dot{C}O} + \mathrm{RCOCO\dot{C}O} \quad \text{or} \quad 2\ \mathrm{RCO\dot{C}O}$$

Calculated bond dissociation energies, as shown below, provide some support for this view. In fact, α-cleavage has been shown to be an important process in vapor phase photolysis of biacetyl and results in formation of carbon monoxide and ethane. The situation in solution photochemistry is much less clear.

	CH_3COCH_3	$CH_3COCOCH_3$	PhCOCOPh
D	81	68	45

For example, the products obtained from irradiation of benzil[1] in cyclohexane solution are shown below. It is immediately apparent that a number of the observed products include benzoyl groups which must be derived from cleavage of benzil. α-Cleavage of benzil

$$\mathrm{PhCOCOPh} + \text{(S)} \longrightarrow \mathrm{PhCHO} + \mathrm{PhCO{-}C_6H_{11}} + \mathrm{PhCOCH(OH)Ph} + \mathrm{PhCOCH(OCOPh)Ph}$$

was accordingly suggested to be the primary reaction of the benzil triplets involved in this reaction. Subsequent free radical reactions could then account for most of the observed products. However, an alternative mechanism can also be suggested[2] in which the primary process is hydrogen atom abstraction from cyclohexane by triplet benzil. The actual cleavage of benzil would then occur in subsequent free radical steps, as illustrated. In particular, the benzoyl radicals would be formed by fragmentation of a species possessing a radical center beta to the carbonyl group:

$$PhCOCOPh \xrightarrow{h\nu} 2\ Ph\dot{C}O$$

$$Ph\dot{C}O + C_6H_{12} \longrightarrow PhCHO + C_6H_{11}\cdot$$

or

$$PhCOCOPh + C_6H_{12} \xrightarrow{h\nu} PhCO\dot{C}(OH)\text{-}Ph + C_6H_{11}\cdot$$

$$PhCOCOPh \longrightarrow PhCO\text{-}C(O\cdot)(Ph)(C_6H_{11}) \longrightarrow PhCO\text{-}C_6H_{11} + Ph\dot{C}O$$

$$\text{-}\overset{O}{\overset{\|}{C}}\text{-}C\text{-}\dot{X}\text{-} \longrightarrow \text{-}\overset{O}{\overset{\|}{C}}\cdot + C{=}X\text{-}$$

Either or both of the above mechanism could be operative; no distinction between them has been made to date.

Careful examination of a number of other cases reported in the literature shows that, with the exception of 1,2-cyclobutanediones, there is no clear-cut evidence for α-cleavage in reactions of diketones in the absence of oxygen. A recent example from the author's laboratory[3] is the photolysis of bicyclo[2.2.2]octanedione (I) in p-xylene solution. In addition to the major product II, a one percent yield of p-methylbenzyl cyclohexyl ketone (III) and carbon monoxide was obtained. Formation of II and III involved the same excited state of I. Similar results were obtained with camphorquinone.

$$\mathrm{I} \xrightarrow[h\nu]{ArCH_3} \mathrm{II} + \mathrm{III}\ (1\%) + CO$$

$$\mathrm{I} + \mathrm{II} \xrightarrow{h\nu} \mathrm{III}$$

$$\mathrm{IV} \xrightarrow[Bz]{h\nu} \text{acetone } (80\text{-}100\%) + \ldots$$

$$\mathrm{V} \xrightarrow{h\nu}$$

The possibility that III was formed via an α-cleavage biradical was eliminated by the observation that III is not formed in the initial stages of reaction but is formed by photoreaction of I with II. Similarly, the high yield of acetone obtained from photolysis of diketone IV in benzene solution[3] does not appear to involve α-cleavage. Rearrangement of unsaturated diketones such as V may involve concerted reaction rather than α-cleavage followed by recoupling at the allylic position.

Investigations of tri- and tetraketones have been performed with diaryl derivatives. The keto-acyl radicals which would be formed by α-cleavage of such molecules would be expected to decarbonylate readily; lower carbonylogs should be observed among the products. In fact, the quantum yield for disappearance of diphenyl triketone in benzene or toluene solution[4] at various wavelengths was less than 1×10^{-3}. A complex mixture of products was obtained as illustrated:

$$PhCOCOCOPh + ArCH_3 \xrightarrow[R.T.]{N_2,\ \lambda > 420\ nm} ArCH_2OH + ArCHO + ArCH_2CH_2Ar + PhCOCH_2Ar + PhCOOPh + PhCOCOPh$$

Photochemistry of a number of tetraketones is being studied.[5] Quantum yields for disappearance are much higher than for triketones; elucidation of product structures is in progress. It has definitely been established, however, that diphenyl triketone or benzil are formed in trace amounts, if at all, in photolysis of diphenyl tetraketone. Interestingly, di-t-butyl tetraketone does not appear to react via intramolecular hydrogen atom abstraction.

In summary, di-, tri-, and tetraketone excited states are remarkable for the absence of α-cleavage as a primary process. A variety of rationalizations can be suggested for this unexpected behaviour.

References:

1) D.L. Bunbury and C.T. Wang, Can. J. Chem., **46**, 1473 (1968).

2) B.M. Monroe in "Advances in Photochemistry", Wiley, New York, Vol. 8, 1971, p. 86.

3) A.L. Gutman, Technion, unpublished results.

4) M. Heller, Technion, unpublished results.

5) M. Weiner, E. Krochmal, Jr., Technion, unpublished results.

FLUORESCENCE FROM UPPER EXCITED SINGLET STATES

Urs P. Wild

Physical Chemistry Laboratory
Swiss Federal Institute of Technology
CH-8092 Zürich, SWITZERLAND

1. INTRODUCTION

Fluorescence from upper excited singlet states in large organic molecules has been considered until recently as a rare and unusual phenomenon. The best known exception to the famous Kasha rule - "Only the lowest excited state of a given multiplicity emits" - has been azulene.

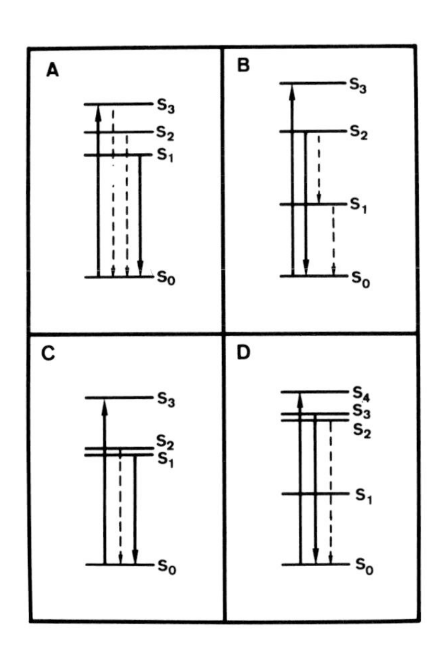

Figure 1
Typical energy level diagrams.

B. Pullman and N. Goldblum (eds.), Excited States in Organic Chemistry and Biochemistry, 387-395.

In Figure 1 some relevant energy level diagrams and the observed radiative transitions are shown. A typical molecule (A) has a rather large energy gap between S_0 and S_1, the higher electronic states being much closer spaced. Such a molecule indeed obeys Kasha's rule to a very good approximation and in order to see fluorescence from upper excited singlet states - which is always present, but extremely weak - high-sensitivity light detection in combination with special techniques is required. Recently Nickel[1] has reported in an elegant study this type of upper state fluorescence. In order to eliminate the disturbing influence from stray light, the upper excited singlet states were populated by triplet-triplet annihilation. The level scheme B may be called the "large gap" case and as examples I will discuss azulene and [18]annulene. The large gap between S_2 and S_1 inhibits fast internal conversion and opens the possibility for a significant radiative deactivation of the S_2 level. Molecules with close lying singlet states S_1 and S_2 (C) may show thermally excited S_2 emission such as the well studied example 3,4-benzpyrene. Finally, small and large energy gap effects can be found in a single molecule such as azulenophenalene (D) which will be discussed in some detail.

2. [18]ANNULENE

Let us start with [18]annulene[2] as a member of the "large gap" family. Assuming D_{6h} symmetry and performing a PPP type calculation the following energies and oscillator strengths for the low lying singlet states are obtained: S_1 12350 cm^{-1}, S_2 17000 cm^{-1} and S_3 28700 cm^{-1} (f = 7.14). The extremely large separation of S_3 and S_2 of about 12000 cm^{-1} could rise hope for a S_3 luminescence. The experimental spectrum (Figure 2) shows excellent agreement with respect to the posi-

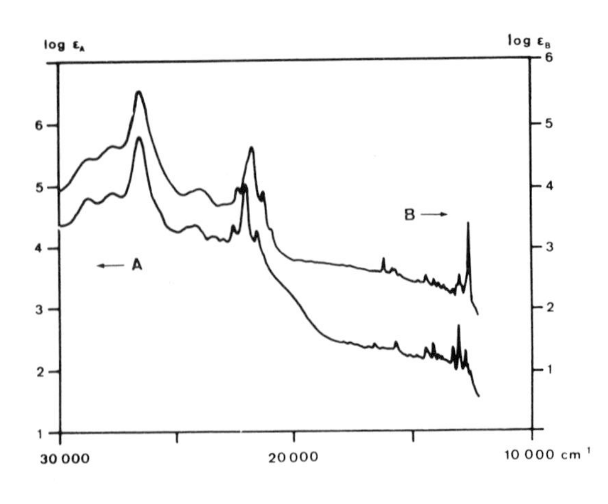

Figure 2
Absorption spectra of [18]annulene (A) and fluoro[18]annulene (B) at 77 K in a 3-methylpentane glass.

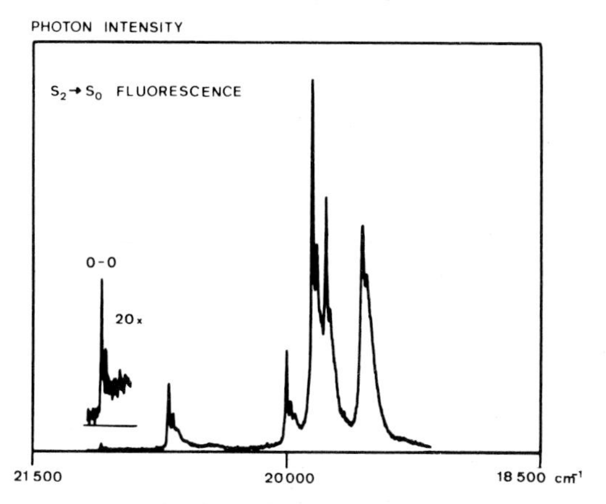

Figure 3
Fluorescence spectrum of the S_2 state of [18] annulene. Excitation at 21 468 cm^{-1} with an argon ion laser.

tions of S_1 and S_3. The S_2 state lies, however, at about 21 000 cm^{-1} and gains appreciable intensity by vibrational borrowing. This opens now the possibility of a weak S_2 luminescence and this is exactly what can be observed (Figure 3). The 0-0 band of the $S_2 \rightarrow S_0$ band is forbidden by symmetry and is very weak. In contrast to the S_1 emission

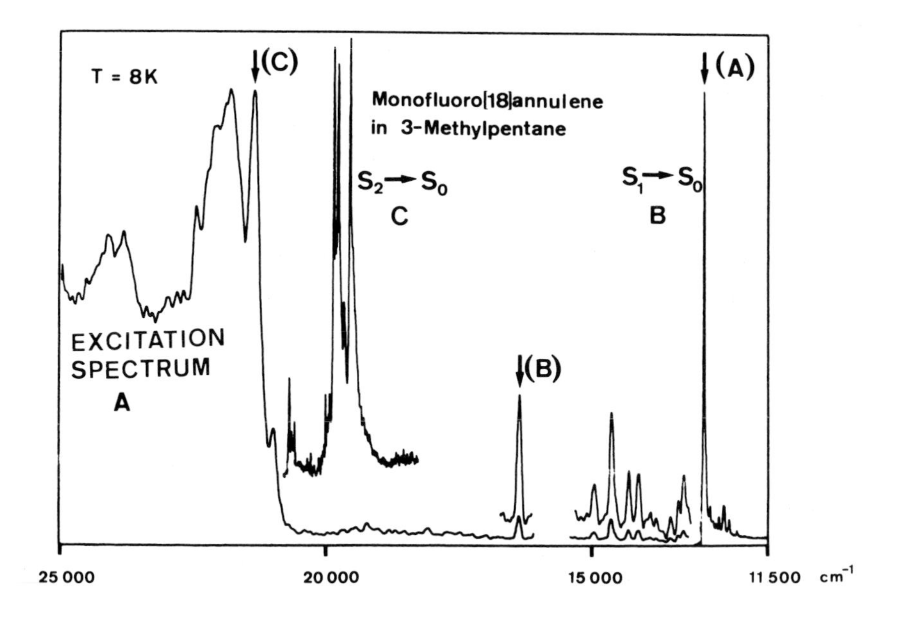

Figure 4
Monofluoro [18] annulene in 3-methylpentane at 8 K.
A. Excitation spectrum monitored at 12 750 cm^{-1} (A).
B. $S_1 \rightarrow S_0$ fluorescence excited at 16 242 cm^{-1} (B).
C. $S_2 \rightarrow S_0$ fluorescence excited at 21 250 cm^{-1} (C).

of benzene, which consists of progressions of totally symmetric modes built on a nontotally symmetric vibration, no progressions can be seen in the $S_2 \rightarrow S_0$ emission of [18] annulene. In the parent compound no emission from S_1 was observed. If the symmetry of the molecule is lowered such as in monofluoro [18] annulene the $S_1 \rightarrow S_0$ transition gets weakly allowed and indeed a very weak $S_1 \rightarrow S_0$ emission was observed (Figure 4). Most of the intensity of this fluorescence is contained in the 0-0 band. The perturbing effect of the fluor atom is much smaller in the S_2 state; the S_2 emissions of [18] annulene and of its monofluoro derivative are almost identical.

3. AZULENE

The classic molecule for observing S_2 emission is azulene. In Figure 5 we show an $S_2 \rightarrow S_0$ emission spectrum which has been recorded at 4 K with a 10^{-3} molar solution of azulene in n-hexane[3]. Some of the molecules are frozen in glass-like regions and by exciting the sample with an argon ion laser at 3 511 Å one is able to achieve site selection. The observed spectrum is a superposition of a diffuse spectrum and a very sharp structure. These sharp lines disappear after prolonged irradiation. Efforts to obtain similar structure in the $S_2 \rightarrow S_1$ region[4, 5] were unsuccessful. The molecule/solvent interaction which leads to the inhomogeneously broadened lines is strongly dependent on the electronic

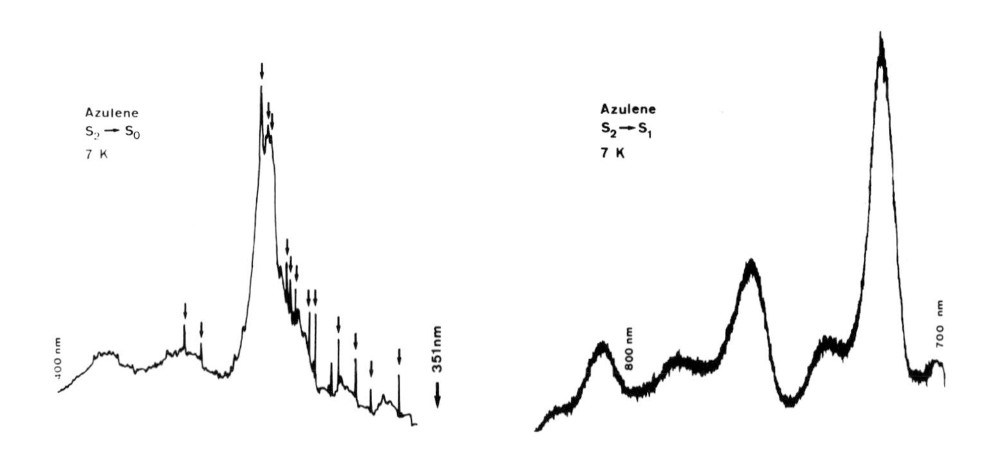

Figure 5
$S_2 \rightarrow S_0$ and $S2 \rightarrow S_1$ fluorescence spectra of a 10^{-3} molar azulene solution in n-hexane at 7 K.

state and almost independent on the vibronic levels. Site selection spectroscopy will thus give sharp spectra only, if the molecule is excited into the emitting level and the emission to the ground state is observed. The $S_2 \rightarrow S_1$ emission which has a different final state is therefore not expected to show sharp features (Figure 5). Similarly, molecules which show a site selection effect in fluorescence - when excited into S_1 - will generally emit only broad phosphorescence.

4. 3,4-BENZPYRENE

An example of a molecule with a small gap is 3,4-benzpyrene. Van den Bogaardt et al.[6] reported a thermally activated S_2 emission and further details about the photophysical properties of this system. If the temperature is decreased the intensity of the S_2 emission decreases also and reaches a limiting value. The remaining part results from a population of the S_2 state which is not in thermal equilibrium with the S_1 state. A fluorescence spectrum of 3,4-benzpyrene in an n-octane Shpolskii matrix at 4 K shows a weak but clearly resolved S_2 emission when excited into higher electronic states. Emissions from states which are not in thermal equilibrium have thus to be considered in high-sensitivity low temperature studies.

5. AZULENO[5,6,7-cd]PHENALENE

The molecule azulenophenalene[7] combines features of small and large gaps and shows some new interesting characteristics. If one looks at this large pentacyclic system one wonders whether it still shows an azulene-type energy level spacing. Calculations using the consistent force field approach of Warshel et al.[8] to estimate the molecular geometry and the electronic absorption spectrum indeed show that azulenophenalene can be considered as a perturbed azulene π system with two double bonds and a benzene ring. PPP calculations have also been performed by Thulstrup, Michl and Jutz[9] in order to understand the polarized absorption spectra and the magnetic circular dichroism. As in most semiempirical calculations on azulene-type molecules the energy of the S_1 state is too high. The absorption spectrum of phenalenoazulene (Figure 6) shows a first-band system at 13 400 cm^{-1} and a second one at 22 600 cm^{-1}. Let us also focus attention to the weak band 1 000 cm^{-1} below the origin of the second system and to the slow increase in absorption in-between. The gap between the second and the first band system is decreased from the azulene value of 14 000 cm^{-1} to about 9 200 cm^{-1} which is still favourable for upper excited state emission. Indeed, a very nicely

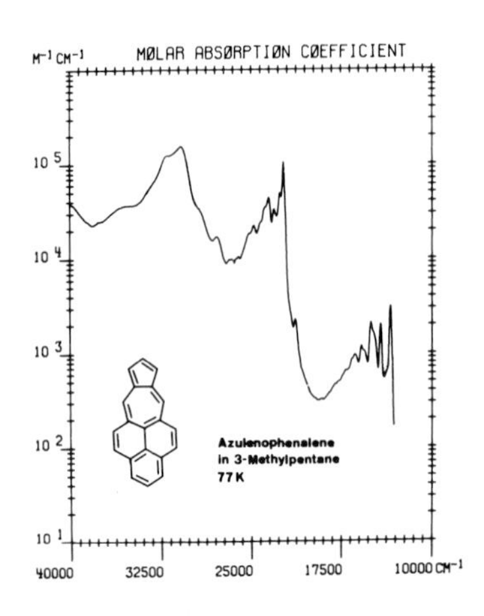

Figure 6
Absorption spectrum of azuleno-[5,6,7-cd]phenalene in 3-methyl-pentane glass at 77 K.

structured fluorescence (Figure 7) is observed in ethanol solution at 4 K. There are strong progressions involving a 1 230 and a 280 cm^{-1} mode. The excitation spectrum (Figure 8) monitored at 21 300 cm^{-1} shows a very good mirror relationship with the fluorescence. If it is compared with the absorption spectrum one notes important differences: the excitation spectrum has the structure expected from a single unperturbed emitting state. In addition to the vibrational structure seen in the excitation spectrum the absorption spectrum shows further bands. The same mirror-image relationship between the emission and excitation spectrum is also observed under conditions of high resolution. Actually the fluorescence spectrum of phenalenoazulene in an n-heptane Shpolskii matrix is one of the clearest spectrum we ever analyzed. Progression in the 280 cm^{-1} mode can be followed up to five quanta. Let us now compare the excitation spectrum monitored at 21 300 cm^{-1} (Figure 8A) with the spectrum monitored at 21 500 cm^{-1} (Figure 8C). The second spectrum shows at least five lines which are not observed in the first spectrum. The first band at 22 350 cm^{-1} - which is buried in the absorption spectrum under the rising edge of the second band system - is clearly resolved. There is, however, no excitation peak at 21 600 cm^{-1} which corresponds to the weak peak seen in the absorption spectrum. We may therefore construct the following energy level scheme: S_1 13 400 cm^{-1}, S_2 21 600 cm^{-1}, S_2^* 22 300 cm^{-1} and S_3 22 600 cm^{-1}. Consistent with such a level diagram are the following observations:

- Excitation into the high energy region of the absorption spectrum opens three pathways for deactivation: a) population of the strongly emitting state S_3 (Figure 7A, B), b) population of the weakly emitting state S_2^* and c) deactivation to the non-emitting states S_2 and S_1 seen in the absorption spectrum only.
- The quantum yield of the S_3 state is temperature dependent and

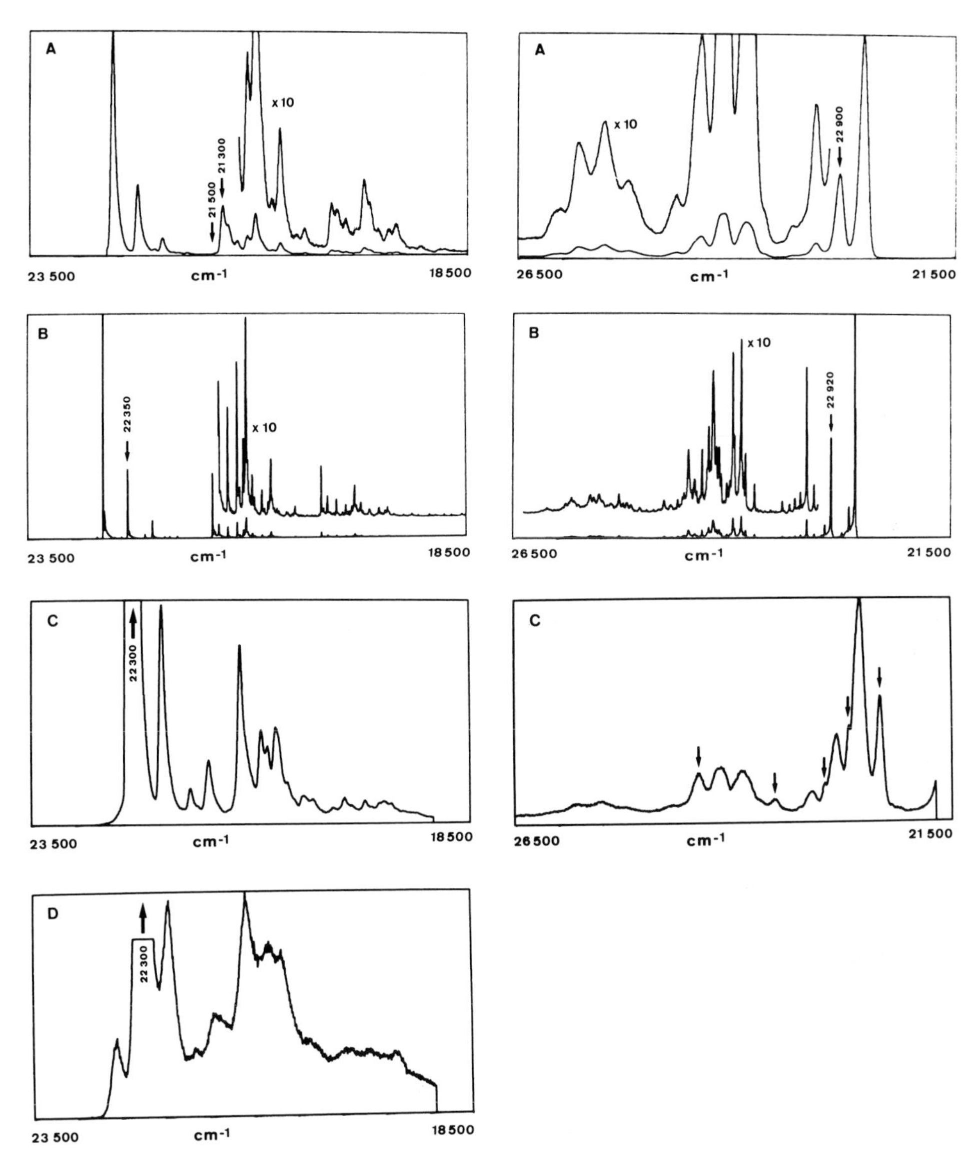

Figure 7
Fluorescence spectra of azuleno[5,6,7-cd]phenalene.
A) c = 2×10^{-6} M, ethanol, 9 K, excited at 22 900 cm^{-1}
B) c = 2×10^{-5} M, n-heptane, 4 K, excited at 22 920 cm^{-1}
C) c = 2×10^{-6} M, ethanol, 9 K, excited at 22 300 cm^{-1}
D) c = 2×10^{-6} M, ethanol, 107 K, excited at 22 300 cm^{-1}

Figure 8
Fluorescence excitation spectra of azuleno[5,6,7-cd]phenalene.
A) c = 2×10^{-6} M, ethanol, 9 K, observed at 21 300 cm^{-1}
B) c = 2×10^{-5} M, n-heptane, 4 K, observed at 22 350 cm^{-1}
C) c = 2×10^{-6} M, ethanol, 9 K, observed at 21 500 cm^{-1}

decreases with an activation energy of about 300 cm^{-1}. Such a behaviour would not be expected if S_3 interacted only with the much lower lying S_1 state in a statistical sense.

- Excitation into the S_2^* level at low temperature leads to the fluorescence spectrum of the vibrationally hot state. The emission spectra of S_3 and S_2^* are compared in Figure 7A and C.
- Excitation into the S_2^* level in the intermediate temperature range (50 - 150 K) gives a superposition of the S_3 and S_2^* emission spectra (Figure 7D). Such an emission is a good indication that the assigned S_2^* emission does not originate from an impurity but from the same molecule which also emits S_3 fluorescence. The activation energy of the $S_2^* \rightarrow S_3$ process is in the order of the observed energy difference between the two states.

The work on [18]annulene and azulene has been performed by H. J. Griesser, the investigations on 3,4-benzpyrene and azulenophenalene by A. Holzwarth as part of their Ph. D. Thesis.

REFERENCES

[1] B. Nickel,
Chem. Phys. Letters 27, 84 (1974)

[2] Urs P. Wild, Hans J. Griesser, Vo Dinh Tuan and Jean F. M. Oth,
Chem. Phys. Letters 41, 450 (1976)

[3] Hans J. Griesser and Urs P. Wild,
to be published.

[4] P. M. Rentzepis, J. Jortner and R. P. Jones,
Chem. Phys. Letters 4, 599 (1970)

[5] Gregory D. Gillespie and E. C. Lim,
J. Chem. Phys. 65, 4314 (1976)

[6] P. A. M. Van den Bogaardt, R. P. H. Rettschnick and J. D. W. Van Voorst,
Chem. Phys. Letters 18, 351 (1972)

[7] Alfred R. Holzwarth, K. Razi Naqvi and Urs P. Wild,
Chem. Phys. Letters 46, 473 (1977)

[8] A. Warshel and M. Levitt,
QCFF/PI: QCPE 247,
Quantum Chemistry Program Exchange, Indiana University (1974)

[9] Erik W. Thulstrup, Josef Michl and Christian Jutz,
J. Chem. Soc. Faraday Trans. II 9, 1618 (1975)

DISCUSSION

MITCHL :

Azuleno [5,6,7-cd] phenalene : It is possible to account for the absorption in the region of 21 600 cm^{-1} without postulating an additional excited state ?

WILD :

We are well aware that the calculations you reported do not support an additional electronic state in this region. We have extended these calculations using QCFF/PI program of Warshel and Levitt to optimize the geometry of the molecule and have -in collaboration with Dr. Baumann, ETH Zürich- performed a CNDO calculation using singly and doubly excited configurations. Still, no new state in this energy region is obtained. From the fluorescence experimental work we must, however, conclude that these bands behave like a new electronic state.

NON-ADIABATIC INTERACTIONS IN THE UNIMOLECULAR DECAY OF POLYATOMIC MOLECULES

J.C. Lorquet, C. Galloy, M. Desouter-Lecomte,

M.J. Decheneux and D. Dehareng.

Institut de Chimie de l'Université de Liège,
Sart Tilman, B - 4000 - Liège 1, Belgium.

INTRODUCTION

It is well-known that intersections, or avoided intersections between potential energy hypersurfaces play an important role in the interpretation of photochemical processes (1). In such regions of space, the Born-Oppenheimer approximation breaks down and the behavior of the system is no longer determined by the usual potential energy surfaces (in the Born-Oppenheimer sense), i.e. by the eigenvalues of the electronic hamiltonian. A coupling between (or among) energy surfaces takes place. Several methods are available to calculate the transition probabilities between them. Since the case of polyatomic molecules is fairly complicated, we shall restrict ourselves to the simplest method, called the semi-classical approximation (2) : the nuclear motion is described by a classical trajectory, whereas the electronic motion is treated quantum-mechanically.

Two choices are then possible. First, one may adopt the adiabatic representation, which is nothing else than the usual Born-Oppenheimer procedure. The wave functions and potential energy surfaces are obtained as eigenfunctions and eigenvectors, respectively, of the electronic hamiltonian.

$$H|i\rangle = E_i \; |i\rangle$$

Each pair of electronic states in coupled by (3N-6) matrix elements.

B. Pullman and N. Goldblum (eds.), Excited States in Organic Chemistry and Biochemistry, 397-407.

$$g_j^{12}(q_j) = \langle 1|\partial/\partial q_j|2\rangle \qquad j = 1,2,\ldots (3N-6) \tag{1}$$

$$g_j^{13}(q_j) = \langle 1|\partial/\partial q_j|3\rangle \qquad \text{etc...}$$

q_j being an internal nuclear coordinate. Some of these quantities are negligible, as will be shown below.

It can be shown that the adiabatic representation is especially useful when the nuclear velocities are vanishingly small. Then, the Born-Oppenheimer surfaces, i.e. those which are calculated in quantum chemistry by the configuration interaction (CI) procedure constitute an adequate potential for the calculation of nuclear forces and trajectories. They obey the non-crossing rule. The main drawback of this adiabatic representation is that it is well-known that in the neighborhood of an avoided crossing, the adiabatic wave functions depend in a very sensitive way on the nuclear coordinates. As a matter of fact,when one goes through such a region, there is usually an inversion of the electronic configurations, i.e., as one goes through the avoided crossing zone, the leading term in the CI expansion changes from Φ_1 to Φ_2 or vice-versa, as shown in Fig.1. This inversion usually takes place in a short range of internuclear coordinates (typically, 0.2 Å, or 20 degrees for an angular coordinate). It is precisely this dependence with respect to the nuclear coordinates which gives rise to the coupling matrix elements of Eq (1). This has important consequences, as will be shown later.

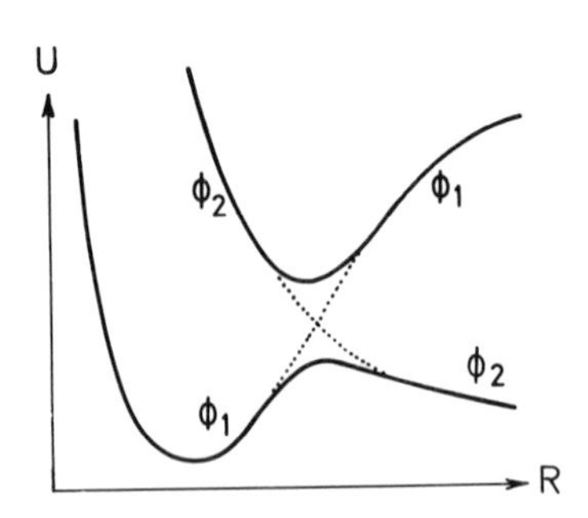

Fig.1.- Inversion of electronic configuration in an avoided crossing.

The alternative possibility (2) is to use the so-called diabatic representation. The diabatic states χ_1 and χ_2 are defined (3) by the property that they should be independent of the internal coordinates.

$$<\chi_1|\partial/\partial q_j|\chi_2> = 0 \qquad (2)$$

A number of variants of this definition have been proposed (3)

These diabatic functions do not diagonalize the electronic hamiltonian. Hence, in the matrix

$$H_{ij} = <\chi_i|H|\chi_j> \qquad (3)$$

the diagonal elements H_{ii} are defined as diabatic energy surfaces. They usually cross, and are coupled by the off-diagonal elements H_{ij}.

It can be shown that these diabatic surfaces provide a faithful description of the nuclear motion when the nuclear velocities become infinitely large, since the electrons do not have time to readjust their motion when the coupling zone is crossed. Hence the nuclei follow the force field determined by the surfaces corresponding to a constant electronic configuration, i.e. the diabatic ones (the dotted lines in Fig.1). This representation is thus useful for the description of high-energy collisions.

If the nuclear velocities are intermediate, as in a chemical reaction, either representation may be adopted as a zero-order description, but coupling effects must be introduced (i.e., the g_j's in the adiabatic representation, or the H_{ij}'s in the diabatic one). Transitions between potential energy surfaces then become possible. The previous ideas essentially go back to a pioneering paper by Hellmann and Syrkin (2).

In principle, the representation leading to the smallest coupling terms is to be preferred. In practice, however, the adiabatic representation is a priori more convenient, since ab initio programs are available which allow the calculation of adiabatic wave functions in a straightforward way. Hence, the following stepwise procedure for a theoretical study of photochemical reactions will be adopted.

(1) Ab-initio calculation of adiabatic potential energy surfaces. Inclusion of configuration interaction is an essential step to ensure proper calculation of crossings. These crossings can be classified as avoided crossings, conical intersections, Renner-Teller touchings, Jahn-Teller intersections, etc...

(2) Coupling matrix elements $g_j(q_j)$ (Eq.(1)) are calculated by numerical differentiation of CI and LCAO coefficients (4).

(3) Transition probabilities between electronic states are calculated by numerical integration of the classical trajectory equations.

Approximate analytical expressions for the transition probability are also available. A well-known one is the Landau-Zener equation; it is valid when the coupling term in the adiabatic representation H_{12} is constant. Also available is the Rosen-Zener approximation, valid when the energy difference between adiabatic states can be considered to be constant. It has been extended and discussed recently (5).

(4) These transition probabilities are then averaged over a Franck-Condon distribution of initial conditions to get an elementary photochemical rate constant.

SELECTION RULES FOR PHOTOCHEMICAL REACTIONS

Let us consider the simple case of an interaction between two electronic states only. The electronic time-dependent wave function is expanded over adiabatic states.

$$\Psi(x;q_1,q_2, \dots q_{3N-6})$$
$$=a_1(t)\ \psi_1(x;q_1, \dots q_{3N-6}) + a_2(t)\ \psi_2(x;q_1, \dots q_{3N-6})$$
$$=a_1(t)\ |1\rangle + a_2(t)\ |2\rangle$$

where x stands for all electron coordinates, $a_1(t)$ and $a_2(t)$ are the time-dependent amplitudes to be in adiabatic state $|1\rangle$ or $|2\rangle$ respectively. The (3N-6) internal nuclear coordinates q_j are classical functions of time.

It is assumed that the initial conditions are well-defined. At time t=0, the system is in adiabatic state k with an amplitude equal to unity. The rate of change of these amplitudes as a function of time is given by the classical trajectory equations (5,6) :

$$\dot{a}_1 = -a_2 \sum_{j=1}^{3N-6} \langle 1|\partial/\partial q_j|2\rangle\ \dot{q}_j\ \exp\left[-(i/h)\int_o^t (E_2-E_1)dt\right]$$
$$\dot{a}_2 = a_1 \sum_{j=1}^{3N-6} \langle 1|\partial/\partial q_j|2\rangle\ \dot{q}_j\ \exp\left[(i/h)\int_o^t (E_2-E_1)dt\right] \qquad (4)$$

The time-derivative of the amplitudes is thus given by a sum of contributions of each of the (3N-6) internal degrees of freedom. Many of them will not contribute to the summation for various reasons. This leads to a great simplification of the problem.

A trivial possibility is to have $\dot{q}_j = 0$. Internal coordinates which remain constant during the course of a photochemical reaction can be ignored in the summation. This will be the case for, e.g., many CH bond lengths, etc...

The second possibility is to have $<1|\partial/\partial q_j|2> = 0$. This brings us to a point of central importance. Do selection rules, or regularities of any kind, determine the value of this matrix element ?

To answer this question, we shall consider a specific problem. Fig.2 represents the potential energy surfaces of two electronic states of the CH_2^+ ion : the third and fourth $^4A_2$ states which were obtained from an ab initio CI calculation(4). The ion is constrained to have C_{2v} symmetry, and energy contours are plotted as a function of the internuclear angle (which is to be read in a polar coordinate system) and of the length of the CH bond (which can be read on the linear scale at the bottom of the figure). That avoided crossings, and hence non-adiabatic interactions, are taking place is evident from two criteria which emerged when considering Fig.1.

First, one notices that to each potential well in the surface of the upper $^4A_2$ state corresponds a ridge, or at least an inflection point in the surface of the lower $^4A_2$ state. In other words, each time the upper surface exhibits a concave structure, the lower surface presents a convex structure at the same internuclear distances.

A second criterion noted earlier is the inversion of electronic configurations as one goes through an avoided crossing. This was found to be the case here. The coefficients of the CI expansion are very sensitive functions of the nuclear geometry. The significance of this will be analyzed below in greater detail.

Let us now go back to the question of possible selection rules affecting the coupling matrix element $g_j = <1|\partial/\partial q_j|2>$. Does it sometimes vanish by symmetry ? Let us first consider internal coordinate q_3, the antisymmetric stretching mode, belonging to the b_2 representation in the C_{2v} point group . If the latter group is used, g_j is expected to vanish by symmetry. However, one may argue that, during the antisymmetric stretching motion, the correct point group to be used is C_s. Then the species of both electronic functions is $^4A''$, that of q_3 becomes a', and the matrix element is no longer expected to vanish.

Fig.3 answers the question. One sees that $<1|\partial/\partial q_3|2>$ vanishes for $q_3 = 0$ (C_{2v} point group) with a zero slope. For larger distortions, the appropriate point group is C_s, and the coupling term may take rather large values. However, since the latter function is antisymmetric, integration of Eqs.(4) yields a low value of the transition probability, of the order of 6.10^{-5}. We therefore conclude that symmetry considerations do play a role in chemical kinetics and may be responsible for low values of particular rate constants.

Such symmetry constraints are absent in the case of matrix elements $<1|\partial/\partial q_1|2>$ and $<1|\partial/\partial q_2|2>$, where q_1 and q_2 are the symmetric stretching and bending modes, respectively. However, other considerations come into play. Only certain trajectories, suitably oriented with respect to the "seam" between diabatic surfaces are able to bring

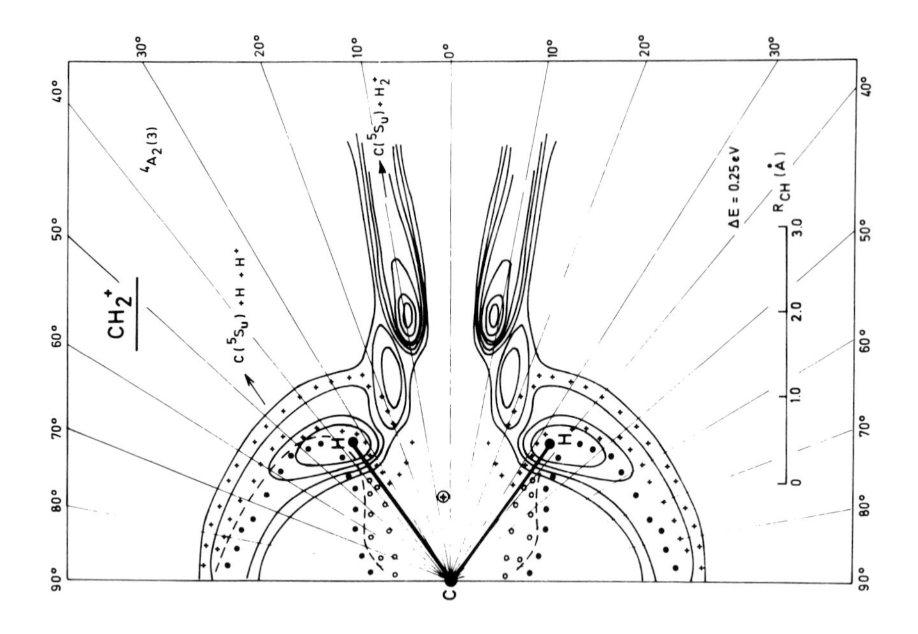
CH$_2^+$
$^4A_2(3)$
C(^5S_u) + H + H$^+$
C(^5S_u) + H$_2^+$
ΔE = 0.25 eV
R$_{CH}$ (Å)
C
H
H

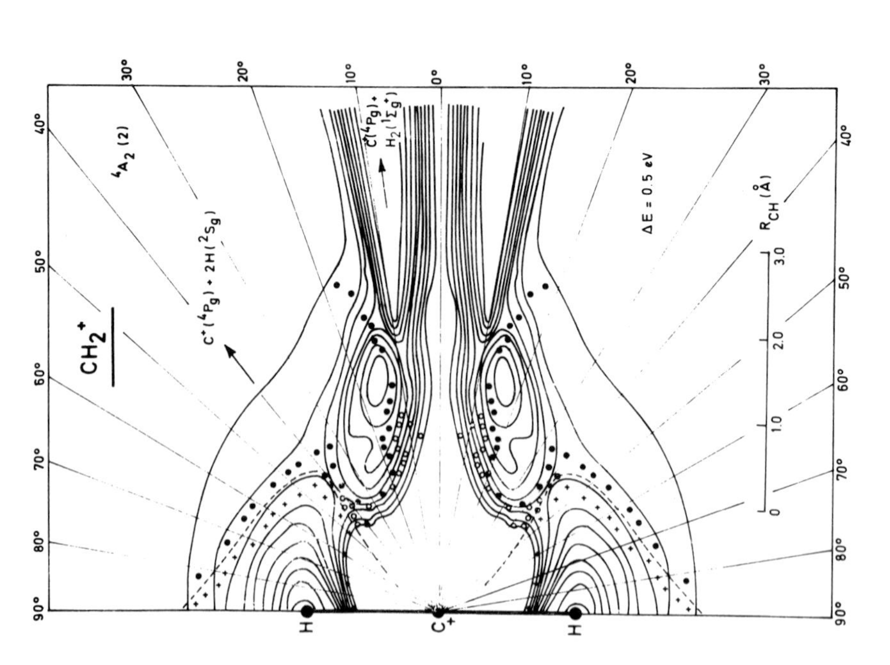
CH$_2^+$
$^4A_2(2)$
C$^+$(^4P_g) + 2H(^2S_g)
C$^+$(^4P_g) + H$_2$($^1\Sigma_g^+$)
ΔE = 0.5 eV
R$_{CH}$ (Å)
H
C$^+$
H

Fig. 2

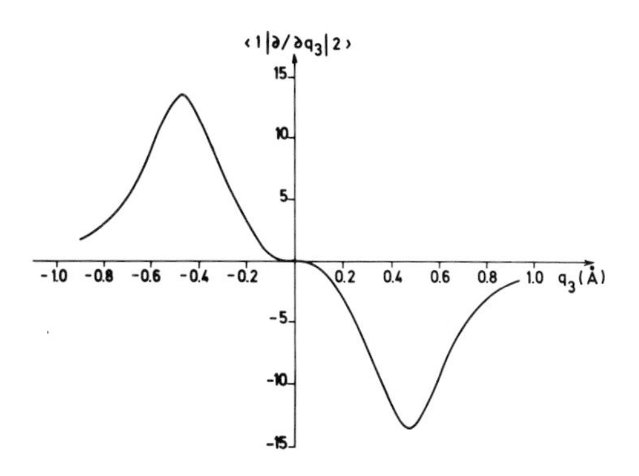

Fig.3.- Non-adiabatic coupling matrix element $<^4A_2(2)|\partial/\partial q_3|^4A_2(3)>$ as a function of q_3, the antisymmetric stretching coordinate.

about non-adiabatic transitions (7). By definition, the seam represents the locus of intersection between diabatic surfaces. A general procedure to locate it will be presented in the next section. However, as a first approximation, it also can be assimilated to a dividing line separating regions of space where the electronic configuration is constant. The CI expansion of the second and third $^4A_2$ states involves three leading configurations :$\Phi_1=(1a_1)^2(2a_1)^2(1b_2)^1(1b_1)^1(4a_1)^1$, $\Phi_2=(1a_1)^2(2a_1)^2(3a_1)^1(1b_1)^1(2b_2)^1$, and $\Phi_3=(1a_1)^2(2a_1)^1(1b_2)^1(3a_1)^2(1b_1)^1$. Their regions of predominance are indicated in Fig.2 by crosses, black dots, and open circles, respectively. The dotted line separates the regions of space where configurations Φ_1 and Φ_2 are respectively dominant. It is readily seen from Fig.2 that, during the stretching mode, the nuclear motion is parallel to the seam, whereas the bending motion takes place perpendicularly to it. In other words, during a symmetric stretching motion, the electronic configuration remains pratically constant, and the matrix element $<1|\partial/\partial q_1|2>$ cannot take very large values. On the other hand, during a bending motion, the system crosses the dividing line between zones of constant configuration. The derivative of the wave function with respect to the q_2 coordinate is large,and so is the matrix element. Thus, in this particular example, we conclude that the bending mode only is apt to bring about non-adiabatic transitions with a substantial probability; the stretching modes are much less efficient. We emphasize that the validity of this conclusion depends on the accuracy with which the seam has been located. Further work is in progress (4). Anyway, we see that, in contradistinction to statistical arguments, the internal energy of a molecule must in general be subdivided into an active part and an inert part.

Non-adiabatic interaction therefore appears as a highly anisotropic process. Eqs.(4) have the form of a scalar product between a coupling vector $\vec{g}$ and a velocity vector $\dot{\vec{q}}$. If these two vectors are perpendicular to each other, the dot product vanishes. In practice, this means that the non-adiabatic transition probability is determined by the component of velocity perpendicular to the seam (7). This is a **useful propensity rule.**

CONSTRUCTION OF DIABATIC STATES.

Once the non-adiabatic coupling matrix elements have been calculated, the obtention of the diabatic quantities (wave functions and matrix elements) is straightforward. The diabatic wave functions χ_1 and χ_2 are related to the adiabatic ones by an orthogonal transformation:

$$\chi_1 = \cos\theta\ |1\rangle - \sin\theta\ |2\rangle$$

$$\chi_2 = \sin\theta\ |1\rangle + \cos\theta\ |2\rangle$$

Substituting in Eq.(2) leads to

$$\partial\theta/\partial q_j = -\langle 1|\partial/\partial q_j|2\rangle = -g_j(q_j)$$

θ can therefore be calculated by numerical integration of the coupling functions. The diabatic functions χ_1 and χ_2 are then obtained at each point as an expansion over slater determinants. The coefficients of this expansion are found to be remarkably insensitive to the internuclear distances (4). The diagonal matrix elements H_{ii} cross and merge into the adiabatic curves at large internuclear distances. The off-diagonal element H_{12} depends only weakly on the internuclear distance. All this is much as expected, except that many coefficients are different from zero in the expansion of the diabatic wave functions. They do not correspond to pure configuration state functions as has often been assumed (3).

CONICAL INTERSECTIONS.

We now briefly present another example, whose study is still under way : the conical intersection between the $\tilde{A}$ and $\tilde{B}$ states of the H_2O^+ ion. When the two OH bond lengths are equal (C_{2v} point group), these two states belong to the $^2A_1$ and $^2B_2$ representations, respectively. Their potential energy curves cross. When the antisymmetrical stretching coordinate q_3 is different from zero, i.e. when the two OH bond lengths are unequal, these two states both belong to the $^2A'$ representation of the C_s point group and avoid crossing. A schematic view is given in Fig.4, where the potential energy surfaces are plotted as a function of the internuclear valence angle α (viz. q_2) and the antisymmetric stretching coordinate $R_1 - R_2$ (viz. q_3).

A region of strong non-adiabatic interaction is centered around the apex of the double cone. The coupling matrix elements have been calculated. It is instructive to use the procedure described in the previous section and to construct the diabatic surfaces from them. The aim is to compare the result of such calculations with the following simple and intuitive model represented on Fig.5.

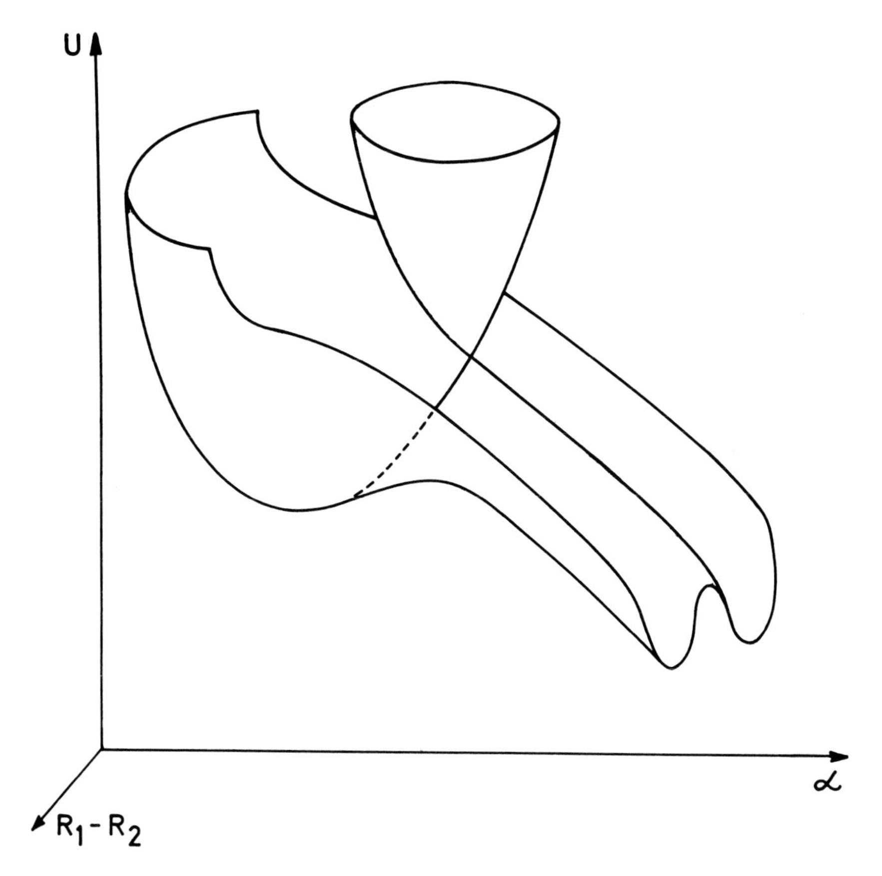

Fig.4.- Schematic view of the conical intersection between the $\tilde{X}^2A_1$ and $\tilde{B}^2B_2$ states of H_2O^+. Internal coordinates are α, the HOH angle (q_2), and R_1-R_2, the antisymmetric stretching coordinate (q_3).

The shape of the adiabatic surfaces around the apex of the double cone has a double origin. Along axis q_2 (viz. $R_1=R_2$) the diabatic and adiabatic potential energy surfaces coïncide by reason of symmetry; the shape is that of the two diabatic surfaces, which in a first approximation can be expected to be ruled. H_{12} vanishes by symmetry along the q_2 axis. Along axis q_3 (α constant), the diabatic surfaces split because of the off-diagonal element H_{12}. At the apex of the double cone, $H_{11}=H_{22}$, $H_{12}=0$, and the two surfaces coïncide. Since the splitting between the two adiabatic surfaces has a different origin in the two directions of space (Fig.5), anisotropic behavior again results. Because the seam between diabatic surfaces is roughly oriented along the q_3 axis, it is to be expected (7) that, the larger the component of the velocity along the q_2 axis, the larger the transition probability. Presently available calculations indicate that

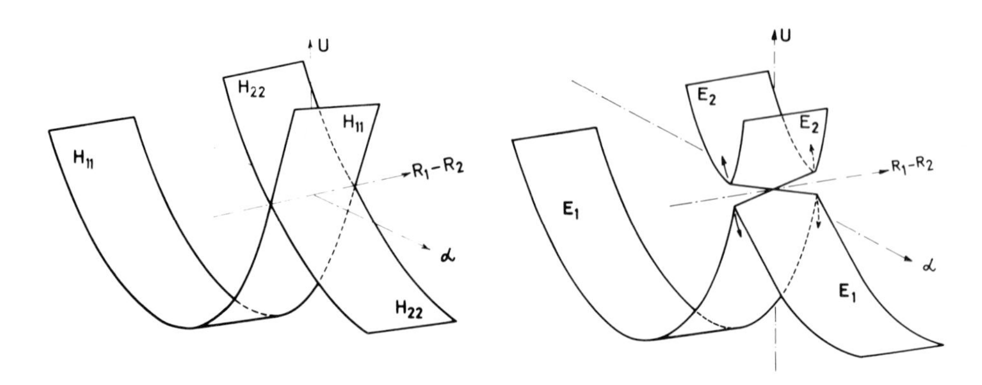

Fig.5.- Diabatic and adiabatic surfaces in a conical intersection.

reality is not that simple, and that deviations with respect to this simple model exist. They are presently under study.

REFERENCES.

1 J.Michl, Topics in Current Chemistry, 46, 1 (1974).
L.Salem, J.Am.Chem.Soc., 96, 3486 (1974).
L.Salem, C.Leforestier, G.Segal and R.Wetmore, J.Am.Chem.Soc., 97,479 (1975).
W.G.Dauben, L.Salem and N.J.Turro, Acc.Chem.Res., 8, 41 (1975).

2 H.Hellmann and J.K.Syrkin, Acta Physicochim.URSS, 2, 433 (1935).
E.E.Nikitin, in Chemische Elementarprozesse, ed.H.Hartmann (Springer, Berlin, 1968).
M.Desouter-Lecomte, J.C.Leclerc and J.C.Lorquet, Chem.Phys., 9, 147 (1975).

3 W.Lichten, Phys.Rev., 131, 229 (1963).
F.T.Smith, Phys.Rev., 179, 111 (1969).
V.Sidis and H.Lefebvre-Brion, J.Phys., B 4, 1040 (1971).
T.F.O'Malley, Adv.Atom.Mol.Phys., 7, 223 (1971).

4 C.Galloy and J.C.Lorquet, to be published.

5 M.Desouter-Lecomte and J.C.Lorquet, J.Chem.Phys., (in press).

6 B.Corrigall, B.Kuppers and R.Wallace, Phys.Rev., A 4, 977 (1971).
J.C.Tully, J.Chem.Phys., 60, 3042 (1974).

7 R.K.Preston and J.C.Tully, J.Chem.Phys., 54, 4297 (1971).
J.C.Tully and R.K.Preston, J.Chem.Phys., 55, 562 (1971).
J.R.Stine and J.T.Muckerman, J.Chem.Phys., 65, 3975 (1976).

DISCUSSION

SALEM :

This is a remarkable piece of work. One question : what is the origin of the avoided crossing between the two $^4A_2$ states, and what is its size ?

LORQUET :

The splitting is due to configuration interaction (i.e., Type C in your nomenclature). The value of the smallest energy gap is the order of 0.3 to 0.5 eV, depending on the basis set which is used in the claculations.

MEASUREMENT OF CIRCULAR DICHROISM IN THE VACUUM ULTRAVIOLET. A NEW CHALLENGE FOR THEORETICIANS.

E. S. Pysh

Department of Chemistry, Brown University, Providence, R. I. 02912 USA

INTRODUCTION

Measurements of vacuum ultraviolet circular dichroism (VUCD)[1] have been made only in recent years. Several prototype instruments have been described[2-20] and reviews have begun to appear in the literature.[20-22] The new results reported here are described from the point of view of illustrating how these and similar data may influence the development of the theories of optical activity and conformational analysis.

VUCD measurements have been reported for a wide variety of compounds. Schnepp[2-4] has described results obtained on small organic molecules in the gas phase, and has recently extended the capability of his instrument to include measurements of vacuum ultraviolet magnetic circular dichroism. Johnson[5-13] has mainly been studying saccharides and nucleic acids. The emphasis of my work [14-21] has been on peptides and polysaccharides. Other applications will undoubtedly be made and additional prototype instruments constructed.

The only novel component in a VUCD instrument is the vacuum ultraviolet quarter wave retarder. Current instruments exploit the invention of an elegantly designed device by J. Kemp.[23] The electrooptic modulators used in commercial instruments do not transmit in the vacuum ultraviolet region, and materials that do, like LiF, CaF_2 and MgF_2, do not have the proper crystal symmetry to display the electroöptic effect. The vacuum uv transmitting materials do, however, exhibit the piezooptic effect, so that when coupled to an electropiezo transducer, such as quartz, the net result is an electropiezo-

B. Pullman and N. Goldblum (eds.), Excited States in Organic Chemistry and Biochemistry, 409-418.

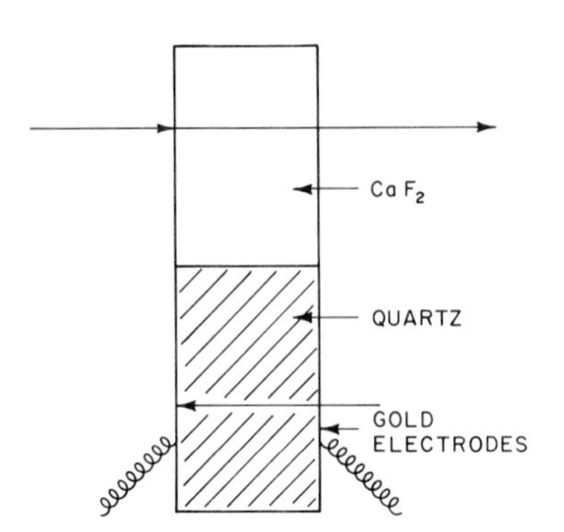

Figure 1. Schematic representation of a vacuum ultraviolet oscillating $\lambda/4$ plate. Modulation is at 50 kHz.

piezooptic modulator (See Figure 1). The other components of the optical train are standard, as is the entire detection system. We have recently reverted from a hot thoriated tungsten filament H_2 discharge lamp to the standard Hinteregger cold cathode mode of operation because of the high maintenance requirements of the hot filament lamp. The optimum sensitivity of all three prototype instruments is approximately the same.

APPLICATIONS

The usefulness of VUCD measurements can be either in extending data on compounds for which near ultraviolet data already exist, or in providing spectra for compounds which absorb only in the vacuum ultraviolet region.

Carboxylic Acids

Figure 2 shows the CD of (-)1-methyl-2,2-dibromocyclopropanecarboxylic acid in HFIP and in cyclohexane. The spectrometer was operated with a spectral width of 1.6 nm, time constant of 10 s, and scan rate of 2 nm/min. The figure shows actual recorder tracings and reflects conditions chosen to optimize the signal/noise ratio in the highly absorbing region near 180 nm; better signal/noise ratios are obtained in the region of low absorbance above 200 nm by increasing sample concentration. The data in these solvents can be extended to slightly lower wavelengths with cells of shorter path length, with an accompanying decrease in signal/noise ratio.

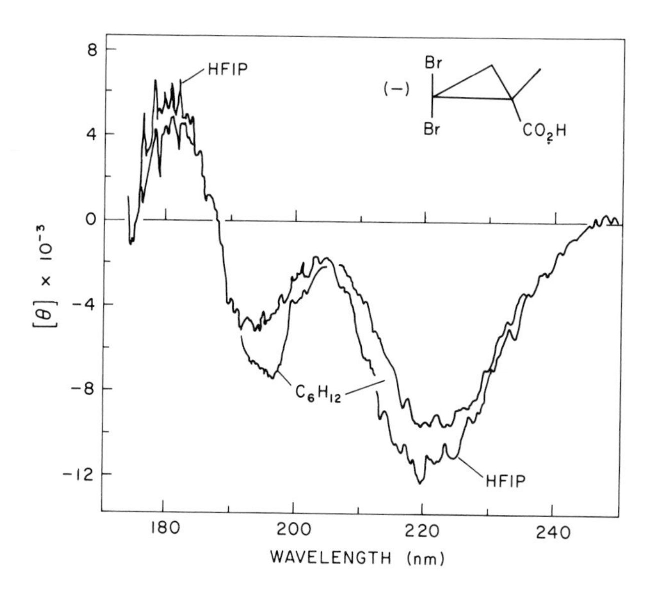

Figure 2. The CD of (-)-methyl-2,2-dibromocyclopropanecarboxylic acid in HFIP (0.062 M) and in cyclohexane (0.070 M) measured in an 0.05 mm path length cell.

A complete analysis of the VUCD of this strained cyclic carboxylic acid and related compounds will be presented elsewhere.[24] The positive band near 181 nm is assigned to the carboxyl π-π^* transition on the basis of the strong absorption in that region. We assign the two negative bands near 194 nm and 220 nm to the n-π^* transition in two rotational isomers. The appearance of two n-π^* bands in this compound is analogous to their appearance in peptides[25] and acid sugars.[26,27]

The ratio of [θ] at 194 nm to that at 220 nm is 0.4 in HFIP and 0.8 in cyclohexane, which indicates that the equilibrium between rotational isomers is strongly solvent dependent. Dimerization is more extensive in cyclohexane than in HFIP and it is the dimerization that alters the equilibrium between the two rotational isomers. Thus the CD solvent effect reflects a dimerization equilibrium as well as a rotational equilibrium. The fact that the total rotational strength of the two negative bands is the same in HFIP as in cyclohexane supports their assignment to two n-π^* bands.

An accompanying large change in the 181 nm band would have cast doubt on such an interpretation since any large change at 181 nm might have caused changes in the apparent intensities of the neighboring band even if its intrinsic rotational strength did not change. The advantage of having data which include the 181 nm band lies in being able to establish clearly that the intrinsic rotational strengths of the 194 nm and 220 nm bands are indeed different in the two solvents.

Oligopeptides and Regular Polypeptides

In collaboration with Dr. C. Toniolo of Padova, we have been measuring the VUCD of protected linear homooligopeptides of the type BOC(L-XXX)$_n$OMe where n =2-7 and XXX =Ala, Cha, Leu, Nva, Phe and Val.[17,18,28] These compounds are known to take up beta sheet conformations under appropriate conditions both in solution and in the solid state. We have concluded that the VUCD in the region 170-200 nm can be used to distinguish between beta sheets of the parallel and antiparallel types. Whereas the parallel sheet VUCD is characterized by rapidly decreasing dichroism below 200 nm leading to a crossover to negative dichroism near 192 nm, the antiparallel sheet VUCD decreases more slowly and includes a positive shoulder below 200 nm so that the crossover to negative dichroism occurs at lower wavelengths, e.g. 178 nm in the alanine heptamer.

By this criterion the VUCD indicates that valine and phenylalanine heptamers take up predominantly parallel sheet forms, while norvaline and leucine take up mixed conformations. We attribute this behavior to the steric effect of the side chains such that bulky side chains and those branched at the β-carbon atom lead to destabilization of the antiparallel form, and a favoring of the parallel form in which these larger side chains can be better accomodated.

Table 1. Polypeptide circular dichroism in the vacuum ultraviolet, showing the sign of the dichroism in three wavelength regions.[14,15,17]

	145 nm	165 nm	180 nm
Alpha helix	+	-	+
Poly-L-proline I	-	-	-
Poly-L-proline II	-	-	-
Antiparallel β-sheet	+	-	+
Parallel β-sheet	+	+	-

A preponderance of valine, phenylalanine and isoleucine residues in the parallel sheets of globular proteins with known conformations supports this interpretation and indicates that the effect is important in the folding of long chains, as well as the aggregation of short chains.

From the point of view of developing a theory for the higher excited states of peptides, VUCD data place very stringent criteria on future calculations of peptide wave functions and peptide optical activity. A successful theoretical treatment of the optical activity of polypeptides would have to include discrete transitions near 145 nm and 165 nm as well as those near 190 nm and 220 nm usually included. The possible role of an additional transition near 175-180 nm should be examined. Polarizability contributions will have to be included in calculations of VUCD in order to account for the several cases of non-conservative CD (i.e. the preponderance of dichroism of one sign). The most extreme case in this regard is poly-L-proline II (Table I) which displays negative dichroism over almost the entire observable wavelength range.

Poly-L-Valine

Figure 3 shows the VUCD of a film of poly-L-valine (Pilot Chemical Co.) cast from TFE. The instrument was operated under the same conditions as described above. There was no dependence of signal on

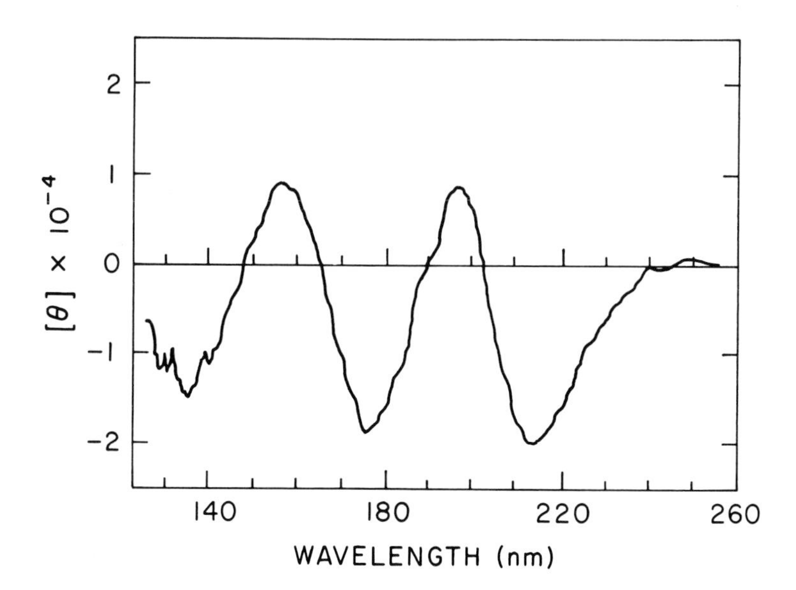

Figure 3. VUCD of a poly-L-valine film cast from TFE.

orientation of the film in the light beam. The reported molar ellipticity is calculated assuming no difference in molar ellipticity between film and solution at 215 nm.

The conformation of beta-form poly-L-valine has been described several times in the literature. It is usually described as being antiparallel or cross-beta. It is known however that valine oligomers do not form beta turns readily, and our earlier work[17] indicates that valine oligomers prefer the parallel rather than antiparallel beta conformation. It is unlikely, therefore, that poly-L-valine is made up entirely of beta turns and antiparallel segments; i.e. the term cross-beta is not appropriate.

Kubota and Fasman[29] obtained a spectrum for $(DL\text{-}Lys\cdot HCl)_{10}$-$(L\text{-}Val)_{20}(DL\text{-}Lys\cdot HCl)_{10}$ in aqueous solution which is similar to our film data above 190 nm except that in our spectrum the positive 195 nm band is somewhat less intense. On the basis of the absence of any CD concentration dependence, Kubota and Fasman accounted for their observation of beta-like infrared absorption, ultraviolet hyperchromicity and CD by proposing the existence of single extended beta-chains.

On the basis of the VUCD data we wish to propose a model for poly-L-proline which reflects its preference for extended chain segments. Fig. 3 shows that the VUCD of poly-L-valine resembles neither the parallel beta sheet CD (e.g. oligovaline) nor the antiparallel beta sheet CD (e.g. oligoalanine). Figure 4 represents schematically a chain conformation consistent with these data. Rather than a well developed cross-beta structure, made up entirely of beta turns and antiparallel segments, the chain is better described as an aggregate of extended chain segments, rather disordered on the whole, but containing some segments aggregated in an antiparallel orientation and some perhaps even in a parallel orientation (See Figure 4).

Figure 4. Schematic representation of poly-L-valine.

Polysaccharides

We have now completed VUCD studies of ι-carrageenan[16], hyaluronic acid[19], agarose[20] and the carob, tara and guar galactomannans.[21] In hyaluronic acid the acetamido and carboxyl substituents are predominantly responsible for the observed dichroism and the analysis of those data are presented elsewhere.[19] In the other cases the observed dichroism arises from the backbone saccharide chromophores; there is no dichroism above 200 nm. The regularity which we have observed in these polysaccharides is summarized in Table 2.

There are two optically active transitions, one near 180 nm and the other near 160 nm. These two transitions are always of opposite sign. The 180 nm band is positive in agarose and in the galactomannans, but negative in ι-carrageenan.

The VUCD is sensitive to primary structure in that the intensity of the 180 nm band differs within the series of galactomannans. The intensity is not, however, a simple function of galactose/mannose ratio as is the optical rotation.

The 180 nm band in agarose is displaced to higher wavelengths during the heat induced gel-sol transition, and therefore reflects some aspects of that complex conformational change, but not the same aspects as optical rotation, the behavior of which displays both cooperativity and hysteresis.

Precisely how the observed VUCD features of these polysaccharides are related to the details of their rich conformational behavior is not yet known. It is worth pointing out, however, that no nonempirical theory of polysaccharide optical activity exists, and there has been relatively little theoretical work done in the area of polysaccharide configurational statistics.

Table 2. Polysaccharide circular dichroism in the vacuum ultraviolet, showing the sign of the dichroism in two wavelength regions.[16,20,21]

	160 nm	180 nm
ι-Carrageenan	+	-
Agarose	-	+
Galactomannan	-	+

CONCLUSIONS

Applications of standard methods of classical configurational analysis can now be made to small molecules with chromophores absorbing in the far ultraviolet and vacuum ultraviolet regions. The fact that alkanes, alkenes, alcohols, esters and acids begin to absorb only in the vacuum ultraviolet region need not inhibit the application of classical configurational analysis by optical activity measurements.

Some VUCD applications will involve empirical correlations of the dichroism with conformation, as in the case of the beta-forming peptides. The maximum amount of detailed information will only be extracted if theoretical developments occur. The useful developments would be in two areas: (a) the extension of spectral calculations into the vacuum ultraviolet region in cases where calculations have in the past been limited to the near ultraviolet region, and (b) the application of current spectral theories to systems heretofore not examined because of the lack of experimental data (e.g. alcohols, esters, saccharides).

ACKNOWLEDGMENTS

The work described here was supported by grants from NSF, USPHS, PRF and NATO.

REFERENCES

1. The following abbreviations are used: CD(circular dichroism), VUCD (vacuum ultraviolet circular dichroism), HFIP (hexafluoroisopropanol), TFE (trifluoroethanol), BOC (*tert*-butyloxycarbonyl), OMe (methoxy), Ala (alanine), Cha (cyclohexylalanine), Leu (leucine), Nva (norvaline), Val (valine).

2. O. Schnepp, E. F. Pearson, and E. Sharman, *J. Chem. Phys.*, 52, 6424 (1970); *Chem. Commun.*, *1970*, 45 (1970).

3. O. Schnepp, S. Allen, and E. F. Pearson, *Rev. Sci. Instrum.*, 41, 1136 (1970).

4. O. Schnepp, in *Chemical Spectroscopy and Photochemistry in the Vacuum Ultraviolet*, Eds. C. Sandorfy, P. J. Ausloos, and M. B. Robin, Reidel Publ. Co. (NATO ASI Series), 1974, p. 211.

5. W. C. Johnson, Jr., *Rev. Sci. Instrum.*, 42, 1283 (1971).

6. H. J. Li, I. Isenberg, and W. C. Johnson, Jr., Biochemistry, 10, 2587 (1971).

7. R. G. Nelson and W. C. Johnson, Jr., J. Am. Chem. Soc., 94, 3343 (1972).

8. W. C. Johnson, Jr. and I. Tinoco, Jr., J. Am. Chem. Soc., 94, 4389 (1972).

9. P. A. Snyder, P. M. Vipond, and W. C. Johnson, Jr., Biopolymers, 12, 975 (1973).

10. W. C. Johnson, Jr., J. Chem. Phys., 59, 2618 (1973).

11. D. G. Lewis and W. C. Johnson, Jr., J. Mol. Biol. 86, 91 (1974).

12. R. G. Nelson and W. C. Johnson, Jr., J. Am. Chem. Soc., 98, 4290 (1976).

13. R. G. Nelson and W. C. Johnson, Jr., J. Am. Chem. Soc., 98, 4296 (1976).

14. M. A. Young and E. S. Pysh, Macromolecules, 6, 790 (1973).

15. M. A. Young and E. S. Pysh, J. Am. Chem. Soc., 97, 5100 (1975).

16. J. S. Balcerski, E. S. Pysh, G. C. Chen, and J. T. Yang, J. Am. Chem. Soc., 97, 6274 (1975).

17. J. S. Balcerski, E. S. Pysh, G. M. Bonora, and C. Toniolo, J. Am. Chem. Soc., 98, 3470 (1976).

18. M. M. Kelly, E. S. Pysh, G. M. Bonora, and C. Toniolo, J. Am. Chem. Soc., 99, xxxx (1977).

19. L. A. Buffington, E. S. Pysh, B. Chakrabarti, and E. A. Balazs, J. Am. Chem. Soc., 99, xxxx (1977).

20. E. S. Pysh, Ann. Rev. Biophys. Bioeng., 5, 63 (1976).

21. E. S. Pysh, Adv. Enzymology, in press.

22. W. C. Johnson, Jr., Adv. Phys. Chem., in preparation.

23. J. C. Kemp, J. Opt. Soc., Am., 59, 950 (1969).

24. T. H. Morton, L. A. Buffington, and E. S. Pysh, in preparation.

25. J. C. Craig and W. E. Pereira, Tetrahedron Lett., 18, 1563 (1970); Tetrahedron, 26, 3457 (1970).

26. E. R. Morris, D. A. Rees, G. R. Sanderson, and D. Thom, J. Chem. Soc., Perkin Trans., 2, 1418 (1975).

27. I. Listowsky, G. Avigad, and S. Englard, J. Org. Chem., 35, 1080 (1970); Trans. N. Y. Acad. Sci., 34, 218 (1972).

28. E. Peggion, M. Palumbo, G. M. Bonora, C. Toniolo, and E. S. Pysh, Proc. Fifth Amer. Peptide Symposium, San Diego, 1977.

29. S. Kubota and G. D. Fasman, Biopolymers, 14, 605 (1975).

30. J. Liang, E. R. Morris, and E. S. Pysh, in preparation.

31. L. A. Buffington, E. R. Morris, and E. S. Pysh, in preparation.

NON EMPIRICAL CALCULATIONS OF EXCITED STATES OF LARGE MOLECULES BY THE METHOD OF IMPROVED VIRTUAL ORBITALS

R. Janoschek

Institut für Theoretische Chemie,
Universität Stuttgart, W. Germany

1. METHOD OF CALCULATION

Within the Hartree-Fock approximation a closed shell system of 2n electrons is described by n occupied orbitals. In addition, a set of unoccupied orbitals (virtual orb.) is obtained. According to the theorem of Koopmans the negative energies of the occupied orbitals approach the ionisation energies.

$$e_i = E^i(2n-1) - E(2n) \quad , \; i \in \{occ\}$$

The energies of the virtual orbitals can be seen to represent the electron affinities.

$$e_k = E^k(2n+1) - E(2n) \quad , \; k \notin \{occ\}$$

Thus the virtual orbitals are determined by the potential of 2n electrons and are therefore unsuitable for the calculation of excited electronic states.

Only one condition must be fulfilled by the virtual orbitals: the orthogonality onto the occupied orbitals.

$$\langle \psi_i \mid \psi_k \rangle = 0 \quad , \; i \in , \; k \notin \{occ\}$$

If virtual orbitals are used for the calculation of excited states, they should come out of a 2n-1 electron potential rather than a 2n one. A modified Fock operator F(j,2n) which yields the restricted HF orbitals as well as virtual orbitals for a 2n-1 particle potential is

$$F(j,2n) = PF(2n)P + (1-P)F(j,2n-1)(1-P)$$

where F(2n) is the common 2n electron Fock operator, F(j,2n-1) is a 2n-1 electron Fock operator in which the

B. Pullman and N. Goldblum (eds.), Excited States in Organic Chemistry and Biochemistry, 419-429.

j-th orbital is singly occupied and P is a projection operator [1,2].

$$F(2n) = h + \sum (2J_i - K_i)$$

$$F(j,2n-1) = F(2n) - J_j + \begin{pmatrix} 2K_j & , \text{Singlet} \\ 0 & , \text{Triplet} \end{pmatrix}$$

$$P = \sum |\psi_i\rangle\langle\psi_i| \;, \qquad i,j \in \{occ\}$$

In practice, open shell virtual orbitals must be generated after the HF calculation. This is represented schematically by

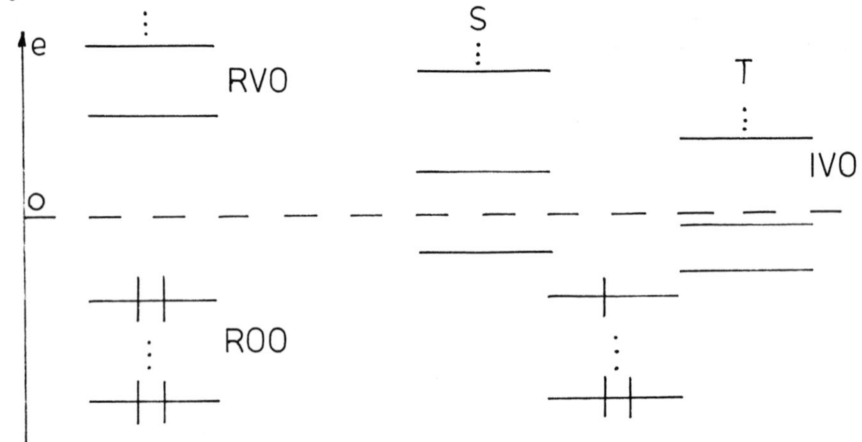

where ROO are the restricted occupied and RVO the restricted virtual orbitals and IVO stands for the improved virtual orbitals which are unrestricted.

The most convenient procedure is to solve the secular equations for the right lower submatrix

$$F(j,2n) = \left(\begin{array}{ccc|c} e_1 & & o & \\ & \ddots & & o \\ o & & e_n & \\ \hline & o & & e(RVO) + F(j,2n-1) - F(2n) \end{array}\right)$$

The results are IVOs and their energies $\bar{e}_k$. The excitation energies are now simply

$$\bar{e}_k - e_j \qquad , \quad j \in , \; k \notin \{occ\}$$

This result can be seen as the 'Koopmans theorem for excitations'.

Two effects are neglected in the Koopmans approximation. These are the reorganisation and the correlation which mutually cancel out to a large extent, due to their opposite signs. This so called Koopmans defect will be discussed and a possibility for reducing it will be given in the next chapter.

2. The KOOPMANS DEFECT

Before the IVO method is applied to molecules some essential features should be discussed by considering small atoms. Firstly, one has to take care, as is necessary for any method, that the basis set is suitable not only for describing the ground state but, moreover, for representing the excited states under consideration.

With the He atom a {10s,5p} basis set [3] yields the following lowest excitation energies in Tab. 1. The calculated singlet excitation energies exceed the observed values by about 0.4 eV, whereas the singlet-triplet separations are reproduced more accurately. The corresponding RVO results can be inaccurate by as much as 2-3 eV and show an incorrect sequence.

		RVO	IVO	EXP [4]
(1s)(2s)	^1S	22.92	21.12	20.61
	^3S	22.91	19.82	19.82
(1s)(2p)	^1P	23.42	21.69	21.22
	^3P	23.42	21.24	20.96
(1s)(3s)	^1S	22.49	23.35	22.92
	^3S	22.37	23.05	22.72
(1s)(3p)	^1P	22.97	23.52	23.09
	^3P	22.96	23.39	23.01
(1s)	^2S	24.97	24.97	24.58

Table 1. Excitation energies (eV) for He.

The discrepancies between calculated and observed excitation energies can be explained by the Koopmans defect. For He the HF energy -2.861, the exact energy -2.903 and the Koopmans ionization energy of 0.918 a.u. yield a reorganisation energy for He+ of 1.5 eV and a correlation energy for He (1s)(1s) of -1.1 eV. Thus the singlet excitations are overestimated by about 0.4 eV. Recently, it has been suggested one should calculate the IVOs in the field of the open shell He+ rather than Koopmans He+ [5].

In so doing, the reorganisation is taken completely into account, but the excitation energies are now too low by an amount equal to the full correlation energy. The more accurate singlet-triplet separations might be explained by the quasi-consideration of the correlation energy due to the Fermi hole in the pair density for equal spin.

The Koopmans defect leads to the suggestion that the predominating reorganisation defect may possibly be reduced by the use of an inner shell core which is more contracted than those of a neutral atom. This reduction can be achieved by using largely small contracted basis sets, which have been obtained by energy variation, for the configurations (1s)(1s)(2s)(2s) for the atoms C, N, O, etc [6]. A first attempt with such a basis set will be undertaken in the following section by means of the C atom.

The C atom has been calculated using the {9s,5p} basis set [3] which was extended by {1s,1p} with the exponents 0.0506 and 0.0382, respectively. The virtual orbitals have been considerably rearranged in their sequence after improvement. Calculated separations of the lowest states are compared with the experiment.

	UHF [7] (4-31 G)	IVO (10s,6p)	IVO (4s,2p)	EXP [4]
$^1D - ^3P$	1.62	1.52	1.39	1.26

Table 2. Separation of the lowest states of the C atom (eV).

For the (4s,2p) basis set the s core has been taken from C++. Exponents: 141.0, 21.6, 4.85; contraction coefficients: 0.07854, 0.41183, 0.63891. A further s function (0.739) was added and two p functions have been taken from [8].

The stage has now been reached where one can apply the truncated basis to large molecules.

3. LOW LYING EXCITATION ENERGIES FOR ETHYLENE, BUTADIENE, BENZENE AND TOLUENE

For the ethylene molecule numerous attempts have been made to calculate the $\pi - \pi^*$ singlet excitation energy. Experience has shown that additional diffuse basis functions are important for the description of the Rydberg-like state. The best results have been obtained by the use of an extended Huzinaga basis set for a CI expansion [9] as well as for the IVO method [10]. The calculated value of 8.3 eV is still far away from the observed one of 7.66 eV.

With the (4s,2p) basis set already described, extended by one diffuse {1s,1p} set (0.04), IVOs have been calculated for the ethylene molecule. The excitation energies will be compared with CI results and with the observed values.

	Excitation	SCF [9]	CI [9]	IVO	EXP
Sing	$1b_{1u}$-$3b_{3u}$	6.67	7.55	7.98	-
	$1b_{1u}$-$4a_{g}$	6.06	6.94	7.51	7.11
	$1b_{1u}$-$1b_{2g}$	7.41	8.32	8.44	7.66
	$1b_{1u}$-$2b_{1u}$	-	8.03	9.13	8.26
	$1b_{1u}$-$5a_{g}$	-	8.44	9.36	8.62
	$1b_{1u}$-$2b_{2u}$	6.64	7.50	8.05	-
	$1b_{1u}$-$2b_{1g}$	7.64	8.44	9.57	-
	$1b_{1u}$-∞	8.96	9.93	10.15	10.51
Trip	$1b_{1u}$-$3b_{3u}$	6.62	7.49	7.81	-
	$1b_{1u}$-$4a_{g}$	5.92	6.79	7.16	-
	$1b_{1u}$-$1b_{2g}$	3.29	4.25	4.18	4.4
	$1b_{1u}$-$2b_{1u}$	6.78	7.73	8.23	-
	$1b_{1u}$-$5a_{g}$	-	8.37	9.08	-
	$1b_{1u}$-$2b_{2u}$	6.62	7.47	7.95	-
	$1b_{1u}$-$2b_{1g}$	7.63	8.43	9.52	-

Table 3. Calculated and observed excitation energies (eV) of the ethylene molecule.

Larger molecules, for which the excitations under consideration are mainly of the $\pi-\pi^*$ type, have been treated with a similar basis. To the C++ s-core only one {1s,1p} set (0.739) is added for the σ framework. Three π-type functions were used again (0.739, 0.175, 0.04). For testing, the ethylene molecule has been investigated once more. The $\pi-\pi^*$ transition energies for singlet and triplet are 7.67 and 3.99 eV, respectively.

In Tab. 4 some $\pi-\pi^*$ excitation energies of the trans-butadiene are reported. For the present the A terms are referred to only one configuration although a configuration mixing would be necessary.

	Excitation	CI [11]	IVO	EXP [11]
Sing	B_u $(\pi_2-\pi_1^*)$	7.67	6.04	5.71-6.29
	A_g $(\pi_2-\pi_2^*)$	7.02	6.47	7.06
	B_u $(\pi_2-3p\pi)$	6.67	7.33	6.657
Trip	B_u $(\pi_2-\pi_1^*)$	3.31	3.79	3.2-3.3
	A_g $(\pi_2-\pi_2^*)$	4.92	5.87	4.8-4.93

Table 4. Calculated and observed excitation energies (eV) of the trans-butadiene molecule.

With the benzene molecule the two-fold degenerate orbitals $1e_{1g}$ and $1e_{2u}$ lead to the terms B_{2u}, B_{1u} and E_{1u} if a suitable mixing of the four configurations is performed. The use of RVOs would give a four-fold degeneracy for these configurations. With the IVO method, however, a two-fold degeneracy in pairs is obtained. For the present these energies are seen to be a first approximation for the $\pi-\pi^*$ excitation energies in Tab. 5.

	Excitation	CI [12]	IVO	EXP [12]
Sing	$1e_{1g}-1e_{2u}$	B_{2u} 5.20		4.67-5.00
		B_{1u} 8.09	(6.63) (7.29)	5.96-6.20
		E_{1u} 9.30		6.72-6.96
Trip	$1e_{1g}-1e_{2u}$	B_{2u} 7.28		5.40-5.76
		B_{1u} 4.21	(4.91) (5.45)	3.65-3.95
		E_{1u} 5.20		4.58-4.75

Table 5. Calculated and observed excitation energies (eV) of the benzene molecule.

As a preliminary result for the $\pi-\pi^*$ excitation energies of the toluene molecule, the single configuration excitations are reported in Tab. 6. The two-fold degeneracy of the configurations, as is present with benzene, is partly avoided with toluene. In order to find more reliable excitation energies a configuration mixing should be performed again owing to the quasi-degeneracy of the configurations. The semi-empirical results have been obtained by CI for the two corresponding configurations.

Excitation	CNDO-CI [13]	IVO	EXP [13]
$\pi_2 - \pi_1^*$	4.6	6.31 (4.94)	4.6
$\pi_1 - \pi_2^*$	6.8	7.23 (5.89)	
$\pi_1 - \pi_1^*$	5.1	6.55 (4.86)	6.0
$\pi_2 - \pi_2^*$	6.8	6.68 (5.03)	

Table 6. Singlet $\pi-\pi^*$ excitation energies (eV) of the toluene molecule (triplet in parentheses).

4. THE SINGLET-TRIPLET SEPARATION FOR MOLECULAR INTERACTIONS INVOLVING THE METHYLENE RADICAL

The energy hypersurface for the systems CH_2 and H_2 leads in its absolute minimum to CH_4. Numerous calculations on the methylene radical predict a triplet ground state due to the quasi-degenerate orbitals $1b_1$ and $3a_1$ [14,15]. The reaction $CH_2 + H_2 \rightarrow CH_4$ leads, however, to a singlet ground state. In principle, the methylene radical would be able to react with a suitable reactand to a product with a triplet ground state.

Starting point for the search for such triplet products are IVO calculations on the low lying electronic states of the methylene radical. The basis set 4-31 G [16] was extended by one diffuse s function and one p set, both with an exponent of 0.05. With both the orbitals $1b_1$ and $3a_1$ singly occupied, the terms $^1B_1$ and $^3B_1$ can be obtained. The HF states with the configurations $...(3a_1)^2$ and $...(1b_1)^2$ are good approximations for $^1A_1$ states only for essentially bent conformations. Thus, a configuration mixing must be performed for nearly linear arrangements. The resulting electronic states are drawn over the HCH angle in Fig. 1. The $^3B_1$ - $^1A_1$ separation results to be 27 kcal/mole. UHF calculations overestimate this separation by 18 kcal/mole due to the unbalanced consideration of electron correlation [7]. The experimental value is recently found to be 19.5 kcal/mole [17].

The singlet-triplet situation before and after the reaction $CH_2 + H_2 \longrightarrow CH_4$ is presented in the following scheme (Fig. 2). The lowest triplet state of methane is found to be unstable with respect to dissociation CH_2+H_2 for the arrangements T_d, D_{4h}, C_{4v} and C_{2v}. All the calculated unstable states show a distinct Rydberg character.

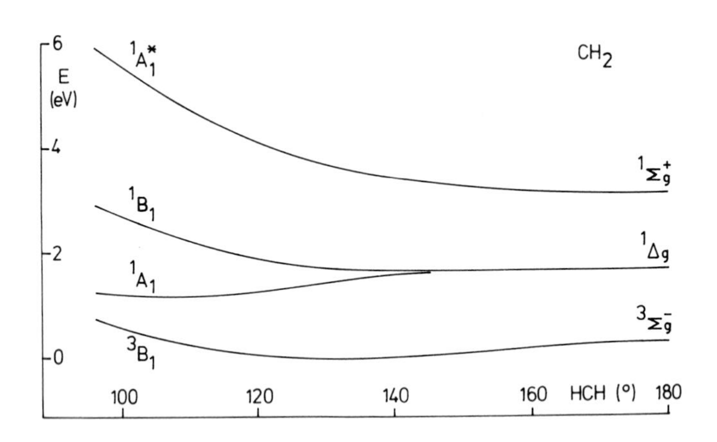

Figure 1. IVO-energy curves for the lowest states of CH_2.

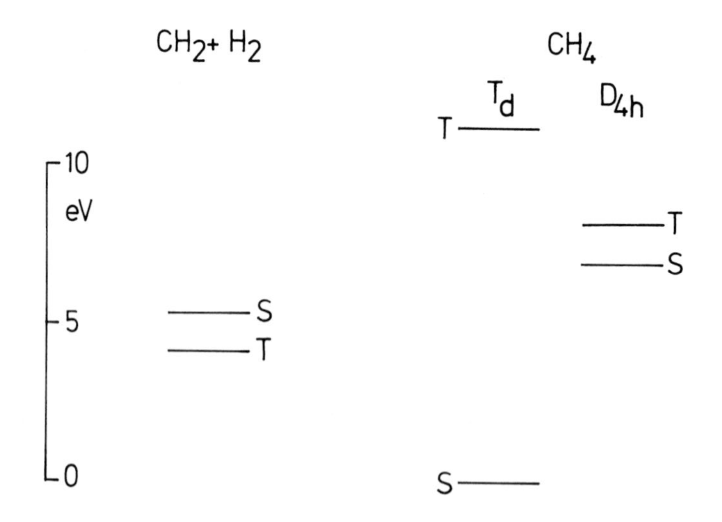

Figure 2. Singlet-triplet separation for the reaction $CH_2 + H_2 \longrightarrow CH_4$.

The H_2 molecule has now been replaced by Li_2. For the arrangements planar singlet, planar triplet, tetrahedral singlet and tetrahedral triplet the corresponding molecular geometries as well as the Li basis set 5-21 G were taken from literature [18,19].

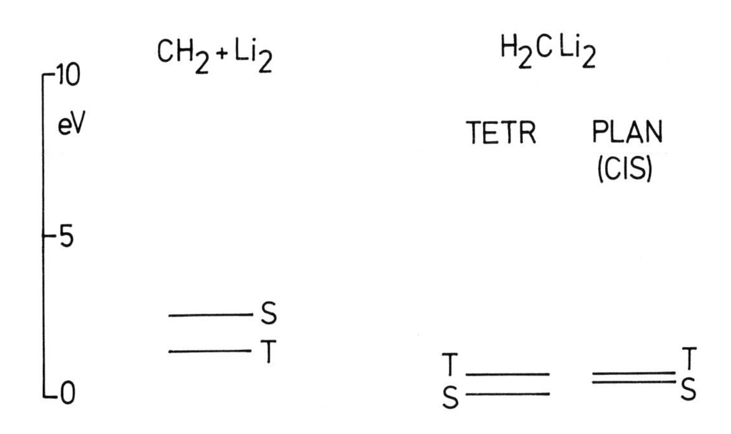

Figure 3. Singlet-triplet separation for CH_2Li_2.

According to Fig. 3 dilithiomethane still has a tetrahedral singlet ground state. The triplet state, however, lies extremely low for both the planar and the tetrahedral arrangement. All these states are found to be stable with respect to dissociation.

Calculations based upon the UHF method involving the 4-31 G basis set yield equal energies for both the planar and the tetrahedral geometry and predict the triplet state below the singlet. However, a comparison between a triplet and a closed shell singlet state is of doubtful value in the UHF approximation because open shell systems are favoured rather than closed shell systems. For instance, the molecules Li-Li and F-F both result to be unstable within the UHF approach. Furthermore, the neglect of additional diffuse basis functions surely provides doubtful results.

The equilibrium geometry of the triplet state of CH_2Li_2 justifies the complex notation $H_2C...Li_2$ (C...Li= 2.0 Å, LiCLi=69°) rather than interpreting the associate as a disubstituted methane as is the case with the singlet ground state.

5. CONCLUSION

Improved virtual orbital (IVO) calculations have been performed on the low-lying excited states for a variety of atoms and molecules. Near-HF investigations provide an agreement between calculated and experimental transition energies which varies by less than 0.5 eV. Even with small basis sets, but still including additional diffuse basis functions, the accuracy of the excitation energies is to within 0.8 eV. Thus, the IVO method, poorly applied until now [20], can be exceeded in accuracy only by large scale CI. All excitations from one occupied orbital can be obtained from one HF calculation; this emphasizes the IVO method as being the most convenient procedure with respect to numerical work. An improvement of the excitation energies by the consideration of the Koopmans defect, as done for ionization energies of closed shell systems, would be desirable. Unfortunately, no perturbation treatment of the Green function has been found until now for open shell systems [21].

Acknowledgement. I would like to thank the Deutsche Forschungsgemeinschaft for financial support and Prof. H. Preuss for his interest in this work. Programming work by F.Fraschio and G.Wagenblast as well as technical help from Miss D.Frampton is gratefully acknowledged. All computations have been performed by the use of the program ELEMOT.

REFERENCES

[1] W.J.Hunt and W.A.Goddard, Chem.Phys.Lett.3,414(1969)
[2] S.Huzinaga and C.Arnau, J.Chem.Phys.54,1948(1971)
[3] S.Huzinaga, J.Chem.Phys.42,1293(1965)
[4] Atomic Energy Levels,Vol I,US Department of Commerce, Washington,1949
[5] K.Hirao and S.Huzinaga, Chem.Phys.Lett.45,55(1977)
[6] H.Preuss, Zeitschr.f.Naturforschg.20a,1290(1965)
[7] W.A.Lathan,W.J.Hehre,L.A.Curtiss and J.A.Pople, J.Am. Chem.Soc.93,6377(1971)
[8] B.Mely and A.Pullman, Theoret.chim.Acta(Berl.)13,278 (1969)
[9] R.J.Buenker,S.D.Peyerimhoff and W.E.Kammer, J.Chem. Phys.55,814(1971)
[10] T.H.Dunning,W.J.Hunt and W.A.Goddard, Chem.Phys.Lett. 4,147(1969)
[11] R.J.Buenker,S.Shih and S.D.Peyerimhoff, Chem.Phys.Lett. 44,385(1976)
[12] S.D.Peyerimhoff and R.J.Buenker, Theoret.chim.Acta (Berl.)19,1(1970)
[13] J.DelBene and H.H.Jaffe, J.Chem.Phys.49,1221(1968)

[14] V.Staemmler, Theoret.chim.Acta(Berl.)31,49(1973)
[15] J.H.Meadows and H.F.Schaefer III, J.Am.Chem.Soc.98, 4383(1976)
[16] R.Ditchfield,W.J.Hehre and J.A.Pople, J.Chem.Phys.54, 724(1971)
[17] P.F.Zittel,G.B.Ellison,S.V.ONeil,E.Herbst,W.C. Lineberger and W.P.Reinhardt, J.Am.Chem.Soc.98,3731 (1976)
[18] J.B.Collins,J.D.Dill,E.D.Jemmis,Y.Apeloig,P.v.R. Schleyer,R.Seeger and J.A.Pople, J.Am.Chem.Soc.98, 5419(1976)
[19] J.D.Dill and J.A.Pople, J.Chem.Phys.62,2921(1975)
[20] W.R.Wadt and W.A.Goddard, Chem.Phys.18,1(1976)
[21] L.S.Cederbaum, Theoret.chim.Acta(Berl.)31,239(1973)

INDEX OF SUBJECTS

Ado-tryptophan phosphorescence spectra, 62
A-T base pairs, 11
adiabatic photoreactions, 221
adventitious biological chemiluminescence, 345, 352
aggregates, 53
alcanes excited state surfaces, 260
amino acid proteins, 53
amino acid residues, 233
amino acid excited states, 113
aniline, 306
anions dinucleoside phosphates, 17
annulene, 1, 388
aqueous solution effect, 373
aromatic amines, 303
azulene fluorescence, 390
azulene(5,6,7-CD) phenalene fluorescence, 391

benzene, 337
benzene epoxide, 337
benzene excitation energies, 422
benzene perepoxide, 337
benzpyrene-3,4 fluorescence, 391
benzylamine, 306
butadiene excitation energies, 422

C-nitroso excited compounds, 323
carbonyls photoelectron spectroscopy, 247
carboxylic acids, 410
carcinogenesis electronic theories, 347
carcinogenic aromatic hydrocarbons, 345
chiral pyrazines excited states, 295
chiral strength, 127, 131
chiroptical properties, 128, 131
chloroplast, 180
circular dichroism, 100

circular dichroism in the vacuum ultraviolet, 409
conformational change, 66
conjugated hydrocarbons 4n-membered ring, 283
conrotatory closure, 169
cross-linking, 79
cycloalkene rings twisted form, 271
cycloheptenone, 278

di(2-anthryl)propane(1.3), 314
diabatic and adiabatic surfaces, 404
diabatic states, 404
dioxetane, 334
disrotatory closure, 170
disulfide group, 56
DNA base anion radicals formation, 17
DNA base π-cation radicals, 15, 20
DNA excited states, 27
DNA ion radicals, 17, 23
DNA luminescence, 9
DNA π-cations, 20
DNA radiation damage, 15
double beam photochemistry, 366

electron donor scaling, 254
electron photoejection, 158
energy transfer, 1, 53, 57, 69, 72
epoxide, 334, 337, 342
ethane vertical excitations, 259
ethylene, 333
ethylene dioxetane, 334
ethylene epoxide, 334
ethylene excitation energies, 422
ethylene perepoxide, 334
euglena, 179
excited conjugated dienes, 199
excited singlet states fluorescence, 387
excited states of saturated molecules, 257

flavin fluorescence, 239
flavin photochemical reactivity, 239
flavin radical, 243
flavin ring system, 233
flavin spectra, 235
flavin triplet, 242
flavin-sensitized photooxidations, 241
fluorescence detected circular dichroism, 101
fluorescence quantum yield, 303
fluorescence quenching, 47
Forster theory, 58

gene product, 32, 99

halouracils(5-), 85
halouracils(5-)radicals, 87, 93
Hg^{2+} cation influence, 27
higher excited states, 361
hydrated benzene, 376
hydration, 48
hydrogen bonding, 69, 71

indole absorption spectra, 153
indole excited states, 151
indole fluorescence, 119, 155
indole nitrogen deprotonation, 157
intra-strand interactions, 102
intramolecular excited state interactions, 313
isotope effect of ^{2}H, 204

Koopmans defect, 421

lisozyme circular dichroism, 128
luminescence, 32

metal ions, 27
methane vertical excitations, 259
method of improved virtual orbitals, 419
methyl pyranosides CD, 212
methyl substitution influence, 204
methyl-4-umbelliferone, 221
mononucleotides, 4

naphtalene, 342
non-adiabatic interactions, 397
nucleic acid bases, 53
nucleic acids excited states, 1
nucleoprotein complexes, 79
nucleosides, 2
nucleotides, 3

O reactions with unsaturated hydrocarbons, 331
oligopeptides, 412
oligosaccharides CD, 18
organic photochemistry in dense media, 361
O2 reactions with unsaturated hydrocarbons, 331

peptides, 81, 137
peptides excited states, 113
peracetylated carbohydrates, 217
perepoxide 334, 337
phenylalanine, 73
phenylcycloheptene, 273
phenylcyclohexene, 273
phenylethylamine, 306

phosphate groups, 68
phosphorescence, 34
photohydration of pyrimidines, 39
photoionisation, 15
photoionisation of amino acids, 113
photolysis of substituted butadiene, 202
photooxidation, 323
photoprocesses, 175
photoreceptors, 175, 177, 187
photosynthesis, 180
phototaxis, 178
phototropism, 176
phycomices blakesleeanus, 176
poly-L-valine, 413
polynucleotides, 5
polypeptides, 412
polysaccharides, 415
protein structure, 123
protein-nucleic acid complexes, 65, 75, 79
Pullman k-indexing, 252
pyrazine CD-spectrum, 298

quenching of amino acid, 65

retinylidene(N-) chromophore, 171
rhodopsin, 182, 187
rhodopsin electron-electron interactions, 194
rhodopsin-like hamiltonian, 191
RNA bases photoelectron spectra, 250
rotatory strength, 124, 143

selection rules for photochemical reactions, 400
sialic acid, 215
single beam photochemistry, 365
singlet transfer, 61
singlet-triplet separation, 425
solvated electrons, 303
solvent effects on excited states, 373
spectral shifts of solvents, 375
stacking interactions, 67, 70
sudden polarisation, 163, 167

tert-butyl-NO, 325
tert-butyl-DO, 325
thymine dimerisation, 35, 53
toluene, 305, 341
toluene excitation energies, 422
tRNA, 11
tryptophan, 54, 116
tryptophan fluorescence spectra, 67, 236
tryptophan transitions, 130

tyrosine, 56, 70, 114
tyrosine quenching reactions, 115

ultraviolet circular dichroism, 209

vibrational relaxation, 59
vicinal polyketones photochemistry, 381

zwiterionic states, 165

INDEX OF NAMES

Abdulner, S. 247, 256
Alagona, G. 379
Alberts, B. 99, 100, 110, 111
Albrecht, A.C. 370
Allen, S. 416
Altar, W. 127, 134
Ambady, G.K. 149
Ames, B. 358
Amesz, J. 357
Amodo, F.J. 110
Amos, A.T. 375, 379
Anbar, M. 122
Anderson, W. 348, 359
Andrews, L.J. 122
Anfinsen, C.B. 77
Apeloig, Y. 429
Apicella, M. 51
Arcos, J.C. 343
Argus, M.F. 343
Armbruster, A.M. 373, 379
Arnau, C. 428
Arnesson, R.M. 358
Arnold, W. 357
Arnone, A. 78
Ashwell, G. 209, 220
Astier, R. 371
Atkinson, T. 359
Ausloos, P. 371
Avery, J. 148
Avigad, G. 418
Ayscough, P.B. 24
Azzi, J. 357

Bacher, A. 302
Badger, G.M. 347, 358
Badruddin, 301
Bailey, J.L. 186
Baker, A.D. 149, 256
Baker, C. 149, 256
Balazs, E.A. 417
Balcerski, J.S. 210, 219, 417
Balke, C.C.F. 134
Balke, D.E. 370
Ballini, J.P. 1, 13
Bambenek, M. 122
Banba, F. 322
Baraldi, I. 281
Bardeen, J. 187, 198
Baremboim, G.M. 358
Bartman, P. 76
Baugher, J.F. 122
Bayley, P.M. 124, 134, 148
Bazin, M. 160
Bebault, G.M. 219
Becker, R.S. 362, 370
Bednar, T.W. 162
Behr, O.M. 293
Beinert, H. 244
Beland, F.A. 343
Bell, A.P. 245
Bent, D.V. 121, 122, 159
Bent, R. 160
Bergene, R. 96
Berger, H. 111
Bergmann, A. 230
Bernardin, J.E. 153, 161
Bernhard, W.A. 96
Berns, D.S. 186
Bernstein, H. 135
Berry, J.M. 219

Bersohn, R. 27, 37
Berthier, G. 169, 174
Berthod, H. 24
Bertinchamps, A. 24
Bethe, H.A. 174
Beychock, S. 135, 148
Beylaev, S.T. 196, 198
Bier, C.J. 78
Bigwood, M. 207
Billups, C. 153, 161
Birks, J.B. 207, 321, 322, 347, 359, 371
Bishop, D.M. 376, 379
Bittner, M. 111
Blauer, G. 148
Blobstein, S.H. 343
Blout, E.R. 148, 149
Bock, S.J. 111
Bodor, N. 24
Blobstein, S.H. 343
Blout, E.R. 148, 149
Bock, S.J. 111
Bodor, N. 24
Boens, N. 313, 322
Bolle, A. 111
Bolsman, T.A.B.M. 323, 328, 329
Bonacic-Koutecky, V. 174
Bonneau, R. 271, 281
Bonora, M. 417, 418
Borkent, J.W. 322
Bouas-Laurent, H. 322
Boue, S. 199, 207
Bovey, F.A. 149
Bowman, M.K. 121, 310
Bownds, D. 186
Box, H.C. 96
Boy de la Tour, E. 111
Brack, C. 111
Brady, R.D. 210, 220
Brahms, J. 24
Brand, L. 76
Breach, 198
Brint, P. 256
Brodsky, L. 174
Brody, M. 186
Brody, S. 186
Broekhoven, F.J.G. 323, 329
Brookes, P. 358
Brown, I.H. 45, 50
Brown, P.K. 186
Broyde, S.B. 186
Bruckmann, P. 174, 197, 198, 207
Bruice, T.C. 50
Brun, F. 64, 76, 77
Brundle, C.R. 149, 256
Bruni, M.C. 174, 281
Bryant, F.D. 119, 122
Budzik, G.P. 84
Budzinski, E.E. 96
Buenker, R.J. 174, 258, 259, 260, 269, 294, 428
Bunbury, D.L. 386
Buffington, L.A. 417, 418
Burger, U. 370
Burr, J.G. 40, 44, 45, 46, 50, 51
Burrows, B.L. 375, 379
Burstein, E.A. 77
Busel, E.P. 77
Bush, C.A. 209, 212, 216, 219, 220
Bushueva, T.L. 77
Butcher, J.A. 370
Butler, A.R. 50
Butmann, E.D. 110
Buu-Hoi, N.P. 359
Byrne, J.P. 302
Byrom, P. 245

Caldwell, J.W. 257, 269
Calvert, J.G. 310
Calvin, M. 245
Campbell, J.D. 293
Cantor, C.R. 111
Carlile, M.J. 186
Carlson, G.L.B. 51
Carlson, T.A. 256
Carmody, D.J. 324, 328
Carr, R.V. 370
Carrell, H.L. 344
Carter, J.G. 302
Caspary, W.J. 358
Castellan, A. 361
Castillan, A. 322
Cech, C.M. 111
Cederbaum, L.S. 429
Cerletti, P. 244
Chakrabarti, 417
Chan, A. 50
Chan, S.I. 24
Chandra, P. 51
Chang, R.L. 343, 358, 359
Charlier, M. 51, 64, 77

Chen, G. 161, 219, 417
Chen, R.F. 37, 60, 64
Cheung, A.S. 253
Chevalley, R. 111
Cho, S. 293
Choy, Y.M. 219
Christensen, H. 310
Christoffersen, R.E. 374, 375, 379
Christophorou, L.G. 302
Chu, E.H.Y. 359
Chung, C.W. 245
Clapp, R.C. 230
Clar, E. 284, 293
Clark, A.D. 198
Clark, C. 24
Close, D.M. 96
Clough, S. 174
Coche, A. 161
Coduti, P.L. 211, 212, 219
Cohen, A. 328
Cohen, B.I. 358
Cohen, C. 148
Collins, J.B. 429
Collins, R.G. 24
Condon, E.U. 127, 134
Conney, A.H. 343, 358, 359
Cooper, L.N. 187, 198
Corey, E.J. 281
Cornelisse, J. 369
Corrigall, B. 406
Cotton, F.A. 78
Coulson, C.A. 293
Coulter, C.L. 149
Couture, A. 370
Craig, D.P. 376, 379, 418
Cramer, F. 76
Cronin, J.R. 245
Cronwall, E. 76
Croy, R.G. 359
Cuatrecasas, P. 77
Cundall, R.B. 309
Curry, G.M. 186
Curtis, M.J. 111
Curtiss, L.A. 269, 428

Daemen, J. 322
Daiker, K.C. 331, 344
Dainton, F.S. 24
Dale, R.E. 59, 64
Daly, J.W. 343
Daniels, M. 2, 12, 13, 39, 50
Danilov, V.I. 47, 50
Danon, A. 186
Dansette, P. 343, 359
Dauben, W.G. 163, 174, 269, 406
Daudel, P. 14, 347, 359
Daudel, R. 347, 359
Daudey, J.P. 174
Davidson, N. 28, 37
Davydov, A.S. 198
Day, V.W. 78
De Boer, G. 49, 50
De Boer, Th. J. 322, 323, 328, 329
De Brackeleire, M. 322
De Busser, R. 328, 329
De Mayo, P. 370
De Schryver, F.C. 313, 322
De Voe, H. 111
Deber, C.M. 149
Decheneux, M.J. 397
Dederen, J.C. 313
Dehareng, D. 397
Delbene, J. 428
Delbruck, M. 186
Delius, H. 99, 103, 111
Dellweg, H. 51
Dement, J. 357
Denhardt, G.H. 111
Dertinger, H. 25
Desouter-Lecomte, M. 397, 406
Devaquet, A. 206
Dewar, M.J.S. 24, 284, 293, 333, 344
Dexter, D.L. 58, 64
Diamond, R. 129, 135
Dickinson, H.R. 214, 216, 219
Dickson, J. 328
Dienes, A. 231
Digros, E.D. 358
Dill, J.D. 429
Dimicoli, J.L. 56, 64, 77, 78
Dipple, A. 359
Ditchfield, R. 269, 343, 429
Djerassi, C. 302
Dobson, G. 310
Dolan, E. 370
Domanskii, A.N. 358
Domingo, E. 293
Donaruma, L.G. 324, 328
Donnelly, R.A. 344
Doty, P. 148
Dougherty, D. 247, 256

Douzou, P. 24, 56, 64, 77
Downing, J.W. 370
Draper, R.O. 244
Dross, C. 161
Druyan, M.E. 149
Dry, L.A. 77
Dulcic, A. 25
Dunne, L.J. 187, 198
Dunning, T.H. 428
Duquesne, M. 14, 122, 311
Durston, W.E. 358
Dutton, G.C.S. 219

Eaton, D.F. 207
Eaton, P.E. 281
Eckhardt, G. 301
Edelhoch, H. 77
Edgar, R.S. 111
Edmondson, D.E. 244, 245
Eftink, M.R. 122
Eggers, J.H. 293
Eglinton, G. 293
Ehrenberg, A. 24
Eisenberg 27, 34, 37
Eisinger, J. 12, 13, 27, 37, 40, 50, 61, 64, 155, 161
El-Bayoumi, M.A. 162
Elad, D. 82, 83, 84
Ellison, G.B. 429
Engel, G. 76
Engelhardt, M. 24
Englard, S. 418
Enwall, C. 51
Epstein, R.H. 111
Epstein, S.S. 359
Evans, R.F. 159, 161
Eversole, R.A. 186
Eyring, H. 127, 134
Ezumi, K. 161

Failor, R. 24
Falk, H.L. 359
Falk, M.C. 245
Farelly, J.G. 76
Farley, R.A. 96
Fasman, G.D. 409, 418
Favre, A. 13, 50
Feitelson, J. 78, 122, 161, 310
Fernandez-Alonso, J.I. 293
Ferris, F.L. 110
Fink, D.W. 230
Finlayson, B.J. 328
Fischer, E. 370
Fischer, G. 370
Fischmann, P.H. 210, 220
Fisher, G.J. 39, 47, 50, 83
Fitts, D.D. 148
Flouquet, F. 258, 269
Fornier de Violet, P. 271, 281
Forster, L.S. 122
Forster, Th. 58, 64
Fory, W. 245
Francois, P. 293
Frey, L. 100, 111
Fridovich, I. 358
Frisell, W.R. 245
Frohlich, H. 192, 198
Fry, K.E. 111
Fuene, T. 344
Fukui, K. 344
Fukutome, H. 344
Funnell, N. 219

Gaibel, Z.L.F. 281
Galbraith, A.R. 293
Galloy, C. 397, 406
Galogaza, V. 25
Galston, A.W. 245
Gardner, P.D. 281
Garratt, P.J. 293
Gefter, M.L. 77
Geiger, R.E. 137, 149
Gelboin, H.V. 358, 359
Gennis, R.B. 111
Gerard, D. 149
Gerhartz, W. 322, 370
Gerlach, H. 149
Getof, N. 121, 303, 309, 310
Ghiron, C.A. 122, 161
Gibson, J.W. 149
Gielen, J.W.J. 370
Giessner-Prettre, C. 24
Gillard, J.M. 245
Gillespie, G.D. 394
Gilligan, C. 50
Glass, J.W. 24
Gleiter, R. 253, 301
Gloor, J. 370
Glusker, J.P. 344
Goddard, W.A. 301, 428, 429
Gohlke, J.R. 76
Goldfarb, T.D. 279, 281

Goldschmidt, C.R. 310
Gordon, E.C. 219
Gordon, M. 198
Gordon, M.S. 257, 269
Goux, W.J. 135
Gowenlock, B.G. 328
Grabner, G. 121, 309, 310
Grabowski, Z. 173, 174
Graf, F. 293
Graslund, A. 24
Gratzer, W.B. 148
Grebow, P.E. 134
Greenlee, L. 358
Greenstock, C.L. 40, 50
Gregoli, S. 24
Gregioriadis, G. 220
Gregory, R.P.V. 186
Greve, J. 99, 108, 111
Griesser, H.J. 394
Griffin, A.C. 344
Grossweiner, L.I. 121, 122, 158, 161, 310
Grover, P.L. 13, 358, 359
Groves, M.R. 231
Gruber, B.A. 77
Gueron, M. 12, 58, 59, 64
Guschlbauer, W. 13, 45, 50
Gutman, A.M. 385

Haar, F. 76
Haas, G. 149
Hackmeyer, M. 301
Hadley, S.G. 371
Hahn, B.S. 50
Haindl, E. 96
Hajos, G. 295, 301
Hall, G.G. 373, 374, 375, 379
Halper, J.P. 132, 135
Halpern, A. 162, 371
Hamaguchi, K. 135
Hamman, J.P. 345, 348, 358
Handler, P. 358
Hanke, T. 76
Hansen, A.E. 148
Harget, A.J. 24
Hartig, G. 25
Harvey, R.G. 343, 344
Haselbach, E. 148
Hasselmann, C. 151, 152, 158, 160, 161
Hastings, R. 45, 245
Hatchard, C.G. 161, 186
Hattaway, E.J. 281
Hatton, M.W.C. 186
Haug, A. 43, 50
Hauswirth, W. 12, 13, 34, 37, 39, 48, 50, 51
Havinga, E. 173, 174, 369, 370
Havron, A. 79, 80, 81, 83, 84
Hayashi, T. 322
Haydon, S.C. 231
Hayon, E. 113, 121, 122, 159, 160, 310
Hayward, L.D. 219
Hazen, E.E. 78
Hehre, W.J. 269, 343, 428, 429
Heicklen, J. 324, 325, 328
Heider, H. 76
Heilbronner, E. 253, 301
Helene, C. 14, 24, 27, 37, 51, 53, 54, 56, 62, 65, 76, 77, 78
Heller, M. 385
Hellmann, H. 406
Hemmerich, P. 244
Henderson, W. 293
Hendriks, R. 245
Herak, J.N. 25
Herbst, E. 429
Hernandez, O. 343, 358, 359
Hewer, A. 359
Hiberty, P. 174
Hickman, J. 220
Hicks, A.A. 302
Himmelreich, J. 301
Hirao, K. 428
Hirsch, R. 76
Hittenhaussen, H. 328
Ho, K. 370
Ho, P.C. 302
Hochstrasser, R.M. 284, 293
Hodgson, E.K. 358
Hoffmann, R. 260, 269
Holler, E. 76
Holroyd, R.A. 24
Holstrom, B. 245
Holzwarth, A.R. 394
Holzwarth, G. 148
Honda, K. 370
Hood, F.P. 149
Hooker, T.M. 123, 134, 135

Hopkins, T.P. 310
Hopp, J.W. 51
Hornung, V. 253, 301
Horsley, J.A. 258, 269
Hoshino, M. 370
Hosoda, J. 99, 100, 111
Howes, R.M. 358
Howard, J.A. 328
Hubbard, R. 186
Huberman, E. 358
Huberman, J. 111
Hug, W. 149
Hui-Bon-Hoa, G. 56, 64, 77
Hunt, J.W. 50
Hunt, W.J. 428
Hush, N.S. 253, 293
Hutchinson, F. 96
Huttermann, J. 85, 96, 97
Huybrechts, J. 313, 322
Huzinaga, S. 428

Iball, J. 343, 344
Ikeda, K. 135
Ikeda, M. 370
Ilan, Y. 310
Inagaki, S. 344
Ingold, K.U. 328
Ingraham, L.L. 244
Innes, K.K. 302
Inoh, M. 322
Irie, M. 245, 313
Iseli, M. 137
Isenberg, I. 417
Ishimoto, M. 51
Itoh, U. 230

Jackowski, S. 322
Jacobs, H.J.C. 370
Jaffe, H.H. 428
Janoschek, R. 419
Jeffrey, A.M. 343
Jemmis, E.D. 429
Jenkins, M. 110
Jennette, K.W. 343
Jensen, D.E. 100, 111
Jensen, L.H. 134
Jerina, D.M. 343, 349, 358, 359
Johansen, J.T. 134
Johns, H.E. 39, 40, 45, 47, 49, 50, 83
Johnson, G.E. 322
Johnson, P.G. 245
Johnson, W.C. 210, 220, 409, 416, 417
Jones, R.P. 394
Jortner, J. 230, 310, 394
Joschek, H.J. 158, 161, 310
Joussot-Dubien, J. 271, 281, 370
Juneau, R.J. 230
Jutz, C. 391, 394

Kadesch, T.R. 135
Kaler, C. 309
Kamm, K.S. 281
Kammer, W.E. 428
Kampas, F.J. 196, 198
Karle, I.L. 149
Karle, J. 149
Kartha, G. 149
Kasai, H. 343
Kasha, M. 42, 50
Kato, M. 281
Katz, S. 37
Kawasaka, T. 209, 220
Kayen, A.H.M. 328
Kearns, D.R. 344
Keene, J.P. 310
Kellenberg, E. 111
Keller, R.A. 371
Kellogg, M.S. 174, 358
Kelly, M.M. 417
Kelly, R.C. 77, 111
Kemp, J.C. 409, 417
Kenney, W.C. 245
Kerby, E.P. 77, 122
Kevan, L. 121, 310
Khalil, O.S. 310
Khattak, M.N. 46, 51
Kim, B. 370
Kindt, T. 221, 231
Kirby, 77
Kirkwood, J.G. 127, 134, 148
Klasinc, L. 255
Klein, M. 77
Klein, R. 122, 310, 311
Klyne, W. 302
Koehler, V.R. 230
Kohler, G. 121, 303, 309, 310
Kohn, R.L. 231
Kolc, J. 361, 370
Konev, S.V. 161
Koopmans, T. 256

Kopczemwski, R.F. 326
Kornberg, A. 111
Kornhauser, A. 51
Koutecky, J. 174
Kraan, G.P. 357
Krochmal, E. 385
Kropf, A. 186
Kropp, P.J. 281
Kubota, S. 414, 418
Kuhn, W. 127, 134
Kuhnle, W. 322
Kumano, Y. 322
Kumei, S.E. 164, 174
Kuntz, E. 162
Kuntz, R.R. 161
Kuppers, B. 406
Kurachi, K. 134
Kurita, M. 281
Kurtin, W.E. 161

Labhart, H. 148
Labrum, J.M. 370
Lacassagne, A. 359
Lachish, U. 119, 122
Lam, S.S.M. 84
Lami, H. 161
Lamola, A.A. 12, 14, 27, 34, 37, 40, 50, 51, 61
Land, E.J. 293
Landry, L.C. 28, 37
Langlet, J. 174
Lapointe, J. 77
Lapouyade, R. 322
Lathan, W.A. 269, 343, 428
Latovitzki, N. 135
Laustriat, G. 151, 158, 160, 161
Lavalette, D. 122, 293
Le Breton, P.R. 253, 256
Leavitt, J.C. 358
Leclerc, J.C. 406
Leclercq, J.M. 371
Lecocq, J. 359
Lee, F.D. 358
Lefebvre-Brion, H. 406
Leforestier, C. 174, 294, 406
Lehr, R.E. 359
Lemahieu, R. 281
Leonard, N.J. 70, 77
Lesclaux, R. 370
Lesko, S.A. 358
Letsinger, R.L. 51
Levin, A. 108, 111
Levin, W. 343, 358, 359
Levitt, M. 394
Lewis, D.G. 417
Li, H.J. 417
Liang, J. 418
Libbit, L. 281
Lichten, W. 406
Lichtin, N.N. 122
Lielaudi, A. 111
Lim, E.C. 394
Lin, E.T. 84
Lin, K. 281
Lineberger, W.C. 429
Lippert, E. 221, 230, 231
Listowsky, I. 418
Little, W.A. 187, 196, 198
Livingston, D.M. 134
Lodemann, E. 51
Lok, C.M. 369
Longuet-Higgins, H. 84, 293
Longworth, J.W. 12, 24, 60, 64, 76, 151, 161
Lorentzen, R.J. 358
Lorquet, J.C. 397, 406
Loxsom, F.M. 148
Lugtenburg, J. 369
Lumry, R. 310
Lumry, R.W. 162
Luria, M. 310

Maassen, J.A. 328, 329
Mace, D. 111
Mackor, A. 328, 329
Madhavan, V. 122
Madison, V. 148
Maelicke, A. 76
Maestre, M.F. 99, 108, 111
Mahanty, J. 375, 379
Mair, G.A. 134
Malling, H.V. 359
Malrieu, J.P. 47, 51, 171, 173, 174
Mandell, N.J. 111
Mantel, N. 359
Mantione, M.J. 190, 198
Manzara, A.P. 370
Marchessault, R.H. 220
Maria, H.J. 256
Martin, D. 374, 379
Martin, M. 359

Martinson, H.G. 84
Mason, S.F. 295, 301
Mataga, N. 53, 64, 155, 161, 313, 320, 322
Mathers, T. 247
Matheson, M.S. 310
Mathies, R. 173, 174
Matsubara, H. 81, 83
Mault, J. 134
Maurizot, J.C. 24, 78
McAlpine, R.D. 293
McCall, M.T. 51
McCormick, D.B. 233, 244, 245
McCorkle, D.L. 302
McCreery, J.H. 374, 375, 379
McGlynn, S.P. 247, 255, 256, 310
McKenzie, R.E. 245
McKenzie, C. 245
McNesby, J.R. 269
McVey, J.K. 370
Meadows, J.H. 429
Medina, V.J. 358
Meeks, J.L. 255, 256
Meinwald, J. 370
Mely, B. 428
Merck, E. 304
Meyer, Y.H. 370
Michelson, A.M. 24, 50
Michl, J. 206, 293, 322, 361, 370, 371, 391, 394, 406
Miller, C.J. 373
Miller, E.C. 358
Miller, J.A. 358
Misumi, S. 322
Mittal, J.P. 40, 51, 121, 310
Mittal, L.J. 121, 310
Miura, I. 343
Moffitt, W. 127, 134, 148
Mohan, P. 24
Moise, H. 100, 111
Molineux, I.J. 77
Momicchioli, F. 174, 281
Montagnani, R. 259, 269
Montenay-Garestier, T. 51, 53, 54, 59, 62, 64, 77
Monroe, B.M. 385
Moore, A.M. 43, 45, 51
Moran, L. 111
Morawetz, H. 322
Morell, A.G. 209, 220
Morgan, D.D. 359
Morita, M. 322
Morita, T. 293
Morokuma, K. 374, 379
Morris, E.R. 418
Morris, F.C. 111
Morrisson, H.A. 328
Morton, T.H. 417
Moscowitz, A. 149, 296, 301
Mosig, G. 111
Muckerman, J.T. 406
Muel, B. 122
Mukherjee, S. 220
Mulac, W.A. 310
Muller, A. 96
Mulliken, R.S. 174, 258, 269
Mumchausen, L.L. 28, 37
Murata, Y. 53, 64
Murrell, J.N. 149
Mutai, K. 77
Myers, L.S. 96
Nagelschneider, G. 244
Nakagawa, C.S. 186
Nakanishi, K. 343
Nakashima, M. 230
Naqvi, K.R. 394
Nathanson, B. 186
Navon, G. 155, 161
Neilson, G.W. 97
Nelson, J.H. 24
Nelson, R.G. 210, 220, 417
Nesta, J.M. 122
Neta, P. 96
Neymann-Spallart, 310
Neumuller, W. 96
Newbold, R.F. 358
Newton, M.D. 343
Nichols, C. 24
Nickel, B. 394
Nielsen, E.B. 124, 134, 148
Nikitin, E.E. 406
Ninham, B.W. 375, 379
Nnadi, J.C. 40, 42, 45, 51
Nobs, F. 370
Noell, J.O. 374, 379
North, A.C.T. 134
Noyori, R. 281
Nozaki, H. 277, 281

O Donnell, J. 256
O Malley, T.F. 406
Oesterhelt, D. 186

Ogilvie, 309
Ogiwara, T. 293
Okabe, H. 269
Okamoto, T. 322
Olast, M. 24
Oloff, H. 96
ONeil, S.V. 429
Oosterhoff, J. 206
Ootacki, T. 186
Ostrowski, J. 358
Oth, J.F.M. 394
Ottolenghi, M. 310
Oxygena, A.G. 304
Owtschinnikow, J.A. 149

Pachmann, U. 76
Padva, A. 174, 253, 256
Pagni, P.G. 180, 186
Pagni, R.M. 370
Pal, K. 359
Palumbo, M. 418
Parfait, R. 77
Pariser, R. 148, 284, 293
Park, E.H. 40, 44, 50
Parker, C.A. 161, 186
Parr, R.G. 148
Pate, C.T. 328
Pauli, A. 77
Pearson, E.F. 416
Peggion, E. 418
Peng, S. 253
Penzer, G.R. 245
Peradejordi, F. 293
Pereira, W.E. 418
Pereyre, J. 281
Petke, J.D. 301
Pettei, M. 24
Peyerimhoff, S.D. 174, 258, 259, 260, 269, 294, 428
Pfeffer, G. 152, 161
Phillips, D.C. 134
Philpott, M.R. 148
Pileni, M.P. 122
Pilling, G.M. 293
Pincock, R.E. 40, 51
Pisanias, M.N. 302
Pitts, J.N. 310, 324, 325, 328
Platt, J.R. 284, 293
Podjorny, A. 134
Politzer, P. 331, 344
Pongs, O. 77
Pople, J.A. 148, 269, 343, 428, 429
Poppinger, D. 259, 269
Porschke, D. 13
Porter, G. 294
Poshusta, R.D. 322
Prangova, L.S. 310
Prelog, V. 149
Preston, R.K. 406
Pretorius, H.T. 77
Preuss, H. 428
Prino, G. 359
Pruchova, M. 310
Ptak, M. 13
Pullman, A. 24, 252, 256, 347, 358, 373, 379, 428
Pullman, B. 190, 198, 252, 256, 347, 358
Put, J. 322
Puza, M. 322
Pysh, E.S. 148, 406, 417, 418

Que Hee, S.S. 358
Quickenden, T.I. 358

Rabani, J. 310
Rabinovich, D. 134
Radda, G.K. 245
Radding, W. 134
Rahn, R.O. 13, 21, 24, 27, 37
Ramamurthy, V. 370
Ramey, B.J. 281
Raphael, R.A. 293
Rapp, W. 231
Rau, H. 302
Raymonda, J.W. 269
Reardon, E.J. 281
Rees, D.A. 418
Rees, Y. 328
Reinhardt, W.P. 429
Rentzepis, P.M. 293, 394
Rettschnick, R.P.H. 328, 329, 394
Riani, P. 259, 269
Ricci, R.W. 122, 161
Richards, F.M. 82, 83
Richardson, D.C. 78
Richardson, J.S. 78
Rigler, R. 76
Riordan, J.F. 134
Ritscher, J.S. 174
Riveiro, C. 358
Rizzuto, F. 245

Robin, M.B. 149
Rommel, E. 293
Rondelez, D. 199, 207
Roothaan, C.C.J. 269
Rosenthal, I. 83
Rosicky, C. 303, 311
Ross, A.B. 122
Ross, I.G. 302
Rothe, M. 149
Rowland, C. 174
Rowlands, J.R. 293
Rubin, M.B. 281, 381
Rudali, G. 359
Rupley, J. 134
Rupprecht, A. 24
Rushbrooke, G.S. 293
Ryan, J.A. 301

Sachs, L. 358
Sakata, Y. 322
Salahub, D.R. 260
Salem, L. 163, 174, 197, 198, 206, 207, 269, 271, 281, 293, 406
Salpeter, E.E. 174
Salvetti, O. 259, 269
Samuelson, G.E. 370
Sanderson, G.R. 418
Sandorfy, C. 260
Santamaria, L. 359
Santry, D.P. 149
Santus, R. 13, 24, 122, 160
Sarko, A. 220
Sarma, V.R. 134
Sasson, S. 83
Sato, K. 322
Saunders, D.S. 322
Saya, A. 134
Scatturin, A. 149
Schabort, J.C. 245
Schaefer, H.F. 429
Schaffner, K. 370
Schang, J. 157, 161
Schechtman, L.M. 358
Schellman, J.A. 124, 134, 148, 212, 302
Schernberg, I.H. 220
Schimmel, P.R. 79, 84
Schleyer, P.V.R. 429
Schmelzer, A. 148
Schmidt, G. 96
Schmidt, O. 358
Schnepp, O. 409, 416
Schoemaker, H.J.P. 84
Scholes, G. 24
Schott, H.N. 84
Schrieffer, J.R. 187, 198
Schuster, O. 302
Schwarz, F.P. 370
Schworer, F. 310
Scoseria, J.L. 358
Scrimgeour, S.N. 343
Scrocco, E. 344, 379
Seeger, R. 429
Seeman, J.I. 174
Segal, G.A. 149, 294, 406
Seliger, H.H. 345, 347, 357, 358, 359
Seliskar, C.J. 310
Selkirk, J.K. 359
Sellini, H. 78
Selzer, R. 51
Sevilla, M.D. 15, 24
Shafferman, A. 122
Shank, C.V. 231
Shapiro, H. 111
Sharman, E. 416
Shelton, J.R. 326
Shetlar, M.D. 84
Shih, S. 428
Shillady, D. 122
Shin, E. 186
Shipman, L.L. 374, 379
Shizuka, T. 290, 293
Shropshire, W. 186
Shugar, D. 45, 51
Shulman, R.G. 21, 24, 27, 37, 64
Sidis, V. 406
Sieker, L.C. 134
Simmons, N.S. 148
Simpson, W.T. 162, 269
Sims, P. 358, 359
Singer, T.P. 244, 245
Sinha, N. 111
Sinsheimer, R.L. 45, 51
Small, M. 359
Smilansky, A. 134
Smith, F.T. 406
Smith, H. 302
Smith, K.C. 79, 80, 84
Smith, C.A. 186
Snatzke, G. 295, 301, 302
Snow, J.W. 134

Snyder, P.A. 417
Solar, S. 310
Soll, D. 77
Sondheimer, F. 293
Song, P.S. 155, 157, 161, 162
Sousa, J.A. 230
Souto, M.A. 361
Sperling, J. 79, 80, 81, 83, 84
Spicer, C.W. 328
Spoinkel, F.M. 122
Srinivasan, R. 207
Staab, H.A. 293
Staemmler, V. 429
Stauff, J. 358
Steele, R.H. 358
Steemers, R.G. 357
Steen, H.B. 120, 121, 122, 158, 161, 310
Steenkampf, D.J. 245
Stein, G. 122, 310
Steinberg, C.M. 111
Steiner, R.F. 77, 122
Stephen, A.M. 219
Stewart, J.C. 186
Stewart, R.F. 269
Stigter, D. 212, 220
Stine, J.R. 406
Stoeckenius, W. 186
Stohrer, W.D. 281
Stone, B.R. 51
Strehler, B. 357
Streitwieser, A. 310
Strickland, E.H. 161
Strickland, R.W. 122, 152
Strom, G. 24
Stryer, L. 161, 173
Stulberg, M.P. 76
Styer, L. 174
Suau, R. 370
Subramanian, L.R. 328
Summers, W.A. 43, 50, 51
Sun, M. 155, 157, 162
Susman, M. 111
Sutherland, B.M. 27, 37
Sutherland, J.C. 27, 37
Suzuki, T. 322
Svartholm, N. 347, 358
Swaisland, A. 359
Symons, M.C.R. 97
Syrkin, J.K. 406
Szent-Gyorgyi, A.G. 148
Szybalski, W. 96

Tada, M. 281
Tait, A.D. 374, 379
Takacs, B. 111
Takakusa, M. 230
Tanaka, H. 370
Tanaka, J. 322
Tatischeff, I. 122, 310, 311
Taylor, M.B. 245
Tazuke, S. 322
Tether, L.R. 245
Tetreau, C. 293
Thiel, W. 333, 344
Thimann, K.V. 186
Thom, D. 418
Thomas, G. 13
Thomson, C.H. 45, 51
Thornber, J.P. 186
Thulstrup, E.W. 293, 391, 394
Tinamari, R.E. 322
Tinoco, I. 111, 148, 155, 162, 417
Tjalldin, B. 24
Tollin, G. 244, 245
Tomasi, J. 344, 378
Toniolo, C. 412, 417, 418,
Torihashi, Y. 161
Totter, J.R. 358
Toulme, F. 77
Toulme, J.J. 69, 77
Tournon, J. 153, 162
Trager, L. 51
Traub, W. 134
Treinin, A. 122
Ts'o, P.O.P. 24, 358
Tu, S.C. 245
Tuan, V.D. 394
Tully, J.C. 406
Turchi, I.J. 344
Turck, G. 51
Turnbull, J.H. 245
Turner, D.W. 111, 149, 256
Turoverov, K.K. 358
Turro, N.J. 269, 370, 406

Ukita, T. 245
Ullman, E.F. 370

Vaidya, V.M. 149
Vaish, S.P. 245
Valeur, B. 153, 162

Vallee, B.L. 134
Van de Vorst, A. 24
Van der Donckt, 155, 162
Van Paemel, C. 24
Van Voorst, J.D.W. 329
Vanderlinden, P. 199, 207
Varghese, A.J. 83, 84
Vaughan, R.A. 96
Vaughn, J.R. 186
Veeger, C. 244
Velthuys, B.R. 357
Verhoeven, J.W. 322
Vertesy, L. 301
Vigny, P. 1, 12, 13, 61, 64, 65
Villa, A. 328
Vipond, P.M. 417
Volkert, W.A. 161
Voll, R. 256
Von Hippel, P. 77, 111

Wacker, A. 51
Wadt, W.R. 301, 429
Wagniere, G. 137, 148, 149
Wajer, Th.A.J.W. 328, 329
Walker, M.S. 155, 162, 175, 245
Wallace, R. 406
Walne, P. 186
Walrant, P. 160
Walter, R. 149
Wang, S.Y. 13, 39, 40, 42, 43, 64
Wang, C.T. 385
Wang, Y.C. 55, 322
Ward, J.F. 96
Ware, W.R. 370
Warren, A.J. 111
Warren, J.G. 123
Warshaw, M.M. 111
Warshawsky, D. 359
Warshel, A. 391, 394
Wawilow, S.J. 310
Weber, G. 77, 153, 162
Wehry, E.L. 186
Weiner, M. 385
Weinstein, I.B. 343
Weller, A. 51
Wendschuh, F.H. 174
Werthemann, J. 281
Wetlaufer, D.B. 148
Wetmore, R. 294, 406
Whitten, D.G. 47, 51
Whitten, J.L. 301
Whittington, S.G. 322
Wiebe, H.A. 328
Wierzchowski, K.L. 51
Wietmayer, N.D. 174
Wild, U.P. 387, 394
Willard, K.F. 281
Williams, G.H. 328
Williams, G.R.J. 231
Williams, O.M. 259, 269
Wilson, T. 245
Windsor, N.W. 294
Winnik, M.A. 322
Wirz, J. 283, 293
Wislocki, P.G. 343, 358, 359
Wismonski-Knittel, T. 370
Witkop, B. 343
Wittel, K. 256
Wolf, H. 358
Wolken, J.J. 186
Wolowsky, R. 293
Wong, H.N.C. 293
Wood, A.W. 343, 358, 359
Woodward, R.B. 260, 269
Woody, R.W. 129, 135, 148
Wu, F. Y-H. 245
Wu, J.R. 111
Wulfman, C.E. 164, 174
Wyckoff, H.W. 82, 83

Yabe, A. 370
Yagi, H. 343, 358, 359
Yakatan, G.J. 230
Yamaguchi, K. 344
Yamane, T. 28, 37
Yamasaki, E. 358
Yanath, A. 134
Yang, J.T. 219, 417
Yang, N.C. 370
Yang, S.K. 358
Yaniv, M. 13, 76, 77
Yarwood, A.J. 281
Yeh, Y.C. 111, 212, 220, 244
Yonath, A. 78
Youmg, M.A. 343
Younathan, E. 256
Young, D.W. 417

Zacharias, D.E. 344
Zachariasse, K. 322
Zachau, H.G. 76
Zechner, J. 121, 309, 310

Zehner, H. 96
Zemb, T. 311
Ziman, J.M. 198
Zimbrick, J.D. 96
Zimmerman, H.E. 51
Zittel, P.F. 429
Zorman, G. 24